Genomes and Genomics of Nitrogen-fixing Organisms

Nitrogen Fixation: Origins, Applications, and Research Progress

VOLUME 3

Genomes and Genomics of Nitrogen-fixing Organisms

Edited by

Rafael Palacios

Nitrogen Fixation Research Center,
National University of Mexico,
Cuernavaca, Morelos, Mexico

and

William E. Newton

Department of Biochemistry,
Virginia Polytechnic Institute & State University, Blacksburg, U.S.A.

Springer

A C.I.P. Catalogue record for this book is available from the Library of Congress.

ISBN 1-4020-3053-3 (HB)
ISBN 1-4020-3054-1 (e-book)
ISBN 1-4020-2510-6 (Divisible set)

Published by Springer,
P.O. Box 17, 3300 AA Dordrecht, The Netherlands.

Sold and distributed in North, Central and South America
by Springer,
101 Philip Drive, Norwell, MA 02061, U.S.A.

In all other countries, sold and distributed
by Springer,
P.O. Box 322, 3300 AH Dordrecht, The Netherlands.

<u>background figure caption</u>:
"A seed crop of clover (*Trifolium hirtum*) in flower near Moora, Western Australia. Photograph courtesy of Mike Davies, Senior Technical Officer, Pasture Research Group of Agriculture WA and reproduced with permission."
<u>Caption for "Volume III Specific Cover Figure"</u>:
"Proteomics applied to *Rhizobium etli* under two different growth conditions demonstrates its regulatory complexity. Proteins present under both conditions are shown as the lighter spots, whereas those appearing under only one of the conditions are shown as darker spots.
Figure courtesy of Sergio Encarnación and Jaime Mora, Programa de Ingeniería Metabólica, Centro de Investigación sobre Fijación de Nitrógeno, Universidad Nacional Autónoma de México, Cuernavaca, Morelos CP62210, México."

Printed on acid-free paper

Printed in the Netherlands.

TABLE OF CONTENTS

Preface to the Series ix

Preface xiii

List of Contributors xv

Chapter 1. Origins of Genomics in Nitrogen-Fixation Research
G. Dávila and R. Palacios 1
1. Introduction 1
2. Symbiotic Organisms 2
3. Free-Living Organism 3
4. Conclusion 5
References 5

Chapter 2. Genomics of Diazotrophic Archaea
J. A. Leigh 7
1. Introduction 7
2. The Core *nif*-gene Cluster 8
3. Other *nif* Genes 9
4. Other Nitrogen Assimilatory Genes 9
5. P_{II} Proteins 10
References 11

Chapter 3. Genomic Aspects of Nitrogen Fixation in the Clostridia
J.-S. Chen 13
1. Introduction 13
2. The Nitrogen-Fixing Clostridia 14
3. The Genome of the Clostridia 15
4. Organization of the Nitrogen-Fixation Gene Cluster 17
5. Regulatory Genes for Nitrogen Metabolism 21
6. Altenative Nitrogen-Fixation (*anf*) Genes 23
7. Genes for Nitrogen Assimilation 23
8. Concluding Remarks 24
References 24

Chapter 4. The Genome of the Filamentous Cyanobacterium *Nostoc punctiforme*
J. C. Meeks 27
1. Introduction 27
2. Phenotypic Traits of *N. punctiforme* 29
3. Overview of the *N. punctiforme* Genome 37
4. N. punctiforme genes Involved in Heterocyst Formation, Nitrogenase Expression, and Ammonia and Nitrate Assimilation 46
5. Summary and Conclusions 63
Acknowledgements 64
References 64

Chapter 5. The *nif* Genes of *Rhodobacter capsulatus*, *Rhodobacter sphaeroides* and *Rhodopseudomonas palustris*
R. Haselkorn and V. Kapatral 71
1. Introduction 71
2. Regulation of the Nitrogen-Fixation System 73
3. Operon Structure and Gene Organization 75
Acknowledgement 80
References 80

Chapter 6. Genomic Architecture of the Multiple Replicons of the Promiscuous *Rhizobium* Species NGR234
P. Mavingui, X. Perret and W. J. Broughton 83
1. Introduction 83
2. Promiscuity of NGR234 84
3. Structural Organization of the NGR234 Genome 85
4. Coding Capacity of eth NGR234 Genome 91
5. Conclusions and Perspectives 92
References 93

Chapter 7. Facets of the *Bradyrhizobium japonicum* 110 Genome
M. Göttfert, H. Hennecke and S. Tabata 99
1. Introduction 99
2. Materials and Methods 100
3. Genome Characteristics 100
4. Perspectives 108
References 108

Chapter 8. pSymA of *Sinorhizobium meliloti*: Nitrogen Fixation and More
M. J. Barnett and M. L. Kahn 113
1. Introduction 113
2. History 114
3. The pSymA Sequence Project 115
4. General Features of pSymA. 115
5. Comparative Genomics 116
6. Plasmid Biology 117
7. Elements of External Origin 117
8. Transfer RNA Genes 119
9. Nodulation Genes 119
10. Nitrogen-Fixation Genes 120
11. Carbon and Nitrogen Metabolism 121
12. Chemotaxis and Pilus Formation 123
13. Transport 123
14. Regulation and Signal Transduction 124
15. Stress Responses 125
16. Sulfur Metabolism 126
17. Orphan Genes 126

18. Genome-Wide Analysis . . . 127
19. A Strategy for Analyzing pSymA of *S. meliloti* . . . 127
References . . . 128

Chapter 9. *Rhizobium etli* Genome Biology
G. Dávila, V. González, M. A. Ramírez-Romero and O. Rodríguez . . . 133
1. Introduction . . . 133
2. *Rhizobium etli* Genome Structure . . . 134
3. *Rhizobium* Genome Plasticity . . . 136
4. *Rhizobium etli* Taxomony and Evolution . . . 138
Acknowledgements . . . 140
References . . . 141

Chapter 10. The Dawn of Functional Genomics in Nitrogen-Fixation Research
S. Encarnación . . . 143
1. Introduction . . . 143
2. Functional Genomics – The Role of Gene-Expression Studies . . . 144
3. The Transcriptome . . . 145
4. Transcriptomics in Nitrogen-Fixation Research. . . . 146
5. Transcriptomics in Plants during Symbiotic Nitrogen Fixation . . . 148
6. The Proteome . . . 150
7. Proteomics and Nitrogen-Fixation Research . . . 152
8. Proteomics in Plants during Symbiotic Nitrogen Fixation . . . 156
9. Proteomics in Concert with Transcriptomics . . . 157
10. Global Approaches to Study the *R. etli-P. vulgaris* Interaction . . . 158
11. Protein-Protein Interactions: Applications of Molecular Maps . . . 160
12. Transcriptomics, Proteomics and Bioinformatics . . . 160
13. Conclusions . . . 162
Acknowledgements . . . 163
References . . . 163

Chapter 11. Transcriptomics in *Sinorhizobium meliloti*
A. Becker and F. J. De Bruijn . . . 169
1. Introduction to Transcriptomics . . . 169
2. Introduction to the Biological System . . . 171
3. *Sinorhizobium meliloti* Microarrays. . . . 172
4. *S. meliloti* Macroarrays . . . 175
5. Conclusions and Perspectives. . . . 178
Acknowledgements . . . 179
References . . . 180

Chapter 12. Genome Dynamics in Rhizobial Organisms
R. Palacios and M. Flores . . . 183
1. Introduction . . . 183
2. Reiterated Sequences . . . 184
3. Genomic Instability . . . 186

4. Natural Gene Amplication 188
5. Artificial Gene Amplification 189
6. Dynamics of Genome Architecture 191
7. Prediction of Genome Rearrangements 193
8. Identification of Genome Rearrangements 194
9. Artificial Selection of Genomic Rearrangements 195
10. Natural Genomic Design 197
11. Concluding Remarks 197
Acknowledgements 197
References 198

Chapter 13. Impact of Genomics on the Reconstruction of Evolutionary Relationships of Nitrogen-Fixing Bacteria and Implications for Taxomony
P. Van Berkum and B. D. Eardly 201
1. Systematics 201
2. Current Reflections for Evolution of Diazotrophy 203
3. Reconstruction of Evolutionary Relationships among Members of the Kingdom Monera 203
4. The Rapid Spread of Antibiotic Resistance: Implications of Reticulate Microbial Evolution 205
5. Mechanisms of Horizontal Gene Transfer in Microbes 205
6. Significance of Horizontal Gene transfer in Nature 206
7. Evidence for Lateral Gene Transfers and Recombination in Microbial Genomes 208
8. Genomic Islands 209
9. Microbial Evolution and Genetic Recombination 210
10. Established Species Concepts Applied to Bacteria 210
11. A Proposed Unified Species fro Bacteria 212
12. Relevant Insights from Recent Genomic Comparisons 212
13. Implications and Future Strategies 213
Acknowledgements 214
References 214

Chapter 14. The Phylogeny and Evolution of Nitrogenases
J. P. W. Young 221
1. Introduction 221
2. The Genetic Organization of Nitrogenase Genes 222
3. Nitrogenase Genes from Genome Sequencing Projects 224
4. Organization of the Nitrogenase Genes 225
5. Evolutionary Relationships of the Nitrogenase Genes 225
6. Nitrogenase Phylogeny *versus* Organism Phylogeny 231
7. Nitrogenase Genes in their Genomic Context 238
8. Conclusions and Prospects 238
References 239

Subject Index. 243

SERIES PREFACE

Nitrogen Fixation: Origins, Applications, and Research Progress

Nitrogen fixation, along with photosynthesis as the energy supplier, is the basis of all life on Earth (and maybe elsewhere too!). Nitrogen fixation provides the basic component, fixed nitrogen as ammonia, of two major groups of macromolecules, namely nucleic acids and proteins. Fixed nitrogen is required for the N-containing heterocycles (or bases) that constitute the essential coding entities of deoxyribonucleic acids (DNA) and ribonucleic acids (RNA), which are responsible for the high-fidelity storage and transfer of genetic information, respectively. It is also required for the amino-acid residues of the proteins, which are encoded by the DNA and that actually do the work in living cells. At the turn of the millennium, it seemed to me that now was as good a time as any (and maybe better than most) to look back, particularly over the last 100 years or so, and ponder just what had been achieved. What is the state of our knowledge of nitrogen fixation, both biological and abiological? How has this knowledge been used and what are its impacts on humanity?

In an attempt to answer these questions and to capture the essence of our current knowledge, I devised a seven-volume series, which was designed to cover all aspects of nitrogen-fixation research. I then approached my long-time contact at Kluwer Academic Publishers, Ad Plaizier, with the idea. I had worked with Ad for many years on the publication of the Proceedings of most of the International Congresses on Nitrogen Fixation. My personal belief is that congresses, symposia, and workshops must not be closed shops and that those of us unable to attend should have access to the material presented. My solution is to capture the material in print in the form of proceedings. So it was quite natural for me to turn to the printed word for this detailed review of nitrogen fixation. Ad's immediate affirmation of the project encouraged me to share my initial design with many of my current co-editors and, with their assistance, to develop the detailed contents of each of the seven volumes and to enlist prospective authors for each chapter.

There are many ways in which the subject matter could be divided. Our decision was to break it down as follows: nitrogenases, commercial processes, and relevant chemical models; genetics and regulation; genomes and genomics; associative, endophytic, and cyanobacterial systems; actinorhizal associations; leguminous symbioses; and agriculture, forestry, ecology, and the environment. I feel very fortunate to have been able to recruit some outstanding researchers as co-editors for this project. My co-editors were Mike Dilworth, Claudine Elmerich, John Gallon, Euan James, Werner Klipp, Bernd Masepohl, Rafael Palacios, Katharina Pawlowski, Ray Richards, Barry Smith, Janet Sprent, and Dietrich Werner. They worked very hard and ably and were most willing to keep the volumes moving along reasonably close to our initial timetable. All have been a pleasure to work with and I thank them all for their support and unflagging interest.

Nitrogen-fixation research and its application to agriculture have been ongoing for many centuries – from even before it was recognized as nitrogen fixation. The Romans developed the crop-rotation system over 2000 years ago for maintaining and improving soil fertility with nitrogen-fixing legumes as an integral component. Even though crop rotation and the use of legumes was practiced widely but intermittently since then, it wasn't until 1800 years later that insight came as to how legumes produced their beneficial effect. Now, we know that bacteria are harbored within nodules on the legumes' roots and that they are responsible for fixing N_2 and providing these plants with much of the fixed nitrogen required for healthy growth. Because some of the fixed nitrogen remains in the unharvested parts of the crop, its release to the soil by mineralization of the residue explains the follow-up beneficial impact of legumes. With this realization, and over the next 100 years or so, commercial inoculants, which ensured successful bacterial nodulation of legume crops, became available. Then, in the early 1900's, abiological sources of fixed nitrogen were developed, most notable of these was the Haber-Bosch process. Because fixed nitrogen is almost always the limiting nutrient in agriculture, the resulting massive increase in synthetic fixed-nitrogen available for fertilizer has enabled the enormous increase in food production over the second half of the 20th century, particularly when coupled with the new "green revolution" crop varieties. Never before in human history has the global population enjoyed such a substantial supply of food.

Unfortunately, this bright shiny coin has a slightly tarnished side! The abundance of nitrogen fertilizer has removed the necessity to plant forage legumes and to return animal manures to fields to replenish their fertility. The result is a continuing loss of soil organic matter, which decreases the soil's tilth, its water-holding capacity, and its ability to support microbial populations. Nowadays, farms do not operate as self-contained recycling units for crop nutrients; fertilizers are trucked in and meat and food crops are trucked out. And if it's not recycled, how do we dispose of all of the animal waste, which is rich in fixed nitrogen, coming from feedlots, broiler houses, and pig farms? And what is the environmental impact of its disposal? This problem is compounded by inappropriate agricultural practice in many countries, where the plentiful supply of cheap commercial nitrogen fertilizer, plus farm subsidies, has encouraged high (and increasing) application rates. In these circumstances, only about half (at best) of the applied nitrogen reaches the crop plant for which it was intended; the rest leaches and "runs off" into streams, rivers, lakes, and finally into coastal waters. The resulting eutrophication can be detrimental to marine life. If it encroaches on drinking-water supplies, a human health hazard is possible. Furthermore, oxidation of urea and ammonium fertilizers to nitrate progressively acidifies the soil – a major problem in many agricultural areas of the world. A related problem is the emission of nitrogen oxides (NO_x) from the soil by the action of microorganisms on the applied fertilizer and, if fertilizer is surface broadcast, a large proportion may be volatilized and lost as ammonia. For urea in rice paddies, an extreme example, as much as 50% is volatilized and lost to the atmosphere. And what goes up must come down; in the case of fertilizer nitrogen, it returns to Earth in the rain, often acidic in nature. This uncontrolled deposition has unpredictable environmental effects, especially in pristine environments like forests, and may also affect biodiversity.

Some of these problems may be overcome by more efficient use of the applied fertilizer nitrogen. A tried and tested approach (that should be used more often) is to ensure that a balanced supply of nutrients (and not simply applying more and more) is applied at the right time (maybe in several separate applications) and in the correct place (under the soil surface and not broadcast). An entirely different approach that could slow the loss of fertilizer nitrogen is through the use of nitrification inhibitors, which would slow the rate of conversion of the applied ammonia into nitrate, and so decrease its loss through leaching. A third approach to ameliorating the problems outlined above is through the expanded use of biological nitrogen fixation. It's not likely that we shall soon have plants, which are capable of fixing N_2 without associated microbes, available for agricultural use. But the discovery of N_2-fixing endophytes within the tissues of our major crops, like rice, maize, and sugarcane, and their obvious benefit to the crop, shows that real progress is being made. Moreover, with new techniques and experimental approaches, such as those provided by the advent of genomics, we have reasons to renew our belief that both bacteria and plants may be engineered to improve biological nitrogen fixation, possibly through developing new symbiotic systems involving the major cereal and tuber crops.

In the meantime, the major impact might be through agricultural sustainability involving the wider use of legumes, reintroduction of crop-rotation cycles, and incorporation of crop residues into the soil. But even these practices will have to be performed judiciously because, if legumes are used only as cover crops and are not used for grazing, their growth could impact the amount of cultivatable land available for food crops. Even so, the dietary preferences of developed countries (who eats beans when steak is available?) and current agricultural practices make it unlikely that the fixed-nitrogen input by rhizobia in agricultural soils will change much in the near-term future. A significant positive input could accrue, however, from matching rhizobial strains more judiciously with their host legumes and from introducing "new" legume species, particularly into currently marginal land. In the longer term, it may be possible to engineer crops in general, but cereals in particular, to use the applied fertilizer more efficiently. That would be a giant step the right direction. We shall have to wait and see what the ingenuity of mankind can do when "the chips are down" as they will be sometime in the future as food security becomes a priority for many nations. At the moment, there is no doubt that commercially synthesized fertilizer nitrogen will continue to provide the key component for the protein required by the next generation or two.

So, even as we continue the discussion about the benefits, drawbacks, and likely outcomes of each of these approaches, including our hopes and fears for the future, the time has arrived to close this effort to delineate what we know about nitrogen fixation and what we have achieved with that knowledge. It now remains for me to thank personally all the authors for their interest and commitment to this project. Their efforts, massaged gently by the editorial team, have produced an indispensable reference work. The content is my responsibility and I apologize upfront for any omissions and oversights. Even so, I remain confident that these volumes will serve well the many scientists researching nitrogen fixation and related fields, students considering the nitrogen-fixation challenge, and administrators wanting to either become acquainted with or remain current in this field. I also acknowledge the many scientists who were not direct contributors to this series of books, but whose contributions to the field are documented in their pages. It would be remiss of me not to acknowledge also the patience and assistance of the several members of the Kluwer staff who have assisted me along the way. Since my initial dealings with Ad Plaizier, I have had the pleasure of working with Arno Flier, Jacco Flipsen, Frans van Dunne, and Claire van Heukelom; all of whom provided encouragement and good advice – and there were times when I needed both!

It took more years than I care to remember from the first planning discussions with Ad Plaizier to the completion of the first volumes in this series. Although the editorial team shared some fun times and a sense of achievement as volumes were completed, we also had our darker moments. Two members of our editorial team died during this period. Both Werner Klipp (1953-2002) and John Gallon (1944-2003) had been working on Volume II of the series, *Genetics and Regulation of Nitrogen-Fixing Bacteria*, and that volume is dedicated to their memory. Other major contributors to the field were also lost in this time period: Barbara Burgess, whose influence reached beyond the nitrogenase arena into the field of iron-sulfur cluster biochemistry; Johanna Döbereiner, who was the discoverer and acknowledged leader in nitrogen-fixing associations with grasses; Lu Jiaxi, whose "string bag" model of the FeMo-cofactor prosthetic group of Mo-nitrogenase might well describe its mode of action; Nikolai L'vov, who was involved with the early studies of molybdenum-containing cofactors; Dick Miller, whose work produced new insights into MgATP binding to nitrogenase; Richard Pau, who influenced our understanding of alternative nitrogenases and how molybdenum is taken up and transported; and Dieter Sellmann, who was a synthetic inorganic chemistry with a deep interest in how N_2 is activated on metal sites. I hope these volumes will in some way help both preserve their scientific contributions and reflect their enthusiasm for science. I remember them all fondly.

Only the reactions and interest of you, the reader, will determine if we have been successful in capturing the essence and excitement of the many sterling achievements and exciting discoveries in the research and application efforts of our predecessors and current colleagues over the past 150 years or so. I sincerely hope you enjoy reading these volumes as much as I've enjoyed producing them.

William E. Newton
Blacksburg, February 2004

PREFACE

Genomes and Genomics of Nitrogen-fixing Organisms

This is Volume 3 of a seven-volume series on all aspects of Nitrogen Fixation. The series aims to be the definitive authority in the field and to act as a benchmark for some years to come. Rather than attempting to cram the whole field into a single volume, the subject matter is divided among seven volumes to allow authors the luxury of writing in depth with a comprehensive reference base. All authors are recognized practicing scientists in the area of their contribution, which ensures the high quality, relevance, and readability of the chapters.

In establishing the rationale for, and the organization of, this book, we realized the need to divide it into two sections. The first section should be organism based and should review our current knowledge of the genomes of nitrogen-fixing organisms and what these nucleotide sequences tell us. The second section should then be technology based. It should review what technologies are available to mine the data inherent in the nucleotide sequences and how they are now being used to produce gene-function data from differential gene expression.

The first section starts with a brief overview of the origins of genomic research in nitrogen fixation and then reviews the current state of our understanding with respect to the application of genomics to various nitrogen-fixing organisms. The following chapters cover the genomes of Archaea (Chapter 2), Clostridia (Chapter 3), Cyanobacteria (Chapter 4), and Rhodobacterales (Chapter 5). Then, the last four chapters (Chapters 6-9) in this section are devoted to the Rhizobiales because of their agricultural importance. In each of the chapters, the review first briefly describes which organisms have had their genomes sequenced, how the data can be accessed, and the relative size and overall structure of the genomes, plus an outline of their physiology and how it changes on interactions with a host. Then, based on the genome sequences, the location and organization of the core *nif*-gene cluster is described and compared for the various member organisms, including any reiterations and whether it is accompanied by either the *vnf*-gene or *anf*-gene system or both. This discussion is usually followed by an outline of which of the other *nif* genes are present and how the system is likely be regulated. Then, important related systems, such as those involving nitrogen-assimilatory genes/proteins are outlined. We believe that the close juxtaposition of these reviews will facilitate ready comparisons among the genomes of these quite different types of bacteria.

In contrast, the second section cuts across organism lines. Chapters 10 and 11 focus on functional genomics, both transcriptomics and proteomics, and the high throughput technologies that allow gene expression to be monitored at the genome level. These chapters review how functional genomics is used to collect either a "gene-expression" profile or a "protein profile" for a particular organism under defined environmental conditions, including involvement in a symbiotic relationship. The technologies and instrumentation and their specific development for these purposes is also briefly described. Next, genome rearrangements, which can involve deletions, amplifications, and co-integrations, in rhizobial organisms are reviewed in Chapter 12 to illustrate that genomes are not simply static structures but are dynamic entities, whose rearrangements have significant biological consequences. Chapter 13 then outlines how genomics has impacted taxonomy and how the evolutionary relationships among nitrogen-fixing bacteria are being reconstructed. The volume closes with a discussion (Chapter 14) of the evolution of the three classical N_2-fixing enzymes, the molybdenum-nitrogenase, the vanadium-nitrogenase, and the iron-nitrogenase, including whether and how their phylogeny correlates with the phylogeny of the organisms that contain them and how such information impacts evolutionary arguments.

Now, as we finalize this volume, we are reminded that it has been in preparation for more than two years and it is really satisfying to have completed the task at last. We thank all of the authors who so willingly committed the necessary effort and so readily agreed to our editorial requirements. We believe that the high quality of the final product is the best thanks we can give to each of them. We also thank the many other experts in these fields, who are not authors but whose contributions are vital to the contents of this volume, and we apologize in advance for any omissions. There is a related volume (volume 2) in this series, which is entitled *Genetics and Regulation of Nitrogen Fixation in Free-Living Organisms*, and it should be of interest to readers of the current volume. It deals with many of the nitrogen-fixing bacteria described or referenced in this volume and provides a considerable amount of information closely related to the topics of this volume.

We've enjoyed working as part of the editorial and production teams from the first planning phase through to the completion of this volume and, in closing, we pause to remember our colleagues who are no longer with us to share the continuing sense of fun and discovery.

Rafael Palacios
Cuernavaca, May, 2004

William E. Newton
Blacksburg, May, 2004

LIST OF CONTRIBUTORS

Melanie J. BARNETT
Department of Biological Sciences, Stanford University, Stanford, CA 94305, USA.
Email: melbar@stanford.edu

Anke BECKER
Lehrstuhl für Genetik, Fakultät für Biologie, Universität Bielefeld, P.O. Box 100131, D-33501 Bielefeld, Germany.
Email: anke.becker@genetik.uni-bielefeld.de

William J. BROUGHTON
Laboratoire de Biologie Moléculaire Des Plantes Supérieures, Université de Genève, 1 Chemin de'Imperatrice, CH-1292 Chambésy, Genéve, Switzerland.
Email: William.Broughton@bioveg.unige.ch

Jiann-Shin CHEN
Department of Biochemistry, Virginia Polytechnic Institute and State University, Blacksburg, VA 24061, USA.
Email: chenjs@vt.edu

Guillermo DÁVILA
Programa de Evolución Molecular, Centro de Investigación sobre Fijación de Nitrógeno, UNAM. Apdo. Postal 565-A, Cuernavaca, Morelos CP 62170, México.
Email: davila@cifn.unam.mx

Frans J. DE BRUIJN,
UMR INRA-CNRS 2594/441 Laboratoire des Interactions Plantes-Microorganismes, Chemin de Borde Rouge, BP27 31326 Castanet-Tolosan cedex, France.
Email: debruijn@toulouse.inra.fr

Bertrand D. EARDLY
Penn State Berks Campus, Tulpehocken Road, P.O. Box 7009, Reading, PA 19610, USA
Email: bde1@psu.edu

Sergio ENCARNACIÓN
Programa de Ingeniería Metabólica, Centro de Investigación sobre Fijación de Nitrógeno, UNAM. Apdo. postal 565-A, Cuernavaca, Morelos CP 62170, México
Email: encarnac@cifn.unam.mx

Margarita FLORES
Programa Dinámica del Genoma Centro de Investigación sobre Fijación de Nitrógeno, UNAM. Apdo. postal 565-A, Cuernavaca, Morelos CP 62170, México
Email: mflores@cifn.unam.mx

Víctor GONZÁLEZ
Programa de Evolución Molecular, Centro de Investigación sobre Fijación de Nitrógeno, UNAM. Apdo. Postal 565-A, Cuernavaca, Morelos CP 62170, México.
Email: vgonzal@cifn.unam.mx

Michael GÖTTFERT
Technische Universität Dresden, Institut für Genetik, Mommsenstrasse 13, D-01062 Dresden, Germany
Email: mgoettfe@rcs.urz.tu-dresden.de

Robert HASELKORN
Dept of Molecular Genetics and Cell Biology, University of Chicago, 920 E. 58th Street, Chicago, IL 60637, USA
Email: r-haselkorn@uchicago.edu

Hauke HENNECKE
Eidgenössische Technische Hochschule, Institut für Mikrobiologie, CH-8092 Zürich, Switzerland.
Email: hennecke@micro.biol.ethz.ch

Michael L. KAHN
Institute of Biological Chemistry, Washington State University, Pullman, WA 99164, USA.
Email: kahn@mail.wsu.edu

Vinayak KAPATRAL
Integrated Genomics, 2201 West Campbell Park Drive, Chicago, IL 60612, USA
Email: vinayak@integrated genomics.com

John A. LEIGH
Department of Microbiology, University of Washington, Box 357242, Seattle, WA 98195, USA
Email: leighj@u.washington.edu

Patrick MAVINGUI
Laboratoire d'Ecologie Microbienne, UMR CNRS 5557, UCBL, 43 boulevard du 11 Novembre 1918, 69622 Villeurbanne Cedex, France.
Email: mavingui@biomserv.univ-lyon1.fr

John C. MEEKS
Section of Microbiology, University of California, Davis, CA 95616, USA
Email: jcmeeks@yellow.ucdavis.edu

William E. NEWTON
Department of Biochemistry, Virginia Polytechnic Institute and State University, Blacksburg, VA 24061, USA.
Email: wenewton@vt.edu

Rafael PALACIOS
Programa Dinámica del Genoma Centro de Investigación sobre Fijación de Nitrógeno, UNAM. Apdo. postal 565-A, Cuernavaca, Morelos CP 62170, México.
Email: palacios@cifn.unam.mx

Xavier PERRET
Laboratoire de Biologie Moléculaire Des Plantes Supérieures, Université de Genève, 1 Chemin de 'Imperatrice CH-1292 Chambésy, Genéve, Switzerland
Email: perret@sc2a.unige.ch

Miguel A. RAMÍREZ-ROMERO
Programa de Evolución Molecular, Centro de Investigación sobre Fijación de Nitrógeno, UNAM. Apdo. Postal 565-A, Cuernavaca, Morelos CP 62170, México.
Email: mramirez@cifn.unam.mx

Oscar RODRÍGUEZ
Programa de Evolución Molecular,
Centro de Investigación sobre
Fijación de Nitrógeno, UNAM.
Apdo. Postal 565-A, Cuernavaca,
Morelos CP 62170, México.
Email: oscar@cifn.unam.mx

Peter VAN BERKUM
Soybean Genomics and Improvement
Laboratory, Agricultural Research
Service, U. S. Department of
Agriculture, 10300 Baltimore Blvd.
Beltsville, MD 20705, USA.
Email: pberkum@ba.ars.usda.gov

Satoshi TABATA
Kazusa DNA Research Institute,
1532-3 Yana, Kisarazu,
Chiba 292-0812, Japan.
Email: tabata@mail.kazusa.or.jp

J. Peter. W. YOUNG
Department of Biology,
University of York, P.O. Box 373,
York, Y010 5YW, UK
Email: jpy1@york.ac.uk

CHAPTER 1

ORIGINS OF GENOMICS IN NITROGEN-FIXATION RESEARCH

G. DÁVILA AND R. PALACIOS

Centro de Investigación sobre Fijación de Nitrógeno, UNAM, P.O. Box 565-A, Cuernavaca, Morelos 62170, México

1. INTRODUCTION

The advent of genomics is changing our perspective of biology. The central paradigm that was focused on the gene is moving to a more integral one based on the whole genome. From a conceptual point of view, genome sciences are placed at the intersection of three disciplines: molecular biology, mathematics, and computational biology. From a practical approach, genomics is the consequence of new high-throughput methodology that allows the nucleotide sequencing of complete genomes and, consequently, study of their integral patterns of expression at the level of both RNA (transcriptome) and protein (proteome). The overall goal of genomics is to link the structure and expression of genetic information with its biological function and evolution.

The first complete genomic sequence obtained was that of *Haemophilus influenzae* (Fleischmann *et al.*, 1995). Since this major achievement, many genomes have been sequenced and analyzed. These include members of the three life domains: Bacteria, Archaea and Eukarya. The major driving force for the development of genomics has been the human genome project that was historically accomplished in 2001 (Lander *et al.*, 2001; Vender *et al.*, 2001). At the time when this chapter was finished (September 2003), 156 sequences of complete genomes from different organisms had been obtained and made public (http://ergo.integratedgenomics.com/GOLD). These comprise 121 bacterial, 16 archaeal, and 19 eukaryal genomes. In addition, there are 388 prokaryotic and 246 eukaryotic genomic projects ongoing.

Nitrogen fixation is only present in some prokaryotes, including members of two kingdoms, Bacteria and Archaea. Among bacteria, the capacity to fix N_2 has a wide taxonomic distribution and includes Proteobacteria, Cyanobacteria, Clostridia,

R. Palacios and W. E. Newton (eds.), Genomes and Genomics of Nitrogen-fixing Organisms, 1-6.

Chlorobi, and Actinobacteria. Some bacteria fix N_2 in the free-living state whereas others require a symbiotic association with a plant. Therefore, very different types of genome may harbor nitrogen-fixation and symbiotic genes (see Chapter 14). In the Archaea domain, nitrogen fixation appears restricted to the methanogens (see Chapter 2). Interestingly, no cryptic homologous of nitrogenase genes have been found in either bacterial or archaeal organisms that do not fix N_2 (see Chapters 2 and 14). Lateral transfer of genetic information may be responsible for the scattered distribution of nitrogen-fixing genes among prokaryotes.

Nitrogen-fixation research has fully entered the era of genomics (Table 1). The attainment of the nucleotide sequence of the cluster of *nif* genes in *Klebsiella pneumoniae* can be considered as the precursor of genomic projects in the field of nitrogen fixation (Arnold *et al.*, 1988). This cluster is located in the chromosome near the histidine operon and comprises the following sets of genes: the structural genes for nitrogenase, which is the enzyme complex responsible for the reduction of atmospheric N_2; genes whose products participate in the assembly of nitrogenase and in the synthesis of the cofactors necessary for its catalytic reaction; genes that encode for proteins related to the provision of electrons; and regulatory elements.

Table 1. Genomic projects in nitrogen-fixing organisms

Complete genomes	Reference
Methanothermobacter thermoautotrophicus	Smith *et al.*, 1977
Mesorhizobium loti	Kaneko *et al.*, 2000
Sinorhizobium meliloti	Galibert *et al.*, 2001
Clostridium aacetobutilicum	Nölling *et al.*, 2001
Anabaena sp	Kaneko *et al.*, 2001
Methanosarcina acetivorans	Galagan *et al.*, 2002
Chlorobium tepidum	Eisen *et al.*, 2002
Methanosarcina mazei	Deppenmeier, 2002
Bradyrhizobium japonicum	Kaneko *et al.*, 2002
Symbiotic genome compartments	
Rhizobium sp NGR234 symbiotic plasmid	Freiberg *et al.*, 1996
Bradyrhizobium japonicum symbiotic region	Göttfert *et al.*, 2001
Rhizobium etli symbiotic plasmid	González *et al.*, 2003

2. SYMBIOTIC ORGANISMS

Formally, the first genomic project with a nitrogen-fixing organism was the complete sequence of the symbiotic plasmid of *Rhizobium* sp. NGR234 (Freiberg *et al.*, 1996; see Chapter 6). Bacteria of the genus *Rhizobium* and related genera, such as *Bradyrhizobium*, *Mesorhizobium* and *Sinorhizobium*, herein referred to as rhizobia, are Gram-negative α-proteobacteria that establish nitrogen-fixing symbioses with the roots of leguminous plants. During the establishment of this association, both partners, the plant and the bacteria, exchange chemical signals that result in the expression of specific genes that participate in the process. In rhizobia, the genes that participate in nodulation and nitrogen fixation are compartmentalized in the genome either as symbiotic regions or islands in the chromosome, as in *Bradyrhizobium japonicum* and *Mesorhizobium loti,* or as independent replicons or symbiotic plasmids, as in *Rhizobium* sp NGR234, *Rhizobium etli*, *Rhizobium leguminosarum*, and *Sinorhizobium meliloti*.

In addition to the symbiotic plasmid of *Rhizobium* sp. NGR234, the symbiotic genomic compartments of other organisms have been sequenced. These include the symbiotic plasmid, pSymA, of *S. meliloti* 1021 (Barnett *et al.*, 2001; see Chapter 8), the symbiotic islands of *Mesorhizobium loti* strains MAFF303099 (Kaneko *et al.*, 2000) and R7A (Sullivan *et al.*, 2002), the symbiotic regions of *Bradyrhizobium japonicum* strain USDA 110 (Göttfert *et al.*, 2001; Kaneko *et al.*, 2002; see Chapter 7), and the symbiotic plasmid of *Rhizobium etli* (González *et al.*, 2003; see Chapter 9).

Important landmarks in symbiotic nitrogen-fixation research have been the completion of the nucleotide sequence of *M. loti* (Kaneko *et al.*, 2000), *S. meliloti* (Galibert *et al.*, 2001; Capela *et al.*, 2001; Barnett *et al.*, 2001; Finan *et al.*, 2001), and *B. japonicum* (Kaneko *et al.*, 2002). The genome of *M. loti* MAFF303099 consists of a chromosome of 7,036,071 bp and two plasmids, pMLa of 351,911 bp and pMLb of 208,315 bp. Both nodulation and nitrogen-fixation genes are located on the chromosome in a symbiotic island of 611 kb (Kaneko *et al.*, 2002). The genome of *S. meliloti* 1021 consists of three replicons, a chromosome of 3,654,135 bp (Capela *et al.*, 2001) and two megaplasmids, pSymA of 1,354,226 pb (Barnett *et al.*, 2001) and pSymB of 1,683,333 bp (Finan *et al.*, 2001). Most of the nodulation and nitrogen-fixation genes are located in pSymA (see Chapter 8). The genome of *B. japonicum* consists of a single chromosome of 9,105,828 bp and both nodulation and nitrogen-fixation genes are located in a region of this chromosome (Göttfert *et al.*, 2001) that constitutes a presumptive symbiotic island of about 681 kb (Kaneko *et al.*, 2002; see Chapter 7).

The comparison of symbiotic genome compartments with complete rhizobial genomes of different organisms has revealed interesting characteristics in regard to the organization and evolution of the genetic information necessary for the establishment of an efficient nitrogen-fixing symbiosis. All the symbiotic genome compartments are heterogeneous regarding gene content; the genes common to most of them are mainly those involved in nodulation and nitrogen fixation. Moreover, there is a lack of synteny between these conserved genes (see Chapter 9). In addition, symbiotic genome compartments contain an unusually high amount of elements related to insertion sequences as compared to the rest of the genome. These findings support the notion that the symbiotic compartments of rhizobial genomes are mosaic structures, presumably assembled from regions derived from diverse genomic contexts and frequently modified as a consequence of transposition and recombination events (Freiberg *et al.*, 1996; González *et al.*, 2003).

3. FREE-LIVING ORGANISMS

The first complete genome of a nitrogen-fixing organism available was that of the methanogen archaea, *Methanothermobacter thermoautotrophicus* (Smith *et al.*,

1997). This organism can synthesize all its components from CO_2, atmospheric N_2, and mineral salts. It reduces CO_2 to methane using hydrogen as the source of energy and electrons. In contrast to its high biosynthetic capacity, its genome is relatively small, only 1,751,377 bp. To date, two other genomes of nitrogen-fixing archaea have been sequenced. These are for *Methanosarcina acetivorans* with a genome of 5,751,492 bp (Galagan *et al.*, 2002) and *Methanosarcina mazei* (Deppenmeier *et al.*, 2002) of 4,096,345 bp (see Chapter 2).

Other sequenced genomes of free-living nitrogen-fixing organisms include those of *Clostridium acetobutylicum* (Nölling *et al.*, 2001), *Anabaena* sp. strain PCC7120, also known as *Nostoc* sp. PCC7120 (Kaneko *et al.*, 2001), and *Chlorobium tepidum* (Eisen *et al.*, 2001).

Organisms of the genus *Clostridium* are Gram-positive, spore-forming, strictly anaerobic bacteria. Both nitrogen-fixing and non-nitrogen-fixing species are found. The genome of the nitrogen-fixing *Clostridium acetobutilicum* ATCC824 has been sequenced (Nölling *et al.*, 2001). It consists of a chromosome of 3,940,880 bp and a plasmid, pSol1 of 192,000 bp. The *nif* genes are clustered on the chromosome near the *dnaA* gene. The nucleotide sequence of the *nif* cluster of other two nitrogen-fixing clostridia, *C. pasteurianum* and *C. beijerinckii*, are also known. In addition, the complete genomic sequence of two non-nitrogen-fixing clostridia, *C. perfringens* (Shimizu *et al.*, 2002) and *C. tetani* (Brüggemann *et al.*, 2003), have been obtained. The availability of nucleotide sequences from different Clostridia has been used for comparative genomics studies (see Chapter 3).

Anabaena is a filamentous, photosynthetic, nitrogen-fixing cyanobacterium. Photosynthesis and nitrogen fixation are not compatible in the same cell due to the production of O_2 during photosynthesis and the high sensitivity to O_2 of the nitrogen-fixation components. Thus, photosynthesis is performed by vegetative cells, whereas nitrogen fixation occurs in differentiated cells called heterocysts. Heterocysts differentiate from vegetative cells at intervals along the filaments under conditions of nitrogen deprivation. *Anabaena* has been extensively used to study both cell differentiation and nitrogen fixation. The genome of *Anabaena* sp. PCC7120 consists of a chromosome of 6,413,771 bp and six plasmids that range from 5,584 bp to 408,101 bp in size. More than 60 genes that participate in heterocyst formation and nitrogen fixation are present in the chromosome.

A close relative of *Anabaena* sp. PCC7120 is *Nostoc punctiforme*, a filamentous cyanobacterium that can fix N_2 in both the free-living and plant-associated states. There is an ongoing genome project on *Nostoc punctiforme* that is about 98% completed (see Chapter 4). Moreover, the genome of a unicellular non-nitrogen-fixing cyanobacterium, *Synechocystis* sp. PCC6803, was one of the first completed bacterial genomic projects (Kaneko *et al.*, 1996). Comparisons between the genomes of *Anabaena* sp. PCC7120, *Nostoc punctiforme*, and *Synechocystis* sp. PCC6803 are discussed in Chapter 4.

Organisms of the phylum Chlorobi are green-sulfur eubacteria that perform anoxygenic photosynthesis by the reductive tricarboxylic-acid cycle. The only member of this phylum whose genome has been completely sequenced is *Chlorobium tepidum* (Eisen *et al.*, 2002). Its genome consists of a single circular chromosome of 2,154,946 bp with the *nif* genes clustered on the chromosome. The *nif* cluster has a similar organization as that of the methanogen archaea, *Methanothermobacter thermoautotrophicus*. It has been suggested that the entire cluster could have been laterally transferred between the two lineages (Eisen *et al.*, 2002).

In addition to the completed genomic projects, there are several ongoing projects that will certainly improve our knowledge of the organization, function, and evolution of nitrogen-fixing organisms. These efforts include, among others: *Methanosarcina barkeri* (see Chapter 2); *Klebsiella pneumoniae*; *Azotobacter vinelandii*; *Rhodobacter sphaeroides*, *Rhodobacter capsulatus*, and *Rhodobacter palustris* (see Chapter 5); *Rhodosporillum rubrum*; *Nostoc punctiforme* (see Chapter 4); *Rhizobium leguminosarum* bv *viciae*; and *Rhizobium etli* (see Chapter 9).

4. CONCLUSION

At this stage, genomic studies in the field of nitrogen fixation are mainly focused on the genome organization and its expression in particular organisms, however, it must be emphasized that genomics is improving and, in some cases, changing our concepts of taxonomy, evolution, and genome dynamics.

REFERENCES

Arnold, W., Rump, A., Klipp, W., Piefer, U., and Pühler, A. (1988). Nucleotide sequence of a 24,206-base-pair DNA fragment carrying the entire nitrogen fixation gene cluster of *Klebsiella pneuminiae*. *J. Mol. Biol., 203*, 715-738.

Barnett, M. J., Fisher, R. F., Jones, T., Komp, C., Abola, A. P., Barloy-Hubler, F., *et al.* (2001). Nucleotide sequence and predicted functions of the entire *Sinorhizobium meliloti* pSymA megaplasmid. *Proc. Natl. Acad. Sci. USA*, *98*, 9883-9888.

Brüggemann, H., Bäumer, S., Fricke, W. F., Wiezer, A., Liesegang, H., Decker, I., *et al.* (2003). The genome sequence of *Clostridium tetani*, the causative agent of tetanus disease. *Proc. Natl. Acad. Sci. USA*, *100*, 1316-1321.

Capela, D., Barloy-Hubler, F., Gouzy, J., Bothe, G., Ampe, F., Batut, J., *et al.* (2001). Analysis of the chromosome sequence of the legume symbiont *Sinorhizobium meliloti* strain 1021. *Proc Natl. Acad. Sci. USA*, *98*, 9877-9882.

Deppenmeier, U., Johann, A., Hartsch, T., Merkl, R., Schmitz, R. A., Martinez-Arias, R., *et al.* (2002). The genome of *Methanosarcina mazei*: Evidence for lateral gene transfer between bacteria and archaea. *J Mol. Microbiol. Biotechnol., 4*, 453-461.

Eisen, J. A., Nelson, K. E., Paulsen, I. T., Heidelberg, J. F., Wu, M., Dodson, R. J., *et al.* (2002). The complete genome sequence of *Chlorobium tepidum* TLS, a photosynthetic, anaerobic, green-sulfur bacterium. *Proc. Natl. Acad. Sci. USA*, *99*, 9509-9514.

Finan, T. M., Weidner, S., Wong, K., Buhrmester, J., Chain, P., Vorholter, F. J., *et al.* (2001). The complete sequence of the 1,683-kb pSymB megaplasmid from the N_2-fixing endosymbiont *Sinorhizobium meliloti*. *Proc. Natl. Acad. Sci. USA*, *98*, 9889-9894.

Fleishman, R. D., Adams, M. D., White, O., Clayton, R. A., Kirkness, E. F., Kerlavage, A. R., *et al.* (1995). Whole genome random sequencing and assembly of *Haemohilus influenzae* Rd. *Science*, *269*, 496-512.

Freiberg, C., Feilla, R., Bairoch, A., Broughton, W. J., Rosenthal, A., and Perret, X. (1997). Molecular basis of symbiosis between *Rhizobium* and legumes. *Nature*, *387*, 394-401.

Galagan, J. E., Nusbaum, C., Roy, A., Endrizzi, M. G., Macdonald, P., FitzHugh, W., *et al.* (2002). The genome of *M. acetivorans* reveals extensive metabolic and physiological diversity. *Genome Res.*, *12*, 532-542.

Galibert, F., Finan, T. M., Long, S. R., Pühler, A., Abola, P., Ampe, F., *et al.* (2001). The composite genome of the legume symbiont *Sinorhizobium meliloti. Science, 293*, 668-672.

González, V., Bustos, P., Ramírez-Romero, M. A., Medrano-Soto, A., Salgado, H., Hernández-González, I., *et al.* (2003). The mosaic structure of the symbiotic plasmid of *Rhizobium etli* CFN42 and its relation to other symbiotic genome compartments. *Genome Biol., 4,* R36.

Göttfert, M., Rothlisberger, S., Kundig, C., Beck, C., Marty, R., and Hennecke, H. (2001). Potential symbiosis-specific genes uncovered by sequencing a 410-kilobase DNA region of the *Bradyrhizobium japonicum* chromosome. *J. Bacteriol., 183*, 1405-1412.

Kaneko, T., Sato, S., Kotani, H., *et al.* (1996). Sequence analysis of the genome of the unicellular cyanobacterium *Synechocystis* sp. strain 6803. II. Sequence determination of the entire genome and assignment of potential protein-coding regions. *DNA Res., 3*, 109-136.

Kaneko, T., Nakamura, Y., Sato, S., Asamizu, E., Kato, T., Sasamoto, S., *et al.* (2000). Complete genome structure of the nitrogen-fixing symbiotic bacterium *Mesorhizobium loti. DNA Res., 7*, 331-338.

Kaneko, T., Nakamura, Y., Wolk, C. P., Kuritz, T., Sasamoto, S., Watanabe, A., *et al.* (2001). Complete genome sequence of the filamentous nitrogen-fixing cyanobacterium *Anabaena* sp. strain PCC 7120. *DNA Res., 8*, 205-213.

Kaneko. T., Nakamura, Y., Sato, S., Minamisawa, K., Uchiumi, T., Sasamoto, S., *et al.* (2002). Complete genomic sequence of nitrogen-fixing symbiotic bacterium *Bradyrhizobium japonicum* USDA110. *DNA Res., 9*, 189-197.

Lander, E. S., Linton, L. M., Birren, B., Nusbaum, C., Zody, M. C., Baldwin, J., *et al.* (2001). Initial sequencing and analysis of the human genome. *Nature, 15*, 860-921.

Vender, J. C., Adams, M. D., Mayers, E. W., Li, P. W., Mural, R. J., Sutton, G. G., *et al.* (2001). The sequence of the human genome. *Science, 291*, 1304-1351.

Nölling, J., Breton, G., Omelchenko, M. V., Makarova, K. S., Zeng, Q. D., Gibson, R., *et al.* (2001). Genome sequence and comparative analysis of the solvent-producing bacterium *Clostridium acetobutylicum. J. Bacteriol., 183*, 4823-4838.

Shimizu, T., Ohtani, K., Hirakawa, H., Ohshima, K., Yamashita, A., Shiba, T., *et al.* (2002). Complete genome sequence of *Clostridium perfringens*, an anaerobic flesh-eater. *Proc. Natl. Acad. Sci. USA, 99*, 996-1001.

Smith, D. R., Doucette-Stamm, L. A., Deloughery, C., Lee, H., Dubois, J., Aldredge, T., *et al.* (1997). Complete genome sequence of *Methanobacterium thermoautotrophicum* deltaH: Functional analysis and comparative genomics. *J Bacteriol., 179*, 7135-7155.

CHAPTER 2

GENOMICS OF DIAZOTROPHIC ARCHAEA

J. A. LEIGH
Department of Microbiology, University of Washington, Seattle, WA 98195, USA

1. INTRODUCTION

Nitrogen fixation was discovered in the domain Archaea in 1984 (Belay, *et al.*, 1984; Murray and Zinder, 1984). Currently diazotrophic species are known in four of the five orders of methanogenic Archaea: *Methanobacteriales* (*Methanobacterium bryantii*), *Methanococcales* (*Methanococcus maripaludis* and *Methanothermococcus thermolithotrophicus*), *Methanomicrobiales* (*Methanospirillum hungatei*), and *Methanosarcinales* (*Methanosarcina barkeri*) (Boone and Castenholz, 2001). In addition to these species, genome sequencing suggests that the ability to fix N_2 exists in *Methanosarcina acetovorans*, *Methanosarcina mazei*, and *Methanothermobacter thermoautotrophicus* and, for the purpose of this chapter, these species are considered to be diazotrophic. The single species in the order *Methanopyrales*, *Methanopyrus kandleri*, is not known to fix N_2 and nitrogenase is not encoded. Despite its widespread distribution in the methanogens, nitrogen fixation is not known within the Archaea outside of the methanogens, and functional homologs of the nitrogenase proteins are not encoded in known non-methanogenic archaeal genome sequences.

The genomes of the following diazotrophic methanogens are available: *Methanococcus maripaudis* LL (still incomplete, see http://www.genome.washington.edu/uwgc/methanococus/), *Methanothermobacter thermoautotrophicus* ΔH (Smith *et al.*, 1997; see http://www.tigr.org/tigr-scripts/CMR2/Genome Page3.spl?database=ntmt01), *Methanosarcina acetivorans* C2A(Galagan *et al.*, 2002; see http://www.genome.wi.mit.edu/annotation/microbes/methanosarcina/), *Methanosarcina barkeri* fusaro (still incomplete, see http://www.jgi.doe. gov/JGI _microbial/html/methanosarcina/methano_homepage.html), and *Methanosarcina mazei* Go1 (Deppenmeier *et al.*, 2002; see http://www.ncbi.nlm.nih.gov/cgi-bin/ Entrez/framik?db=Genome&gi=239). The genomes of several non-diazotrophic Archaea have also been sequenced (see http://wit.integratedgenomics.com/GOLD/).

R. Palacios and W. E. Newton (eds.), Genomes and Genomics of Nitrogen-fixing Organisms, 7-12.

2. THE CORE *nif* GENE CLUSTER

All five available genome sequences of diazotrophic methanogens contain one or more copies of a common *nif*-gene cluster (Figure 1). The clusters encode dinitrogenase reductase (NifH or the Fe protein) and the dinitrogenase (the MoFe protein) α and β subunits (NifD and NifK) as well as the NifE and NifN proteins that function in the synthesis of the nitrogenase cofactor. Sometimes, NifX is encoded at the end of the cluster. In addition, the *nif*-gene clusters of methanogens consistently contain genes encoding $NifI_1$ and $NifI_2$. These are P_{II} homologs (see below) that were shown in *Methanococcus maripaludis* to regulate switch-off of nitrogenase activity (Kessler, *et al.*, 2001; Kessler and Leigh, 1999). All the genes of the core *nif*-gene cluster are contained in a single operon in *M. maripaudis* (Kessler, *et al.*, 1998). In contrast, these genes (excluding $nifI_1$ and $nifI_2$) are typically divided into several operons in bacterial diazotrophs, although the overall gene order is often the same.

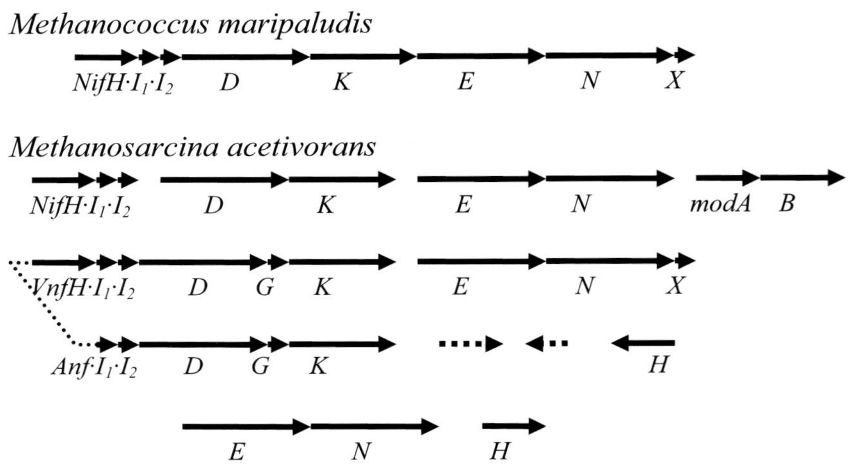

Figure 1. nif *gene clusters of two methanogenic Archaea.*
The nif *cluster of* M. acetivorans *is adjacent to the* modA *and* B *genes possibly encoding molybdate uptake for the purpose of nitrogenase cofactor synthesis. The* vnf *and* anf *clusters of* M. acetivorans *are adjacent and in opposite orientation. There are two genes possibly encoding* AnfH *in* M. acetivorans, *one near* anfK *but separated by two genes and in opposite orientation, and one elsewhere in the genome linked to apparent* anfE *and* anfN *genes.*

M. maripaludis contains a single *nif* gene cluster (Figure 1). This cluster apparently encodes a molybdenum-nitrogenase, consistent with the observation that molybdate is required for diazotrophic growth (Kessler, *et al.*, 1997). *Methano-*

thermobacter thermoautotrophicus also contains a single *nif* cluster as does *Methanosarcina mazei*. In contrast, *Methanosarcina acetivorans* (Figure 1) and *Methanosarcina barkeri* each contains three *nif* clusters. These species appear to have alternative (vanadium- and iron-based) nitrogenases as well as the molybdenum-nitrogenase. In addition to the *vnf* and *anf* homologs of *nif* genes, *vnfG* and *anfG*, which encode the δ subunit of the alternative nitrogenases, are present. Each alternative nitrogenase cluster contains, as a minimum, *vnf/anfH*, *D*, *G*, and *K*. A separate set of genes corresponding to $nifI_1$ and $nifI_2$ is also present in each cluster. Both molybdate and vanadate have been shown to stimulate diazotrophic growth in *Methanosarcina barkeri* 227 (Chien, et al., 2000; Lobo and Zinder, 1988; Scherer, 1989).

Phylogenetic analyses on NifH, NifD, and NifK across a wide spectrum of diazotrophs (Chien *et al.*, 2000; Chien and Zinder, 1996; Leigh, 2000) support a scenario in which nitrogen fixation had an ancient origin that preceded the divergence between Archaea and Bacteria. Ancient gene duplication or horizontal transfer evidently participated in the evolution of the alternative nitrogenases.

3. OTHER *NIF* GENES

Many *nif* genes in bacteria function in electron delivery to the nitrogenase complex, maturation and stabilization of nitrogenase proteins, molybdate transport, and synthesis of homocitrate, a component of the nitrogenase cofactor. Clear orthologs of these genes cannot be identified in diazotrophic Archaea. However, some of these functions can be surmised. For example, *M. maripaludis* has three sets of genes homologous to molybdate ABC transporters. Two of these gene sets have inverted repeat sequences in their promoter regions that share nucleotide identity to a repressor-binding sequence, which regulates *nif* and *glnA* expression (Kessler and Leigh, 1999). If these genes are indeed nitrogen-regulated, they may function in the transport of molybdate for the nitrogenase cofactor. Molybdate ABC transporters are also encoded adjacent to *nif*-gene clusters in the diazotrophic *Methanosarcina* species. Another function, that of homocitrate synthesis, may be performed by 2-oxosuberate synthase (AksA), in the pathway leading to biotin and coenzyme B synthesis (Howell *et al.*, 1998). The dehydrated precursor of homocitrate is an intermediate in that pathway.

Some Archaea contain *nif* gene homologs that are unlikely to have any function in nitrogen fixation. Homologs of *nifH*, *nifS*, *nifB*, and *nifU* are present in non-diazotrophic methanogens.

4. OTHER NITROGEN ASSIMILATORY GENES

4.1. Glutamine synthetase

Most methanogens, and Euryarchaeota (the kingdom within the domain Archaea containing methanogens) in general, contain a single glutamine synthetase, belonging to the α subdivision of type I glutamine synthetases that are typical of the low mol % G+C Gram-positive bacteria (Brown *et al.*, 1994). *M. acetivorans* has a second glutamine synthetase that is divergent and harder to place phylogenetically. It also appears closest to members of the type I α subdivision, but may contain an insertion typical of the β subdivision found in Proteobacteria and other Bacteria.

4.2. Glutamate synthase

Bacterial glutamate synthases are composed of a large chain and a small chain. Archaea typically encode the domains corresponding to the large chain as three separate subunits. The first subunit contains an amidotransferase domain, the second subunit an iron-sulfur cluster and a flavin-binding domain, and the third subunit a domain of unknown function. These three domains are contiguous in the genomes of both *M. maripaludis* and *M. acetivorans*. A gene corresponding to the bacterial small chain is inconsistently present in archaeal genomes.

4.3. Glutamate dehydrogenase

Genes homologous to glutamate dehydrogenases are variably present in Archaea. A glutamate dehydrogenase gene is present in *M. acetivorans* but no homolog is found in *M. maripaludis*.

4.4. Alanine dehydrogenase

M. maripaludis possesses the unusual ability to use alanine as a nitrogen source. This ability may be conferred by the presence of an alanine dehydrogenase, which would produce ammonia from alanine. The gene encoding alanine dehydrogenase in *M. maripaludis* has a highest BLAST hit to a gene in Gram-positive bacteria, and may have been acquired by horizontal gene transfer. The alanine dehydrogenase gene is adjacent to genes for alanine racemase and alanine transferase, also apparently acquired from bacteria.

5. P_{II} PROTEINS

P_{II} proteins are nitrogen transducers that are encoded by *glnB* and *glnK* in bacteria (Arcondeguy, *et al.*, 2001). The P_{II} protein of *E. coli*, encoded by *glnB*, has been extensively characterized. *E. coli* P_{II} is covalently modified by uridylylation depending on the level of glutamine, and apparently serves as an indicator of nitrogen sufficiency. *E. coli* P_{II} also binds 2-oxoglutarate, apparently as an indicator of nitrogen deficiency.

Widespread in bacteria, P_{II} proteins are also common in some of the Archaea. Archaea containing P_{II} proteins include both the methanogens (whether diazotrophic or not) and some non-methanogenic Euryarchaeota. Other Euryarchaeota lack P_{II} proteins as do the Crenarchaeota. Work in the diazotrophic methanogens led to the realization that there are three subfamilies of P_{II} proteins, distinguished from one another in amino-acid alignments (Figure 2). One subfamily contains nearly all P_{II} proteins (GlnB and GlnK) of bacteria and many of Archaea, whereas the second and third subfamilies contain the $NifI_1$ and $NifI_2$ proteins that are encoded in the *nif*-gene clusters of the diazotrophic methanogens and regulate nitrogenase acitvity (see above).

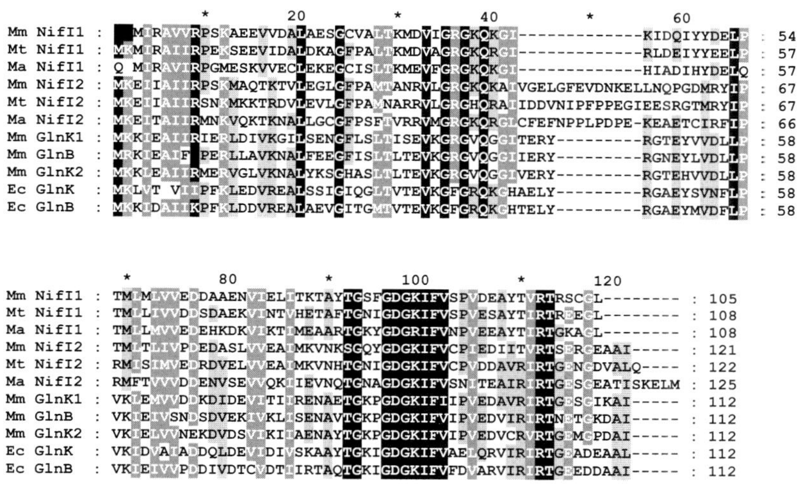

Figure 2. Alignment of $NifI_1$, $NifI_2$, and GlnB-GlnK subfamilies of P_{II} proteins. The T-loop is designated, with an arrow indicating the tyrosine residue that is uridylylated in enteric bacteria. Organism designations: Mm, Methanococcus maripaludis; Mt, Methanobacterium thermoautotrophicum; Ma, Methanosarcina acetovorans; Ec, Escherichia coli.

The three subfamilies of P_{II} proteins have different T-loops (Figure 2), which are domains that protrude from the rest of the protein and are thought, in the GlnB-K subfamily, to mediate interactions with other proteins. The T-loop is also the site of uridylylation in the GlnB-K subfamily, which usually occurs on a conserved tyrosine residue. *M. maripaludis* contains three P_{II} homologs of the GlnB-K subfamily in addition to $NifI_1$ and $NifI_2$. As is often the case in bacteria, at least some of these genes (*glnK* genes) appear to be in operons with *amtB* genes encoding ammonia transporters.

REFERENCES

Arcondeguy, T., Jack, R., and Merrick, M. (2001). PII signal transduction proteins, pivotal players in microbial nitrogen control. *Microbiol. Mol. Biol. Rev., 65*, 80-105.

Belay, N., Sparling, R., and Daniels, L. (1984). Dinitrogen fixation by a thermophilic methanogenic bacterium. *Nature, 312*, 286-288.

Boone, D. R., and Castenholz, R. W. (Eds.). (2001). *The Archaea and the deeply branching phototrophic bacteria* (Second Edition, Vol. 1). New York: Springer-Verlag.

Brown, J. R., Masuchi, Y., Robb, F. T., and Doolittle, W. F. (1994). Evolutionary relationships of bacterial and archaeal glutamine synthetase genes. *J. Mol. Evol., 38*, 566-576.

Chien, Y. T., Auerbuch, V., Brabban, A. D., and Zinder, S. H. (2000). Analysis of genes encoding an alternative nitrogenase in the archaeon *Methanosarcina barkeri* 227. *J. Bacteriol., 182*, 3247-3253.

Chien, Y. T., and Zinder, S. H. (1996). Cloning, functional organization, transcript studies, and phylogenetic analysis of the complete nitrogenase structural genes (*nifHDK*2) and associated genes in the archaeon *Methanosarcina barkeri* 227. *J. Bacteriol., 178*, 143-148.

Deppenmeier, U., Johann, A., Hartsch, T., Merkl, R., Schmitz, R. A., Martinez-Arias, R., *et al.* (2002). The genome of *Methanosarcina mazei*: Evidence for lateral gene transfer between bacteria and archaea. *J. Mol. Microbiol. Biotechnol., 4*, 453-461.

Galagan, J. E., Nusbaum, C., Roy, A., Endrizzi, M. G., Macdonald, P., FitzHugh, W., *et al.* (2002). The genome of *M. acetivorans* reveals extensive metabolic and physiological diversity. *Genome Res, 12*, 532-542.

Howell, D. M., Harich, K., Xu, H., and White, R. H. (1998). Alpha-keto acid chain elongation reactions involved in the biosynthesis of coenzyme B (7-mercaptoheptanoyl threonine phosphate) in methanogenic Archaea. *Biochemistry, 37*, 10108-10117.

Kessler, P. S., Blank, C., and Leigh, J. A. (1998). The nif gene operon of the methanogenic archaeon *Methanococcus maripaludis*. *J. Bacteriol., 180*, 1504-1511.

Kessler, P. S., Daniel, C., and Leigh, J. A. (2001). Ammonia switch-off of nitrogen fixation in the methanogenic archaeon *Methanococcus maripaludis*: Mechanistic features and requirement for the novel GlnB homologues, $NifI_1$ and $NIfI_2$. *J. Bacteriol., 183*, 882-889.

Kessler, P. S., and Leigh, J. A. (1999). Genetics of nitrogen regulation in *Methanococcus maripaludis*. *Genetics, 152*, 1343-1351.

Kessler, P. S., McLarnan, J., and Leigh, J. A. (1997). Nitrogenase phylogeny and the molybdenum dependence of nitrogen fixation in *Methanococcus maripaludis*. *J. Bacteriol., 179*, 541-543.

Leigh, J. A. (2000). Nitrogen fixation in methanogens--the archaeal perspective. In E. Triplett (Ed.), *Prokaryotic nitrogen fixation: A model system for analysis of a biological process*. Wymondham, UK: Horizon Scientific Press.

Lobo, A. L., and Zinder, S. H. (1988). Diazotrophy and nitrogenase activity in the archaebacterium *Methanosarcina barkeri* 227. *Appl. Environ. Microbiol., 54*, 1656-1661.

Murray, P. A., and Zinder, S. H. (1984). Nitrogen fixation by a methanogenic archaebacterium. *Nature, 312*, 284-286.

Scherer, P. (1989). Vanadium and molybdenum requirement for the fixation of molecular nitrogen by two *Methanosarcina* strains. *Arch. Microbiol., 151*, 44-48.

Smith, D. R., Doucette-Stamm, L. A., Deloughery, C., Lee, H., Dubois, J., Aldredge, T., *et al.* (1997). Complete genome sequence of *Methanobacterium thermoautotrophicum* deltaH: Functional analysis and comparative genomics. *J. Bacteriol., 179*, 7135-7155.

CHAPTER 3

GENOMIC ASPECTS OF NITROGEN FIXATION IN THE CLOSTRIDIA

J.-S. CHEN

Department of Biochemistry, Virginia Polytechnic Institute and State University Blacksburg, Virginia 24061, USA

1. INTRODUCTION

The genus *Clostridium* is a diverse collection of rod-shaped, spore-forming, obligately anaerobic bacteria, which do not carry out dissimilatory sulfate reduction and usually stain gram positive in young cultures (Collins *et al.*, 1994). At present, more than 160 named species are included in the traditional genus *Clostridium* (according to the database maintained at the National Center for Biotechnology Information or NCBI, USA). However, it has been proposed that, on the basis of phenotypic criteria and the results of phylogenetic analyses, the diverse species in the traditional genus *Clostridium* be rearranged into different genera (Collins *et al.*, 1994). In this proposed rearrangement, the genus *Clostridium* is reserved for species that belong to the rRNA group I of Johnson and Francis (1975). The redefined genus *Clostridium*, based on *Clostridium butyricum*, retains the better known nitrogen-fixing species of the traditional genus *Clostridium*.

The genus *Paenibacillus* contains nitrogen-fixing species that form spores under anaerobic conditions (Rosado *et al.*, 1997). The former *Clostridium durum* or *Paenibacillus durum*, the dominant organism found in a sediment core from the Black Sea, has been reclassified as a member of the species *Paenibacillus azotofixans* (Rosado *et al.*, 1997). Criteria used in the reclassification include the DNA relatedness at the genome level as measured by the DNA-DNA reassociation technique. Thus, some of the spore-forming, nitrogen-fixing rods do not belong to the genus *Clostridium*.

The genus *Clostridium* has been closely associated with the advancement of our knowledge about biological nitrogen fixation. Modern biochemical studies on nitrogen fixation became possible when consistently active cell-free extracts were

R. Palacios and W. E. Newton (eds.), Genomes and Genomics of Nitrogen-fixing Organisms, 13-26.

prepared from *Clostridium pasteurianum* (Carnahan *et al.*, 1960; Burris, 1988), which was the first free-living nitrogen-fixing organism isolated in pure culture (Winogradsky, 1895; McCoy *et al.*, 1928). Because *C. pasteurianum* is an anaerobe, the study of cell-free nitrogen fixation with this organism was logically conducted under O_2-free conditions to circumvent possible inactivation by O_2 of cellular components that support *in vitro* nitrogen-fixing activity. The use of an obligate anaerobe in early nitrogen-fixation research and the perceived need to conduct *in vitro* studies under O_2-free conditions were momentous circumstances for this field because it turned out that, regardless of the source organism, nitrogenase itself is extremely sensitive to O_2.

The purification of the component proteins of nitrogenase from *C. pasteurianum* was followed by the determination of the amino-acid sequences of the component proteins (Tanaka *et al.*, 1977; Hase *et al.*, 1981; 1984). That the structural genes *nifH*, *nifD*, and *nifK* encode the three polypeptides of nitrogenase is definitive because the amino-acid sequences of these polypeptides have been determined with the purified and active nitrogenase component proteins. Therefore, the early biochemical studies conducted on the nitrogenase of *C. pasteurianum* provided an important base for the ensuing genetic studies with nitrogen-fixing organisms, whereas the later genetic studies provided the essential information for the presumptive identification of other nitrogen-fixation and related genes in the genome of clostridia.

2. THE NITROGEN-FIXING CLOSTRIDIA

Winogradsky in 1895 reported the properties of the first free-living nitrogen-fixing organism in isolation and named it *Clostridium pasteurianum*. The culture of *C. pasteurianum* from Winogradsky was subjected to single-cell isolation in the laboratory of McCoy at the University of Wisconsin; however, this further purification did not alter the nitrogen-fixing activity or efficiency of the culture (McCoy *et al.*, 1928). *C. pasteurianum* strain W5 (= ATCC 6013), which was from the laboratory of McCoy, represents the organism isolated by Winogradsky. Among the nitrogen-fixing clostridia, *C. pasteurianum* is the most thoroughly characterized in terms of the nitrogen-fixing system.

In 1949, Rosenblum and Wilson reported the fixation of $^{15}N_2$ by several species of *Clostridium*, including *C. acetobutylicum*, *C. beijerinckii*, and *C. pasteurianum*. *Clostridium madisoni* also displayed nitrogen-fixing activity, but this organism is now recognized as a member of *C. beijerinckii* (Keis *et al.*, 2001). Besides these mesophilic and saccharolytic species, the nitrogen-fixing clostridia include other species that grow under more extreme conditions: the thermophilic *C. thermosaccharolyticum* or *Thermoanaerobacterium thermosaccharolyticum* (Bogdahn and Kleiner, 1986; Collins *et al.*, 1994), the cellulytic *C. hungatei* (Monserrate *et al.*, 2001), and the acid-tolerant *C. akagii* and *C. acidisoli* (Kuhner *et al.*, 2000).

Among the nitrogen-fixing clostridia, the nucleotide sequences of the *nif* genes encoding the nitrogenase component proteins and those genes required for the synthesis of the iron-molybdenum cofactor (FeMo-cofactor) have been determined, either completely or partially, in *C. acetobutylicum*, *C. beijerinckii*, *C. pasteurianum* and *C. hungatei*. In addition, the complete genome sequence of *C. acetobutylicum* has been determined, which allows a glimpse of the genomic distribution of *nif* and related genes in a phylogenetically ancient obligate anaerobe.

3. THE GENOME OF THE CLOSTRIDIA

Several species of *Clostridium* have been subjected to genome sequence analysis. They include the nitrogen-fixing species *Clostridium acetobutylicum* ATCC 824 (Nölling *et al.*, 2001). Although the genome of the nitrogen-fixing *C. beijerinckii* and *C. pasteurianum* has not been sequenced, the nucleotide sequence of the *nif* cluster of these two species has been determined. The flanking regions of the *nif* cluster from these species may contain information about the origin and propagation of the *nif* cluster among these species. Besides the genes present in the *nif* cluster, other genes must be required for the regulation of the expression of the *nif* genes, for the biosynthesis of the iron-sulfur clusters of the nitrogenase component proteins, and for the assimilation of ammonia produced by nitrogenase. At present, little is known about the latter genes, and a comparison of the genomes of both nitrogen-fixing and non-nitrogen-fixing clostridia may facilitate the identification of the ancillary genes for nitrogen fixation in the clostridia.

The complete genome sequences of two non-nitrogen-fixing clostridia, *C. perfringens* and *C. tetani*, are now available, and several others are nearly complete. These human pathogens are proteolytic or amino acid-fermenting organisms, and their use of nitrogenous compounds for carbon and energy metabolism precludes a need for these organisms to obtain nitrogen via nitrogen fixation. It is thus useful to examine their genomes for the possible presence of any remnants of *nif* genes, if *nif* genes had been present in these species or in their common ancestors before these species became either pathogens or proteolytic.

3.1. *The genome of* Clostridium acetobutylicum

The genome of *C. acetobutylicum* consists of a chromosome of about 4 Mb and a megaplasmid of about 200 kb (Cornillot *et al.*, 1997). The genome sequence of *C. acetobutylicum* ATCC 824 has been determined (Nölling *et al.*, 2001). The chromosome is 3,940,880 bp in length (GenBank accession number AE001437), with a total of 3,740 polypeptide-encoding ORFs and 107 stable RNA genes having been identified and accounting for 88% of the chromosomal DNA. The average length of the intergenic regions is about 121 bp. The megaplasmid, pSOL1, is 192,000 bp in length (GenBank accession number AE001438) and appears to encode 178 polypeptides. There appear to be two unrelated cryptic prophages in the chromosome. The first spans about 90 kb and includes about 85 genes (CAC1113 to CAC1197; genes and open-reading frames are assigned consecutive CAC numbers), whereas the second spans about 60 kb and includes about 79 genes (CAC1878 to CAC1957). Genes for three distinct insertion sequence-related proteins are present on the chromosome, but only one of these is intact. It is believed that no active insertion-sequence elements are present in the *C.*

acetobutylicum genome. The *nif* genes of *C. acetobutylicum* ATCC 824 are clustered between CAC0253 (n*ifH*) and CAC0261 (*nifV*α), which are near the *dnaA* gene (CAC0001) that encodes the DNA replication initiator protein.

3.2. The genome of Clostridium beijerinckii

The presence of nitrogen-fixing activity in *C. beijerinckii* was first reported by Rosenblum and Wilson (1949). More recently, nitrogen-fixing activity was demonstrated in *C. beijerinckii* strains NRRL B592 and NRRL B593, and the sequence of the *nif* cluster of the two strains has been determined (Chen *et al.*, 2001; J. Toth, M. Kasap, and J.-S. Chen, unpublished data). The genome of these *C. beijerinckii* strains has not been sequenced. However, the genome of *C. beijerinckii* strain NCIMB 8052, whose nitrogen-fixing capacity is unknown, has been studied, and its circular chromosome has an estimated size of 6.7 Mb, which is over 50% longer than the genome of *C. acetobutylicum* (Wilkinson and Young, 1995).

3.3. The genomes of non-nitrogen-fixing clostridia

The genomes of the non-nitrogen-fixing *Clostridium perfringens* strain 13 (3.03 Mb chromosome, 54.3 Kb plasmid; Shimizu *et al.*, 2002) and *Clostridium tetani* (2.8 Mb chromosome, 74 Kb plasmid; Brüggemann *et al.*, 2003) have been sequenced, and the genomes of several other non-nitrogen-fixing clostridia, including *C. botulinum* and *C. difficile*, are being sequenced. *C. perfringens*, *C. tetani*, and the nitrogen-fixing clostridia (*C. acetobutylicum*, *C. beijerinckii*, and *C. pasteurianum*) belong to Cluster I - the redefined *Clostridium* group on the basis of 16S rRNA sequence as proposed by Collins *et al.* (1994). *C. perfringens* is phylogenetically closer to *C. beijerinckii* than to either *C. acetobutylicum* or *C. pasteurianum*. *C. tetani*, although a member of Cluster I, is phylogenetically distant to *C. perfringens* and the diazotrophic *Clostridium* species.

No *nif*-specific genes have been identified by BLAST searches of the genomes of both *C. perfringens* and *C. tetani*; however, open-reading frames related to the non-*nif* genes, *nirJ-1* and *nirJ-2*, which are present in the *nif* cluster of *C. beijerinckii*, can be found in the genomes of *C. perfringens* and *C. tetani* (Toth and Chen, unpublished results). The genomes of these non-nitrogen-fixing clostridia may be analyzed for structure, function, and regulation of genes involved: (i) in ammonia assimilation; (ii) in the electron-transport pathways toward the reduction of the low-potential electron carriers (ferredoxin and flavodoxin); and (iii) in the assimilation of iron, sulfur, and molybdenum, which are relevant to the nitrogen-fixation process. These aspects may be compared between the nitrogen-fixing and non-nitrogen-fixing *Clostridium* species to reveal any evolutionary events that may have altered the organization and effectiveness of the *nif* clusters and thereby suggesting approaches that may enhance the nitrogen-fixation activity of either the clostridia or other diazotrophs.

4. ORGANIZATION OF THE NITROGEN-FIXATION GENE CLUSTER

The complete or partial amino acid sequence of the *C. pasteurianum* iron protein (Tanaka *et al.*, 1977) and the molybdenum-iron protein α subunit (Hase *et al.*, 1981) and β subunit (Hase *et al.*, 1984) were determined with the purified protein. Therefore, the assignment of the nitrogenase structural genes, *nifHDK*, in *C. pasteurianum* is definitive, although there is a lack of data from genetic tests. The other *nif* genes, which occur downstream to the *nifHDK* genes, are identified by the conserved amino-acid sequences they encode and by their positional relationships to the *nifHDK* genes.

A cluster of *nif* genes has been identified in three species of clostridia: *C. acetobutylicum*, *C. beijerinckii*, and *C. pasteurianum* (Figure 1). The presumed boundaries for the *nif* cluster are defined by long intergenic regions and the presence of flanking genes that are unrelated to nitrogen fixation or nitrogen metabolism. The proposed gene products or their functions of the *nif*-cluster genes are listed in Table 1.

C. acetobutylicum ATCC824

(553) *nifH*▶ (143) $nifI_1$▶ $nifI_2$▶ *nifD*▶ *nifK*▶ (96) *nifE*▶ *nifN-B*▶ *nifVω*▶ *nifVα*▶

C. beijerinckii NRRL B593

(459) *nifH*▶ (261) $nifI_1$▶ $nifI_2$▶ *nifD*▶ *nifK*▶ (1,123) *nifE*▶ *nifN-B*▶ (328) *fdxA*▶ (370) *nirJ1*▶ *nirJ2*▶ *nirD*▶ *nirH*▶ (236) *nifVω*▶ *nifVα*▶

C. pasteurianum W5

(501) *nifH2*▶ (409) *nifH1*▶ *nifD*▶ *nifK*▶ (284) *nifE*▶ *nifN-B*▶ *modA*▶ (349) *modB*▶ *nifVω*▶ *nifVα*▶

Figure 1. Organization of the nif *cluster of* C. acetobutylicum, C. beijerinckii, and C. pasteurianum.

Each nif *cluster is composed of* nif-*specific genes as well as putative genes that may play a role in nitrogen regulation or nitrogenase synthesis. The direction of transcription is indicated by arrowheads. The length of an intergenic region longer than 50 bp, which may signify a regulatory region, is indicated by bp number in parenthesis.*

4.1. Clostridium pasteurianum

The organization of the *nif* genes in *C. pasteurianum* has been reviewed (Chen and Johnson, 1993; Chen *et al.*, 2001). The major *nif* cluster (Figure 1) spans a region of 13.4 kb (between *nifH2* and *nifVα*), and it is 12.2 kb between *nifH1* and *nifVα*. The orientation of all genes in the cluster is from *nifH2* toward *nifVα*.

Table 1. The proposed gene products or their function of the nif *cluster genes of* C. acetobutylicum, C. beijerinckii, *and* C. pasteurianum

C. acetobutylicum	*C. beijerinckii*	*C. pasteurianum*	Proposed gene product or function	Reference
		nifH2	Homologue of nitrogenase Fe protein	Chen et al., 1986
nifH	*nifH*	*nifH1*	Nitrogenase Fe protein; synthesis of FeMo-cofactor	Chen et al., 1986, 2001
nifI1	*nifI1*		Regulation of nitrogenase activity (switch-off)	Chen et al., 2001
nifI2	*nifI2*		Regulation of nitrogenase activity (switch-off)	Chen et al., 2001
nifD	*nifD*	*nifD*	Nitrogenase MoFe protein, α subunit	Wang et al., 1988b; Chen et al., 2001
nifK	*nifK*	*nifK*	Nitrogenase MoFe protein, β subunit	Wang et al., 1988b; Chen et al., 2001
nifE	*nifE*	*nifE*	Synthesis of FeMo-cofactor	Wang et al., 1988b
nifN-B	*nifN-B*	*nifN-B*	Synthesis of FeMo-cofactor	Chen and Johnson, 1993
		modA	Molybdate transport	Chen et al., 2001
		modB	Molybdate transport	Wang et al., 1990; Chen et al., 2001
	fdxA		2Fe-2S ferredoxin	Toth and Chen, unpublished
	nirJ1		Synthesis of heme d_1 or coenzyme PQQ*	Toth and Chen, unpublished
	nirJ2		Synthesis of heme d_1 or coenzyme PQQ	Toth and Chen, unpublished
	nirD		Synthesis of heme d_1	Toth and Chen, unpublished
	nirH		Synthesis of heme d_1	Toth and Chen, unpublished
nifVω	*nifVω*	*nifVω*	Homocitrate synthase; synthesis of FeMo-cofactor	Wang et al., 1991; Toth and Chen, unpublished
nifVα	*nifVα*	*nifVα*	Homocitrate synthase; synthesis of FeMo-cofactor	Wang et al., 1991; Toth and Chen, unpublished

** PQQ: pyrroloquinoline quinone*

The *nifH1* gene encodes the purified nitrogenase iron protein (273 amino acids) that has been sequenced (Chen *et al.*, 1986), whereas the *nifH2* gene encodes a polypeptide of 272 amino acids, which differs from the *nifH1* product in 23 amino acids (8%). The novel features of the *nifD*- and *nifK*-encoded polypeptides include the presence in the *C. pasteurianum* NifD of an extra stretch of about 50 amino acids (in comparison to the NifD polypeptide of non-clostridial species) and the shortened amino terminal region of about 50 amino acids in the *C. pasteurianum* NifK polypeptide (Wang *et al.*, 1988b).

The fused *nifN-B* and the split *nifVω* and *nifVα* genes are the other novel features that were first discovered in *C. pasteurianum* (Chen and Johnson, 1993). The presence of the *modA* and *modB* genes within the *nif* cluster is rare. Similar *mod* genes are present in other nitrogen-fixing organisms (Lee *et al.*, 2000), but they are located either elsewhere in the genome or at the boundary of the *nif* cluster, such as in *Acetobacter diazotrophicus* (Lee *et al.*, 2000). The proposed function for the putative *modA* and *modB* genes is molybdenum transport.

In addition to *nifH1*, *nifH2*, and *nifH3* (*anfH*; see Section 6), *C. pasteurianum* has three unlinked *nifH*-like genes, *nifH4* through *nifH6* (Wang *et al.*, 1988a). The lengths of the polypeptides encoded by *nifH1* through *nifH6* vary between 272 and 275 amino acids. Except for NifH3, which is about 65% identical to NifH1 at the amino-acid level, the other five NifH polypeptides are between 91.6 and 99.6% identical among themselves.

The postulated NifH2 and NifH6 polypeptides differ in only one amino acid (Wang *et al.*, 1988a). A recent study detected the presence, in nitrogen-fixing cells of *C. pasteurianum*, of a polypeptide that is the product of either the *nifH2* or the *nifH6* gene (Kasap, 2002). The presence of this polypeptide in nitrogen-fixing cells, but not in ammonia-grown cells, suggests a role related to nitrogen fixation for this protein, however, its actual function remains to be determined.

4.2. Clostridium acetobutylicum

The *nif* cluster of *C. acetobutylicum* ATCC 824 spans 10.7 kb (Nölling *et al.*, 2001; Chen *et al.*, 2001) and is the shortest *nif* cluster that has been described to date (Figure 1). The orientation of all genes in the cluster is from *nifH* toward *nifVα*. The assigned gene number is as follows: *nifH*, CAC0253; $nifI_1$, CAC0254; $nifI_2$, CAC255; *nifD*, CAC256, *nifK*, CAC0257; *nifE*, CAC0258; *nifN-B*, CAC0259; *nifVω*, CAC0260; *nifVα*, CAC0261.

The $nifI_1$ and $nifI_2$ genes correspond to the homologous genes that are present at the same location in the *nif* cluster of several methanogens, and these genes were previously designated as the $glnB_1$ and $glnB_2$ genes, respectively, for their sequence relatedness to the nitrogen-regulatory gene *glnB*, which encodes the P_{II} protein (Arcondéguy *et al.*, 2001). The $nifI_1$ and $nifI_2$ genes of *Methanococcus maripaludis* are required for the ammonia-induced switch-off of nitrogen fixation (Kessler *et al.*, 2001).

In *C. acetobutylicum* ATCC 824, *nifS*-like genes are found outside the *nif* cluster as CAC2234 and CAC2972 (Toth and Chen, unpublished results). Open-reading frames similar to *nifS* are also found in the genomes of *C. perfringens* and *C. tetani*, although the genomes of these two clostridia do not contain any other *nif* genes (Toth and Chen, unpublished results). These observations suggest that the *nifS* genes found in these *Clostridium* species are not *nif*-specific.

4.3. Clostridium beijerinckii

The *nif* cluster of *C. beijerinckii* NRRL B593 (Figure 1) resembles those of *C. acetobutylicum* and *C. pasteurianum*, both in the structure and order of the *nif* genes and in the location of the non-*nif* genes in the cluster. However, the identity of these non-*nif* genes is species-specific for these three clostridial species. The direction of transcription for the *nif*-cluster genes is from *nifH* toward *nifV*α.

The deduced amino-acid sequences of the seven core *nif* genes (*nifH, D, K, E, N-B,* V_ω, and V_α) are highly conserved among *C. acetobutylicum*, *C. beijerinckii*, and *C. pasteurianum*, whereas the deduced amino-acid sequences of the two *nifI* genes are conserved in *C. acetobutylicum* and *C. beijerinckii*. The five non-*nif* genes, from *fdxA* to *nirH*, are situated between *nifN-B* and *nifV*ω, the same location where the *modA* and *modB* genes are situated in *C. pasteurianum*. The *C. acetobutylicum* genome does not have a gene similar to *fdxA* of *C. beijerinckii*. On the other hand, the [2Fe-2S] ferredoxin of *C. pasteurianum* (Meyer, 1993) is highly related to the *fdxA*-encoded sequence, although the gene encoding the *C. pasteurianum* [2Fe-2S] ferredoxin is not part of the *nif* cluster. In *C. pasteurianum*, the level of the [2Fe-2S] ferredoxin is greatly increased in nitrogen-fixing cells compared to ammonia-grown cells, but there is no evidence that the [2Fe-2S] ferredoxin is involved in nitrogen fixation.

The gene cluster, CAC2796 through CAC2793 of *C. acetobutylicum*, corresponds to *nirJ1*, *nirJ2*, *nirD*, and *nirH*, respectively, of *C. beijerinckii*. The *nir* cluster of *C. acetobutylicum* is about 1.3 Mb away (across the proposed origin of replication) from the *nif* cluster (Toth and Chen, unpublished results). The *nirJ, D, H* genes are postulated to play a role in the synthesis of either heme d_1 or coenzyme PQQ (pyrroloquinoline quinone), but it is not known if either heme d_1 or coenzyme PQQ is present in the clostridial cell.

Another distinctive feature of the *nif* cluster of *C. beijerinckii* NRRL B593 is the presence of an unusually long intergenic region (1,123 bp) between the *nifK* and the *nifE* genes (Toth and Chen, unpublished results). In *Escherichia coli* K-12, the average distance of the intergenic region is 118 bp, and the longest intergenic region is 1,730 bp (Blattner *et al.*, 1997). There are only 55 intergenic regions in *E. coli* that are over 600 bp in length. A long intergenic region is believed to contain an independent regulatory sequence for the succeeding gene(s). Except for those with the base-pair number indicated in parentheses in Figure 1, all other intergenic regions for the *nif* genes of the three *Clostridium* species are either shorter than 50 bp or have an overlap between the two contiguous genes. These longer intergenic regions are presumed promoter regions, and results of transcriptional analyses of the *C. pasteurianum nif* genes substantiate this prediction (Wang *et al.*, 1988a). The unusually long intergenic region between the *nifK* and *nifE* genes of *C. beijerinckii* perhaps indicates a splicing event(s) accompanied by deletions and base changes.

4.4. Characteristics of the nif *clusters of the clostridia*

The *nif* cluster of *C. acetobutylicum* and *C. pasteurianum* is each composed of nine genes, whereas the *nif* cluster of *C. beijerinckii* is composed of 14 genes. Seven of these genes are common for the three species; these are the *nif*-specific genes arranged in the order of $H < D < K < E < N\text{-}B < V\omega < V\alpha <$. These seven *nif* genes are conceivably the minimum required for nitrogen fixation. The fused *nifN-B* gene and the split $nifV_{\omega}$ and $nifV_{\alpha}$ genes are characteristics of these three *Clostridium* species. The intergenic region preceding the *nif* cluster of these three *Clostridium* species is between 459 and 553 bp. The length of this intergenic region seems to mark a clear upstream boundary for the *nif* cluster.

The genes in the *nif* cluster of *C. acetobutylicum*, *C. beijerinckii*, and *C. pasteurianum* have two other general features. One is an overlap between the *nifD* and *nifK* genes (Wang *et al.*, 1988b; Toth and Chen, unpublished results). The other is a biased codon usage pattern (Chen and Johnson, 1993). The third position of the codons for each amino acid is predominantly or exclusively A and U for the *nifHDK* genes, whereas the other *nif* genes and the *nir* genes use codons with G or C at the third position more frequently than the *nifHDK* genes. The very biased codon usage pattern for the *nifHDK* genes probably reflects the high level of synthesis of nitrogenase component proteins but may also indicate a different origin for these genes.

The *nif* cluster of *C. acetobutylicum*, *C. beijerinckii*, and *C. pasteurianum* can be differentiated by the non-*nif* genes that are present in each cluster (Figure 1). The *nif* cluster of *C. acetobutylicum* does not contain any non-*nif* gene and is thus the simplest. The *nif* cluster of *C. pasteurianum* does not have the *nifI* genes but has the *modA* and *modB* genes, instead. The *nif* cluster of *C. beijerinckii* has the *nifI* genes and, instead of the *mod* genes, it has five other non-*nif* genes (*fdxA* and *nirJ1, J2, D,* and *H*) occupying the location of the *mod* genes in *C. pasteurianum*.

Although the concise *C. acetobutylicum nif* cluster lacks the *mod* genes and the *nir* genes, these genes are present in the *C. acetobutylicum* genome. They are CAC0281 (*modA*), CAC0280 (*modB*), CAC2796 (*nirJ1*), CAC2795 (*nirJ2*), CAC2794 (*nirD*), and CAC2793 (*nirH*). A gene similar to the *fdxA* gene of *C. beijerinckii* is not found in *C. acetobutylicum*; however, *C. pasteurianum* has a homologous gene that encodes a [2Fe-2S] ferredoxin that is synthesized in nitrogen-fixing cells but not in ammonia-grown cells.

The organization of the *nif* and non-*nif* genes in the *nif* clusters of *C. acetobutylicum*, *C. beijerinckii*, and *C. pasteurianum* suggests that the seven core *nif* genes have existed in these species as a cluster in the present order prior to the insertion of other genes into the cluster. The insertion of the *mod* genes would result in the *nif* cluster of *C. pasteurianum*, whereas the insertion of the *nifI* genes would result in the *nif* cluster of *C. acetobutylicum*. The insertion of the five genes (*fdxA* to *nirH*) into the *nif* cluster of *C. acetobutylicum* would then result in the *nif* cluster of *C. beijerinckii*.

5. REGULATORY GENES FOR NITROGEN METABOLISM

Because of the sensitivity of the nitrogenase component proteins to molecular oxygen and the high energy requirement for the nitrogen-fixation process, nitrogen-fixation activity is regulated in response to both the redox and nitrogen status of the cell (Arcondéguy *et al.*, 2001). For the clostridia, the regulatory genes for nitrogen fixation are yet to be identified. However, the discovery of the *nifH*-linked *nifI* genes in some clostridia should accelerate the understanding of this important aspect of nitrogen fixation in this group of diazotrophs.

In the proteobacteria, the expression and activity of the transcriptional regulator (NifA) for the *nif* genes are generally modulated by proteins that respond to the nitrogen status (NtrB, NtrC, and the P_{II} protein) and the O_2 level (NifL). In addition to its role in the transcriptional control of nitrogen fixation, the P_{II} protein may also regulate the nitrogenase activity at a posttranslational level in response to ammonia and other fixed nitrogen sources. The P_{II} protein (named for its chromatographic behavior) is encoded by the *glnB* gene and represents a family of conserved signal-transduction proteins that play a significant role in the coordination of nitrogen metabolism in a wide variety of bacteria.

Because the clostridia are obligately anaerobic bacteria and are evolutionarily ancient, one may not expect to find a *nifL*-like gene in the clostridia, especially if the *nif* genes or their ancestral forms have existed in the clostridia before the earth's atmosphere became aerobic. Indeed, a *nifL*-like gene has not been reported in the clostridia. In addition, a gene similar to *nifA* is not present in the *nif* cluster of *C. acetobutylicum*, *C. beijerinckii*, or *C. pasteurianum* or elsewhere in the genome of *C. acetobutylicum*. The presumed promoter regions for the *nif* genes of *C. pasteurianum* do not have the motif of the *nifA*-regulated promoters (Wang *et al.*, 1988a). Therefore, the *nif*-specific transcriptional regulators of the clostridia may differ from the well-characterized proteins of the proteobacteria.

Interestingly, two *glnB-like* genes are present between the *nifH* and *nifD* genes in both *C. acetobutylicum* and *C. beijerinckii*. These two *glnB*-like genes are referred to as the $nifI_1$ and $nifI_2$ genes here because the gene designation has been proposed for similar genes found in *Methanococcus maripaludis* (Kessler *et al.*, 2001). The $nifI_1$ and $nifI_2$ genes are required for ammonia switch-off of nitrogen fixation in *M. maripaludis*; this switch-off is reversible and does not seem to involve ADP-ribosylation or any other covalent modification of the iron-protein component of nitrogenase. Also, switch-off does not affect *nif*-gene transcription, *nifH* mRNA stability, or the stability of the iron protein (Kessler *et al.*, 2001).

A search of the genome of two non-diazotrophic clostridia, *C. perfringens* and *C. tetani*, did not reveal any open-reading frames related to the *nifI* gene, which further suggests a *nif*-specific role for the *nifI* genes of *C. acetobutylicum* and *C. beijerinckii* (Toth and Chen, unpublished results). The *nif* cluster of *C. pasteurianum* does not have any *glnB*-like genes and the *in vivo* nitrogenase activity of *C. pasteurianum* does not show switch-off on the addition of ammonia, although the synthesis of nitrogenase is repressed by ammonia (Daesch and Mortenson, 1972). In contrast, the *in vivo* nitrogenase activity of *C. beijeirinckii* falls rapidly following the addition of ammonia, but the *in vitro* nitrogenase activity shows a much smaller decrease. Furthermore, there is no change in the mobility of the *C. beijerinckii* iron protein on SDS-PAGE during the period that the *in vivo* nitrogenase activity exhibits a decrease (Kasap, 2002). The nature of the switch-off of nitrogenase activity by ammonia in *C. beijerinckii* is yet to be determined.

6. ALTERNATIVE NITROGEN-FIXATION (*anf*) GENES

Besides the *nif* genes that encode the molybdenum-nitrogenase system, putative *anf* genes, which encode the iron-only nitrogenase, have been identified in two nitrogen-fixing clostridia. Separate from the major *nif* cluster, *C. pasteurianum* has a putative *anf* cluster that consists of five open reading frames: *anfH*< *ORF*> *anfD*< *anfG*< *anfK*< (Zinoni *et al.*, 1993). The *C. pasteurianum anfH* gene is synonymous to the previously reported *C. pasteurianum nifH3* gene (Chen *et al.*, 1986). The ORF between *anfH* and *anfD* is in an opposite orientation relative to the *anf* genes (Wang *et al.*, 1988a; Zinoni *et al.*, 1993). The separation of the *anfH* and *anfD* genes by an ORF in *C. pasteurianum* resembles the organization of the *vnf* genes in *Azotobacter* (Bishop and Premakumar, 1992). A cluster of *anfHDGK* genes is present in the nitrogen-fixing, cellulytic *Clostridium hungatei*; however, the *anfH* and *anfD* genes of *C. hungatei* are not separated by an ORF (GenBank accession number U59415).

On the basis of 16S rRNA sequences, *C. pasteurianum* and *C. hungatei* are phylogenetically distant. *C. pasteurianum* (with a G+C content of 26-28 %) belongs to Cluster I of the clostridia (Collins *et al.*, 1994), whereas *C. hungatei* (with a G+C content of 40-42 %) clusters with species belonging to Cluster III of the clostridia (Monserrate *et al.*, 2001). Further sequence analysis of the putative *anf* genes and the flanking regions in the two clostridia may provide clues on the path of propagation of the *anf* genes.

7. GENES FOR NITROGEN ASSIMILATION

Among the clostridia, *C. pasteurianum* has been the primary organism used in biochemical studies on nitrogen fixation. The *in vivo* and *in vitro* nitrogen-fixing activities of *C. pasteurianum* have been well characterized. In comparison, the assimilation of fixed nitrogen in this organism has received much less attention, and genes for nitrogen assimilation have not been cloned from *C. pasteurianum*. However, several genes for nitrogen assimilation have been cloned from or identified in the clostridia. At present, there is a rudimentary understanding of the biochemistry and genetics of nitrogen assimilation in the clostridia.

In the genome of *C. acetobutylicum*, putative genes for glutamine synthetase (CAC2658), glutamate synthase (CAC0764, CAC1673, CAC1674, CAC2398, CAC3020), and glutamate dehydrogenase (CAC0737) have been identified (annotated genome sequence; GenBank accession number AE001437). Nucleotide sequences for the glutamate synthase genes of the non-nitrogen-fixing *C. perfringens* are also available from the GenBank.

Genes for ammonia assimilation in *Clostridium saccharobutylicum* NCP 262 (formerly *Clostridium acetobutylicum* NCP 262) have been studied, although it remains to be shown whether or not *C. saccharobutylicum* is a diazotroph. From *C. saccharobutylicum* NCP 262, the *glnA* gene for glutamine synthetase (Usdin *et al.*, 1986) and the genes for the large and small subunits of glutamate synthase (GenBank accession numbers: AAD41675 and AAD41676) have been cloned and sequenced. Regulation of glutamine-synthetase activity in *C. saccharobutylicum* NCP 262 appears to involve an antisense RNA (Fierro-Monti *et al.*, 1992). There is no evidence for a global *ntr* system, and the glutamine synthetase of *C. saccharobutylicum* is not regulated by adenylylation (Janssen *et al.*, 1988; Fierro-Monti *et al.*, 1992). The regulation of the glutamine synthetase of *C. pasteurianum* was also reported not to involve adenylylation (Kleiner, 1979). How the diazotrophic clostridia regulate nitrogen assimilation remains to be determined.

8. CONCLUDING REMARKS

Both the structure and the organization of the *nif* genes of the clostridia display distinct features when compared with those of the proteobacteria and the other diazotrophs. The most prominent structural features include: (i) the extended *nifD* gene and the shortened *nifK* gene; (ii) the fused *nifN-B* gene; (iii) the split *nifV* genes; and (iv) a biased codon usage pattern that is consistent with organisms with a low mol% G+C. These structural features and the organization of these genes in the *nif* cluster are conserved in *C. acetobutylicum*, *C. beijerinckii*, and *C. pasteurianum*. It is apparent that the *nif* cluster of these three *Clostridium* species originated from a common ancestral cluster that differed substantially from the ancestral *nif* clusters of the proteobacteria.

The *nif* clusters of *C. acetobutylicum*, *C. beijerinckii*, and *C. pasteurianum* are the most concise among known *nif* clusters and they may represent the minimal set of *nif*-specific genes that are required for diazotrophic growth in the absence of O_2. Despite the small number of genes that constitute the *nif* clusters of these three *Clostridium* species and the conserved organization of *nif* genes in these *nif* clusters, there is significant diversity in the non-*nif* genes that are present between *nifN-B* and *nifVω* genes in these three species. They range from the absence of any non-*nif* gene in this region in *C. acetobutylicum* to the presence of five non-*nif* genes in this region in *C. beijerinckii*. The presumed insertion of these non-*nif* genes into the otherwise highly conserved *nif* cluster manifests the pliability of the genome. Whether or not these non-*nif* genes influence the nitrogen-fixation process remains to be determined.

The presence of the *nifH*-linked genes, $nifI_1$ and $nifI_2$, in *C. acetobutylicum*, *C. beijerinckii*, *C. cellobioparum* as well as in other anaerobic bacteria and archaea suggests that the cluster *nifH*-$nifI_1$-$nifI_2$-*nifD*-*nifK* is ancient in origin and may resemble an ancestral nitrogen-fixation gene cluster. Future research on the function of the $nifI_1$ and $nifI_2$ genes may shed new light on the evolution of the P_{II} protein family and the presumed advantage for an anaerobic diazotroph to possess this regulatory capacity.

REFERENCES

Arcondéguy, T., Jack, R., and Merrick, M. (2001). P_{II} signal transduction proteins, pivotal players in microbial nitrogen control. *Microbiol. Mol. Biol. Rev., 65*, 80-105.

Bishop, P. E., and Premakumar, R. (1992). Alternative nitrogen fixation systems. In G. Stacey, R. H. Burris, and H. J. Evans (Eds.), *Biological nitrogen fixation* (pp. 736-763). New York: Chapman and Hall.

Blattner, F. R., Plunkett, III., G., Bloch, C. A., *et al.* (1997). The complete genome sequence of *Escherichia coli* K-12. *Science, 277*, 1453-1462.

Bogdahn, M., and Kleiner, D. (1986). N_2 fixation and NH_4^+ assimilation in the thermophilic anaerobes *Clostridium thermosaccharolyticum* and *Clostridium thermoautotrophicum*. *Arch. Microbiol., 144*, 102-104.

Brüggemann, H., Bäumer, S., Fricke, W. F., Wiezer, A., Liesegang, H., Decker, I., *et al.* (2003). The genome sequence of *Clostridium tetani*, the causative agent of tetanus disease. *Proc. Natl. Acad. Sci. USA, 100*, 1316-1321.

Burris, R. H. (1988). 100 years of discoveries in biological N_2 fixation. In H. Bothe, F. J. de Bruijn, and W. E. Newton (Eds.) *Nitrogen fixation: Hundred years after* (pp. 21-30). Stuttgart: Gustav Fischer.

Carnahan,J. E., Mortenson, L. E., Mower, H. F., and Castle, J. E. (1960). Nitrogen fixation in cell-free extracts of *Clostridium pasteurianum*. *Biochim. Biophys. Acta, 38*, 188-189.

Chen, J.-S., and Johnson, J. L. (1993). Molecular biology of nitrogen fixation in the clostridia. In D. R. Woods (Ed.), *The clostridia and biotechnology* (pp. 371-392). Boston: Butterwoth-Heinemann.

Chen, K. C.-K., Chen, J.-S., and Johnson, J. L. (1986). Structural features of multiple *nifH*-like sequences and very biased codon usage in nitrogenase genes of *Closttridium pasteurianum*. *J. Bacteriol., 166*, 162-172.

Chen, J.-S., Toth, J., and Kasap, M. (2001). Nitrogen-fixation genes and nitrogenase activity in *Clostridium acetobutylicum* and *Clostridium beijerinckii*. *J. Ind. Microbiol. Biotechnol., 27*, 281-286.

Collins, M. D., Lawson, P. A., Willems, A., Cordoba, J. J., Fernandez-Garayzabal, J., Garcia, *et al.* (1994). The phylogeny of the genus *Clostridium*: Proposal of five new genera and eleven new species combinations. *Int. J. Syst. Bacteriol., 44*, 812-826.

Cornillot, E., Croux, C., and Soucaille, P. (1997). Physical and genetic map of the *Clostridium acetobutylicum* ATCC 824 chromosome. *J. Bacteriol., 179*, 7426-7434.

Daesch, G., and Mortenson, L. E. (1972). Effect of ammonia on the synthesis and function of the N_2-fixing enzyme system in *Clostridium pasteurianum*. *J. Bacteriol., 110*, 103-109.

Fierro-Monti, I. P., Reid, S. J. and Woods, D. R. (1992). Differential expression of a *Clostridium acetobutylicum* antisense RNA: Implications for regulation of glutamine synthetase. *J. Bacteriol., 174*, 7642-7647.

Hase, T., Nakano, T., Matsubara, H., and Zumft, W. G. (1981). Correspondence of the larger subunit of the MoFe protein in clostridial nitrogenase to the *nifD* gene product of other N_2-fixing organisms. *J. Biochem. (Tokyo), 90*, 295-298.

Hase, T., Wakabayashi, S., Nakano, T., Zumft, W. G., and Matsubara, H. (1984). Structural homologies between the amino acid sequence of *Clostridium pasteurianum* MoFe protein and the DNA sequences of *nifD* and *K* genes of phylogenetically diverse bacteria. *FEBS Lett., 166*, 39-43.

Janssen, P. L., Jones, W. A., Jones, D. T., and Woods, D. R. (1988). Molecular analysis and regulation of the *glnA* gene of the gram-positive anaerobe *Clostridium acetobutylicum*. *J. Bacteriol., 170*, 400-408.

Johnson, J. L., and Francis, B. S. (1975). Taxonomy of the clostridia: Ribonucleic acid homologies among the species. *J. Gen. Microbiol., 88*, 229-244.

Kasap, M. (2002). Nitrogen metabolism and solvent production in *Clostridium beijerinckii* NRRL B593. *Ph.D. dissertation*. Virginia Polytechnic Institute and State University, Blacksburg, Virginia, U.S.A.

Keis, S., Shaheen, R., and Jones, D. T. (2001). Emended description of *Clostridium acetobutylicum*, *Clostridium beijerinckii*, and descriptions of *Clostridium saccharobutylacetonicum* sp. nov. and *Clostridium saccharobutylicum* sp. nov. *Int. J. Syst. Evol. Microbiol., 51*, 2095-2103.

Kessler, P. S., Daniel, C., and Leigh, J. A. (2001). Ammonia switch-off of nitrogen fixation in the methanogenic archeon *Methanococcus maripaludis*: Mechanistic features and requirement for the novel GlnB homologues, $NifI_1$ and $NifI_2$. *J. Bacteriol., 183*, 882-889.

Kleiner, D. (1979). Regulation of ammonium uptake and metabolism by nitrogen fixing bacteria III. *Clostridium pasteurianum*. *Arch. Microbiol., 120*, 263-270.

Kuhner, C. H., Matthies, C., Acker, G., Schmittroth, M., Gößner, A. S., and Drake, H. L. (2000). *Clostridium akagii* sp. nov. and *Clostridium acidisoli* sp. nov.: Acid-tolerant, N_2-fixing clostridia isolated from acidic forest soil and litter. *Int. J. Syst. Evol. Microbiol., 50*, 873-881.

Lee, S., Reth, A., Meletzus, D., Sevilla, M., and Kennedy, C. (2000). Characterization of a major cluster of *nif, fix,* and associated genes in a sugarcane endophyte, *Acetobacter diazotrophicus. J. Bacteriol., 182*, 7088-7091.

McCoy, E., Higby, W. M., and Fred, E. B. (1928). The assimilation of nitrogen by pure cultures of Clostridium pasteurianum and related organisms. *Zentralblatt für Bakteriologie Parasitenk II*, 76, 314-320.

Meyer, J. (1993) Cloning and sequencing of the gene encoding the [2Fe-2S] ferredoxin from *Clostridium pasteurianum. Biochim. Biophys. Acta, 1174*, 108-110.

Monserrate, E., Leschine, S. B., and Canale-Parola, E. (2001). *Clostridium hungatei* sp. nov., a mesophilic, N_2-fixing cellulytic bacterium isolated from soil. *Int. J. Syst. Evol. Microbiol., 51*, 123-132.

Nölling, J.,Breton, G., Omelchenko, V., *et al.* (2001). Genome sequence and comparative analysis of the solvent-producing bacterium *Clostridium acetobutylicum. J. Bacteriol., 183*, 4823-4838.

Rosado, A. S., van Elsas, J. D., and Seldin, L. (1997). Reclassification of *Paenibacillus durun* (formerly *Clostridium durum* Smith and Cato 1974) Collins *et al.* 1994 as a member of the species *P. azotofixans* (formerly *Bacillus azotofixans* Seldin *et al.* 1984) Ash *et al.* 1994. *Int. J. Syst. Bacteriol., 47*, 569-572.

Rosenblum, E. D., and Wilson, P. W. (1949). Fixation of isotopic nitrogen by *Clostridium. J. Bacteriol., 57*, 413-414.

Shimizu, T., Ohtani, K., Hirakawa, H., Ohshima, K., Yamashita, A., Shiba, T., *et al.* (2002). Complete genome sequence of *Clostridium perfringens*, an anaerobic flesh-eater. *Proc. Natl. Acad. Sci. USA, 99*, 996-1001.

Tanaka, M, Haniu, M., Yasunobu, T., and Mortenson, L. E. (1977). The amino acid sequence of *Clostridium pasteurinum* iron protein, a component of nitrogenase. *J. Biol. Chem., 252*, 7093-7100.

Usdin, K. P., Zappa, H., Jones, D. T., and Woods, D. R. (1986). Cloning, expression, and purification of glutamate synthetase from *Clostridium acetobutylicum. Appl. Environ. Microbiol., 52*, 413-419.

Wang, S.-Z., Chen, J.-S., and Johnson, J. L. (1988a). The presence of five *nifH*-like sequences in *Clostridium pasteurianum*: sequence divergence and transcription properties. *Nucleic Acids Res., 16*, 439-454.

Wang, S.-Z., Chen, J.-S., and Johnson, J. L. (1988b). Distinct structural features of the α and β subunits of nitrogenase molybdenum-iron protein of *Clostridium pasteurianum*: An analysis of amino acid sequences. *Biochemistry, 27*, 2800-2810.

Wilkinson, S. R., and Young, M. (1995). Physical map of the *Clostridium beijerinckii* (formerly *Clostridium acetobutylicum*) NCIMB 8052 chromosome. *J. Bacteriol., 177*, 439-448.

Winogradsky, S. (1895). Recherches sur l'assimilation de l'azote libre de l'atmosphère par les microbes. *Archives des Sciences Biologiques, 3*, 297-352.

Zinoni, F., Robson, R. M., and Robson, R. L. (1993). Organization of potential alternative nitrogenase genes from *Clostridium pasteurianum. Biochim. Biophys. Acta, 1174*, 83-86.

CHAPTER 4

THE GENOME OF THE FILAMENTOUS CYANOBACTERIUM *NOSTOC PUNCTIFORME*

What can we learn from it about free-living and symbiotic nitrogen fixation?

J. C. MEEKS

Section of Microbiology, University of California, Davis, CA 95616, USA

1. INTRODUCTION

Nostoc punctiforme is a filamentous cyanobacterium, typically found in soil habitats. It has the unusual property of fixing nitrogen in both free-living and plant-associated symbiotic growth states. The shotgun sequencing phase of its approximately 9.25 Mb genome has been completed, with about 98% of the genome provisionally assembled into 203 contiguous units (contigs), computationally annotated, and is publicly available at http:www.jgi.doe.gov. A preliminary analysis has been published (Meeks *et al.*, 2001) and the finishing of the complete sequence is in progress. *N. punctiforme* is amenable to genetic manipulations, including exogenous random transposon mutagenesis (Cohen *et al.*, 1994), complementation in trans (Summers *et al.*, 1995), and targeted gene replacement (Campbell *et al.*, 1998; Hagen and Meeks 1999), such that the physiological role of any gene product can be defined by phenotypic analyses. This chapter examines the genome of *N. punctiforme* in the context of nitrogen fixation.

Cyanobacteria are uniformly characterized by their oxygenic photoautotrophic mode of energy and carbon metabolism for growth. Simultaneous photosynthetic production of O_2 and light-dependent nitrogen fixation in an uncompartmentalized prokaryotic cell appears to be incompatible due to the O_2 sensitivity of the nitrogenase enzyme complex. The evolutionary solution to this incompatibility dilemma in certain cyanobacteria resulted in the differentiation of a specialized, nearly anoxic, nitrogen-fixing cell, called the heterocyst (see Volume 5 for details). Consequently, a spatial separation is established between the production of O_2 and reduced carbon in vegetative cells and nitrogen fixation in heterocysts. Heterocyst differentiation occurs only in filamentous cyanobacteria and is a taxonomic characteristic. The heterocyst-forming species constitute subsections IV (formerly

R. Palacios and W. E. Newton (eds.), Genomes and Genomics of Nitrogen-fixing Organisms, 27-70.

order Nostocales, filamentous genera with a single plane of vegetative cell division) and V (formerly order Stigonematales, filamentous genera with more than one plane of vegetative cell division) of the Phylum Cyanobacteria (Castenholz, 2001). However, based on 16S rRNA sequences, representatives from both subsection IV and V clearly associate as one phylogenetic cluster (Wilmotte and Herdman, 2001). Except for a very few isolates, most notably *Anabaena variabilis* ATCC 29413, which express more than one nitrogenase enzyme complex (see Volume II), synthesis and activity of the single molybdenum-nitrogenase is localized to the heterocyst in heterocyst-forming cyanobacteria. Moreover, transcription of the nitrogenase structural genes appears to be dependent on developmental signals within the heterocyst differentiation cascade and not on environmental signals, except indirectly through those signals inducing heterocyst differentiation (Elhai and Wolk, 1990). Therefore, genomic analysis of the regulated expression of nitrogen fixation is intricately and inescapably linked to analysis of heterocyst differentiation in these organisms.

The genomes of *N. punctiforme* and its close relative, *Anabaena* sp. strain PCC 7120 (also known as *Nostoc* sp. strain PCC 7120), provide information about the copy number, organization, and relative similarity of genes previously identified as involved in nitrogen fixation and heterocyst differentiation (Kaneko *et al.*, 2001). However, they provide little information about transcriptional regulation of those genes and nothing about the identity of previously unknown genes involved in cellular differentiation. The genes and gene products of most core metabolic systems found in many, if not all, organisms provide motifs and organizational structure from which to model newly identified genes and predict functional roles of the gene products. The nitrogenase genes and their products are an excellent example, albeit with a limited organismal distribution, as are the more widely distributed DNA-binding proteins, protein kinases, ABC transporters, electron-transfer proteins, to name but a few, and other core metabolic proteins. Conversely, in predicting genes that could be involved in the heterocyst-differentiation cascade, there appear to be no common models of bacterial developmental pathways that one can follow (Shimkets and Brun, 2000). We hypothesized that the regulatory elements and circuits governing the cellular differentiation alternatives expressed by *N. punctiforme* uniquely evolved in the cyanobacterial lineage (Meeks et al., 2002). Implicit in this hypothesis is an answer to the question posed in the subtitle of this chapter; gene sequences present in the genomes of *N. punctiforme* and *Anabaena* 7120 provide little specific information about the overall process of heterocyst-localized nitrogen fixation in an oxic environment that was not already known. Nor do sequences in the genome of *N. punctiforme* allow one to identify genes that could be responsible for its broad symbiotic competence (see below). Nevertheless, the genomes do provide a vast amount of exciting, and often bewildering, information upon which to formulate additional, more detailed and specific questions.

An objective of this chapter is to consider the information that leads to the above hypothesis. The organizational approach will be to first briefly describe the extraordinarily wide range of ecological niches, physiological properties, and vegetative cell developmental alternatives of *N. punctiforme* that make it a versatile experimental organism. This will be followed by a survey of the currently analyzed status of its genome. The genome is large, complex and most likely in a state of flux; consequently, the survey will not be comprehensive, but it will include selected gene families of unexpected diversity. A comparison will then be made between genes found in *N. punctiforme* and *Anabaena* 7120 that are established as involved in heterocyst differentiation and maturation into a nitrogen-fixing cell and of assimilation of other sources of nitrogen. In general, I will briefly describe what is known of the process and then link that information to what may or may not be predicted from the genome sequence.

2. PHENOTYPIC TRAITS OF *N. PUNCTIFORME*

2.1. Free-living growth state

Many of the phenotypic traits of *N. punctiforme* enhance its photosynthetic competence and contribute to its competitive survival in a terrestrial habitat. *N. punctiforme* can alter its photosynthetic pigment complement in response to light quality (type II complementary chromatic adaptation; Rippka *et al.*, 1979) and produce UV-light-absorbing compounds in response to UV irradiation (Hunsucker *et al.*, 2001). These properties are important in surface terrestrial habitats subject to both unshaded and changing shaded illumination. *N. punctiforme* can assimilate ammonium and nitrate/nitrite, in addition to dinitrogen (N_2). Nitrogen fixation clearly provides a survival advantage in habitats lacking combined nitrogen. *N. punctiforme* is among a limited number of cyanobacteria that can grow in the dark as a respiratory heterotroph when supplied with sucrose, fructose or glucose, although the growth rate is less than half of that supported by light (Summers *et al.*, 1995). The ability to switch to a heterotrophic metabolic mode may be essential in the symbiotic competence of *N. punctiforme*.

The vegetative cells of *N. punctiforme* can mature in four developmental directions (Figure 1) (Meeks *et al.*, 2002). The first development mode occurs when nutrients are unlimited. The cells grow and divide, primarily in a plane transverse to the filament, to perpetuate the vegetative-cell cycle and yield an elongating unbranched filament (Figure 1A). Cell divisions can deviate from a purely transverse plane, resulting in a "kinky" appearing filament. When such "kinky" filament segments are confined within a rigid sheath, the filaments tend to loose their filamentary appearance, leading to aseriate morphology. The aseriate growth stage is a characteristic part of the life cycle of many *Nostoc* species, but it is not obligatory and is rapidly lost in laboratory culture.

The other three vegetative-cell developmental alternatives are induced by environmental signals that specify either nutrient limitation or stress. The second developmental mode involves nitrogen limitation. The differentiation of 5-10% of the vegetative cells into nitrogen-fixing heterocysts is the most highly studied developmental event in cyanobacteria. The environmental signal to induce heterocyst differentiation is clearly deprivation of combined nitrogen (Fogg, 1949). Heterocysts are morphologically distinguished from vegetative cells by their generally larger size and different shape, their pale color resulting from a loss of phycobiliproteins, and the presence of refractile bodies at each pole. Heterocysts are present in a semi-regular spacing pattern in the filaments, with singly spaced heterocysts separated by 10-20 vegetative cells, depending on the heterocyst frequency (Figure 1B) (Golden and Yoon, 1998; Haselkorn, 1998; Adams, 2000; Wolk, 2000; Meeks and Elhai, 2002). Therefore, mechanisms regulating heterocyst differentiation must include not only gene products involved in initiating the differentiation event, but also those in defining which cells will initiate and/or continue differentiation upon receiving the signal. Heterocyst differentiation is a terminal event, although the heterocyst physiological life span has yet to be precisely defined. Terminal differentiation is a basic form of programmed cell death, or apoptosis, and the process may have initially evolved in different developmentally competent bacteria (*e.g.*, the *Bacillus* spore mother cell; Levin and Losick, 2000) and emerged in eukaryotic organisms by acquisition from bacteria (Koonin and Aravind, 2002). Known heterocyst-related gene products are compared in Section 4.

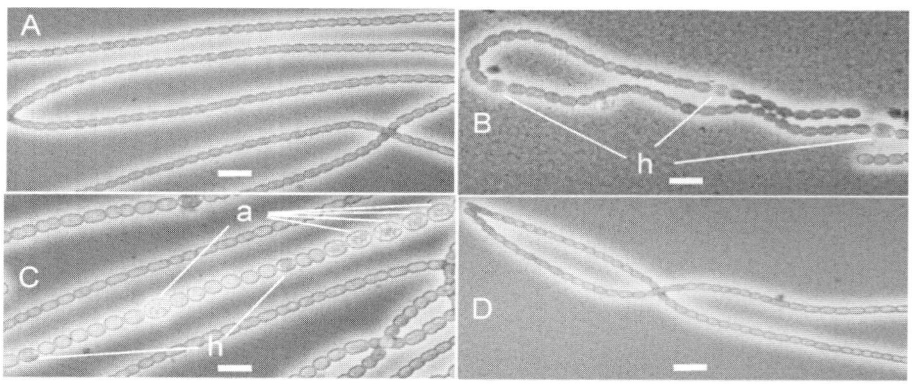

Figure 1. Phase contrast photomicrographs of N. punctiforme.
A, ammonium-grown culture consisting of unbranched, undifferentiated filaments; B, *a dinitrogen-grown culture showing three heterocysts (h) present in a nonrandom spaced pattern;* C, *a dinitrogen-grown culture from an early stationary phase culture showing akinete (a) differentiation, initiating in a cell in the interval between heterocysts (h) and spreading from there;* D, *hormogonium culture, illustrating the smaller size and different shape of the cells. Bars equal 10 μM. Reformatted from Meeks* et al., *2002, with permission of the publisher.*

The third developmental mode occurs under conditions of cellular energy limitation, most often experimentally imposed by phosphate starvation, when some or all vegetative cells can transiently differentiate into spore-like cells called akinetes (Figure 1C). Akinetes are generally larger, have more granulation than vegetative cells, but with similar pigmentation, and they lack the distinct polar bodies of heterocysts. Akinetes are more resistant than vegetative cells to cold. Only, but not all, heterocyst-forming cyanobacteria differentiate akinetes. Initiation of akinete differentiation occurs in one of two patterns within a filament. In some species, akinetes differentiate from vegetative cells adjacent to existing heterocysts, while in others, such as *N. punctiforme* (Figure 1C), the vegetative cells begin to differentiate about midway in the interval between adjacent heterocysts (Meeks *et al.*, 2002). In both cases, akinete differentiation continues in a sequential pattern progressing away from the first cell to differentiate. Akinetes and heterocysts share a unique envelope polysaccharide layer. These, and other properties, contribute to the hypothesis that akinetes were the evolutionary precursors of heterocysts (Wolk *et al.*, 1994). A heterocyst regulatory-gene product, HetR, may (Leganés *et al.*, 1994) or may not (Wong and Meeks, 2002) be involved in the induction of akinete differentiation (Meeks *et al.*, 2002). A gene, *avaK,* whose gene product was found to be enriched in akinetes of *A. variabilis*, has homologues in both the *N. punctiforme* and *Anabaena* 7120 genomes (Zhou and Wolk, 2002).

The fourth developmental mode has non-motile *N. punctiforme* vegetative filaments transiently differentiating into filaments that are motile by gliding and called hormogonia. Hormogonium filaments are characterized by the smaller size and different shape of their cells, relative to cells of vegetative filaments, and the lack of either heterocysts or akinetes (Figure 1D). Hormogonia differentiate in response to a variety of environmental changes that may be either positive or negative for growth. They function in short-distance dispersal (Tandeau de Marsac, 1994) and are the infective units of cyanobacterial symbiotic associations (Meeks, 1998; Meeks and Elhai, 2002). Hormogonium differentiation has been a major phenotypic characteristic used to distinguish the morphologically similar genera *Nostoc* (positive) and *Anabaena* (negative) (Rippka *et al.*, 1979). The smaller size of the hormogonium cells results from cell divisions that are not accompanied by an increase in cell biomass. There is no significant increase in DNA (Herdman and Rippka, 1988), protein, or chlorophyll (Campbell and Meeks, 1989) in a culture differentiating hormogonia. The extensive uncoupling of cell division (enhanced) from DNA replication (repressed) appears to be unique to hormogonia of cyanobacteria (Meeks *et al.*, 2002). Because cyanobacteria contain multiple chromosomal copies per cell (Herdman *et al.*, 1979), hormogonium cells are likely to receive one or more copies of the chromosome in the absence of DNA replication. Hormogonia remain in a motile state for about 72 h, after which they cease to glide, the filament-end cells differentiate into heterocysts and macromolecular synthesis resumes. Transcription of genes encoding a sigma subunit of RNA polymerase and a carboxyl terminal protease are enhanced in hormogonia of *N. punctiforme* (Campbell *et al.*, 1998), as are genes in the related organism, *Calothrix* sp. strain PCC 7601, encoding gas-vesicle proteins (Damerval *et al.*, 1991), cell division proteins, and pili (Doherty and Adams, 1999) However, none of these gene products appear to be involved in the initiation of hormogonium differentiation.

2.2. Symbiotic growth state

Cyanobacterial symbiotic associations are comprehensively described in Volume 5. Therefore, the discussion here will be restricted to *N. punctiforme* and its symbiotic properties that are relevant to genome analyses. *N. punctiforme* has exceptionally broad symbiotic competence with members of three of the major phylogenetic groups of terrestrial plants and a fungus. *N. punctiforme* strain PCC 73102 was isolated from a coralloid root of the gymnosperm cycad, *Macrozamia* sp. (Rippka *et al.*, 1979). Strain PCC 73102 and its sibling culture, strain ATCC 29133, individually establish an association with the angiosperm *Gunnera manicata* (Johansson and Bergman, 1994), the bryophyte hornwort *Anthoceros punctatus* (Enderlin and Meeks, 1983), and the liverwort *Blasia pusilla* (Joseph and Adams, 2000). *N. punctiforme* is identified as the intracellular symbiont of the unique mycorrhizal-like fungus *Geosiphon pyriforme* (Mollenhauer *et al.*, 1996) and strain PCC 73102 is competent as a symbiotic partner in this association (M. Kluge, personal communication).

The broad spectrum of partners that form a symbiotic association with *N. punctiforme* implies that *N. punctiforme* is not the specialized type of symbiont exemplified by the rhizobia. Not only are rhizobia largely restricted to association with leguminous plants, but rhizobia also express a nitrogen-fixing phenotype in symbiosis that they do not express during free-living growth. In contrast, symbiotically associated *Nostoc* species alter only the magnitude of the response of targeted developmental and metabolic systems, all of which are typically expressed also in the free-living state. Thus, the fundamental contrast between the two groups in the consequences of their symbiotic interactions can be described in terms of the degree of the response (*Nostoc*) as opposed to the kind of response (rhizobia).

In cyanobacterial associations, the plant partners regulate the differentiation and behavior of hormogonia, enhance the frequency of heterocyst differentiation, and alter *Nostoc* metabolism, such that the *Nostoc* assume a heterotrophic metabolic mode, their rate of nitrogen fixation is elevated with the excess fixed nitrogen being released to the plant partner (Meeks, 1998; 2003; Meeks and Elhai, 2002). These observations led to the suggestion that, in contrast to the sophisticated exchange of signals between rhizobia and leguminous plants which is essential for establishment of a microoxic environment for nitrogenase expression, the interactions in cyanobacterial symbioses are primarily unidirectional from photosynthetic plant to the cyanobacterium (Meeks, 1998). Perhaps because *N. punctiforme* and its relatives provide their own O_2-protection mechanism by differentiating heterocysts, such an elaborate signal exchange-dependent development sequence is not necessary in the formation of cyanobacterial nitrogen-fixing symbiotic associations.

2.3. Genetic targets in hormogonium differentiation and behavior

The plant partner controls the induction of hormogonium differentiation and, in the case of the bryophyte hornwort, *Anthoceros punctatus*, and liverwort, *Blasia pusilla*, the direction of hormogonium gliding during the infection process. *A. punctatus* also represses hormogonium differentiation once colonization has occurred. There is considerable experimental evidence for production of a hormogonium-inducing factor (HIF) by various plant partners (Meeks and Elhai, 2002; Meeks 2003), but no factor has been isolated and chemically characterized. Only two genes have been shown experimentally to be transcriptionally induced by exposure to HIF; *sigH*, which encodes a class two alternative sigma subunit of RNA polymerase, and its 5' gene, *ctpH*, which encodes a carboxyl terminal protease (Campbell *et al.*, 1998). SigH appears to influence hormogonium behavior, whereas the symbiotic phenotype of CtpH is unknown.

The phenotype of the *sigH* mutant is a higher frequency of infection, relative to the parental *N. punctiforme*, in the *A. punctatus* association. The increased infection by the mutant was not a consequence of either differentiation of more hormogonia, or a prolonged gliding period, but rather that the formed hormogonia appeared to be more efficient in the infection process. *Nostoc* chemotaxis to bryophyte exudate was clearly established by Knight and Adams (1996) and contributes to the efficiency of infection. The *N. punctiforme* genome has multiple copies of genes encoding putative chemotaxis proteins (Meeks *et al.*, 2001). Nothing is known of their expression, cellular localization, signal-transduction pathway, or ultimate target. We speculate that the genes are specifically expressed in hormogonia, which consist of the only motile cells in *N. punctiforme*. Whether SigH has a positive or negative role in their expression is unknown. Mutation of the global nitrogen regulator, NtcA (Herrero *et al.*, 2001), in *N. punctiforme* resulted in a lower frequency of HIF-dependent hormogonium formation and the hormogonia that were formed failed to infect *A. punctatus* (Wong and Meeks, 2002). These results imply positive (NtcA) and negative (SigH) regulation of hormogonium behavior by two transcriptional regulators. Global transcriptional assays will be required to identify the downstream regulated genes.

Hormogonium differentiation is accompanied by fragmentation of the vegetative filament at the junctions between the heterocysts and adjacent vegetative cells, leading to detachment of heterocysts and elimination of their reductant supply (Campbell and Meeks, 1989). Therefore, populations of *N. punctiforme* in the hormogonium stage do not fix N_2. Associated *N. punctiforme-A. punctatus* tissue produces HIF (Campbell and Meeks, 1989). The continual production of the HIF could induce the hormogonium cycle in the symbiotic *N. punctiforme* colonies, which would decrease the rate of nitrogen fixation and counter the selective advantage of a nitrogen-fixing association. *A. punctatus* appears to produce a hormogonium-repressing factor (HRF) that overrides the HIF and suppresses hormogonium differentiation (Cohen and Meeks, 1997). A target of the HRF is an 8.3-kb cluster of 8 genes in the *N. punctiforme* genome that we have termed the *hrm* locus (Campbell *et al.*, 2003). Genes in the *hrm* locus have a sequence similarity and organization that is analogous to the genes encoding enzymes of hexuronic acid metabolism in heterotrophic bacteria (Campbell *et al.*, 2003). We suggest that the gene products synthesize a metabolite inhibitor of hormogonium differentiation. Mutation of the *hrmU* gene, which encodes a product with similarity to 2-keto-3-deoxygluconate dehydrogenase, resulted in continued reentry into the hormogonium cycle in the presence of either HIF or HIF plus HRF and a symbiotic phenotype of high infection frequency (Cohen and Meeks, 1996).

The related genes in the heterotrophic bacteria are organized in an operon and transcription is regulated by a repressor that is encoded by the first gene in the operon. Although co-located in *N. punctiforme*, the *hrm* genes are not transcribed as a unit. Transcription of *hrmU* and its cotranscribed gene, *hrmA*, are induced by HRF and by the plant flavenoid naringin (Cohen and Yamasaki, 2000). The homologous repressor in *N. punctiforme*, HrmR, regulates its own transcription and also the transcription of a nearby gene, *hrmE*, encoding an aldehyde reductase, by binding to specific sequences in their respective operator regions. *In vitro* binding to the operator regions is eliminated by the presence of galacturonic acid and extracts of a *N. punctiforme hrmR* mutant induced by HRF, but not by extracts induced by naringin (Campbell *et al.*, 2003). The *hrmR* mutant is unable to differentiate hormogonia in the presence of HIF, presumably due to unregulated synthesis of the *hrmE*-gene product. Thus, transcriptional regulation in the *hrm* locus involves at least two mechanisms.

The sequences of the genes encoding putative chemotaxis proteins allows one both to predict a physiological role for them in hormogonium behavior and symbiotic interactions and to design specific experiments to test that role(s). However, the sequences of the other known genes identified thus far with an involvement in hormogonium differentiation and behavior provide no clues as to their function in the absence of a phenotype. This is particularly true of the genes of the *hrm* locus. There is strong evidence that the gene products are not involved in the catabolic process that is predicted from their gene organization, sequence similarities, and biochemical characterization in heterotrophic bacteria (Campbell *et al.*, 2003).

2.4. Genetic targets of growth, metabolic and developmental control

The growth of symbiotically associated *Nostoc* is at least 5-fold slower, and considerably more depending on the plant partner, than that of free-living nitrogen-fixing cultures in the laboratory (Meeks, 1998). Photosynthetic carbon dioxide assimilation by *Nostoc* is depressed in proportion to the slower growth and ammonium assimilation is similarly depressed (Meeks, 1998), except for cycad associations (Pate *et al.*, 1988). There is no obvious causal relationship between the slow growth rate and lower carbon- and nitrogen-assimilatory capacities and the mechanisms of growth control remain unidentified. The relatively low rate of photosynthetic CO_2 assimilation, in proportion to the relatively high rate of nitrogen fixation, implies that the plant partner may supply reduced carbon for reductant generation in the symbiotic *Nostoc*. Such a relationship was supported by experiments measuring nitrogen fixation by *Nostoc* mutants that were resistant to photosynthetic inhibitors and dependent on either plant metabolism or an exogenous supply of sucrose, fructose or glucose (Steinberg and Meeks, 1991). The heterotrophic rates of symbiotic nitrogen fixation were equal to light-dependent rates and exceeded all measured heterotrophic rates by free-living cultures. These results imply transition to a very robust heterotrophic metabolism. Neither the genetic basis of the transition nor the metabolic end products (CO_2 and/or an organic acid) in this situation of limited demand for carbon skeletons in the assimilation of ammonium have been investigated.

In the *A. punctatus* association, the lower rates of *Nostoc* CO_2 and NH_4^+ assimilation are a consequence of irreversible inhibition of ribulose bisphosphate carboxylase/oxygenase and glutamine synthetase catalytic activities, respectively (Meeks, 1998). However, irreversible inhibition of enzyme activity does not appear to contribute to the lower assimilatory activities in the *Gunnera* spp. and cycad associations (Meeks and Elhai, 2002; Meeks, 1998, 2003). Direct *in situ* measurements have established that symbiotic *Nostoc* does carry out a low rate of complete photosynthesis in the *A. punctatus* association (Steinberg and Meeks, 1991), verifying that light penetrates to the *Nostoc* embedded in the gametophyte tissue. Therefore, the transition from photoautotrophic to heterotrophic metabolism is not a simple consequence of light deprivation and may depend on signals from the plant partner. Because there appears to be little transcriptional control of the target enzymes in modulation of the metabolic activities that have been measured thus far in symbiotic cyanobacterial associations, identification of genetic mechanisms is not possible. Based on the observed irreversible inactivation of catalytic activity, one would predict that either synthesis or activation of protein-modification enzymes is important in metabolic control in symbiosis. The *N. punctiforme* genome has approximately 150 ORFs with predicted protein protease, phosphorylation, and sulfhydryl-modification activities that could function in modulation of growth and metabolism. Identification of specific modulators will require functional analyses.

Symbiotically associated *Nostoc* species show two morphological changes relative to free-living populations. First, the vegetative cells are larger with a more irregular shape, relatively weak cell-cell connections, and an aseriate appearance. Consequently, the size difference between vegetative cells and heterocysts is minimized and physically isolated *Nostoc* tend to appear microscopically as unicells, bicells, or very short filaments. Because the peptidoglycan layer defines the shape of a bacterial cell, there appear to be modifications in the peptidoglycan of symbiotic *Nostoc*. Putative penicillin-binding proteins involved in peptidoglycan synthesis and/or assembly are predicted in the *N. punctiforme* genome, but their activities have not been documented to be different in the symbiotic growth state.

The most dramatic morphological change is the increase in symbiotic heterocyst frequency to a range of from 25% to more than 65% of the total cells (Meeks and Elhai, 2002). Genes involved in heterocyst differentiation and maturation will be discussed later. The two points to be considered here are: What is the signal that initiates heterocyst differentiation and when is the signal perceived by the vegetative cells that differentiate into heterocysts? The environmental signal for initiation of heterocyst differentiation in free-living populations has long been known to be deprivation of combined nitrogen (Fogg, 1949). However, several lines of evidence imply that limitation of combined nitrogen is not the signal to initiate heterocyst differentiation in symbiosis (Meeks, 1998; 2003; Meeks and Elhai, 2002). Vegetative cells of symbiotic *Nostoc* contain cyanophycin granules, carboxysomes, and phycobiliproteins (Meeks and Elhai, 2002); thus, they do not show the ultrastructural characteristics of nitrogen-limited vegetative cells in the free-living growth state. Yet some of these vegetative cells differentiate into heterocysts. Moreover, symbiotic *Nostoc* are exposed to a concentration of N_2-derived ammonium of *ca.* 0.55 mM in the *A. punctatus* association, which is at least

37-fold higher than that required to repress heterocyst differentiation in free-living populations (Meeks, 2003). Nevertheless, vegetative cells continue to differentiate into heterocysts. A *N. punctiforme* mutant, which is defective in nitrate assimilation, differentiated heterocysts in the presence of nitrate in free-living culture, but heterocyst differentiation was repressed by nitrate when the mutant was in symbiotic association with *A. punctatus* (Campbell and Meeks, 1992). These results are consistent with combined nitrogen acting indirectly, by repressing either synthesis or release of a plant-derived signal of heterocyst differentiation, rather than directly on the symbiotic *Nostoc*.

The global nitrogen regulator NtcA is the first gene product activated during induction of heterocyst differentiation following combined nitrogen-deprivation in free-living cultures. NtcA activates transcription of *hetR*, the primary positive regulator of heterocyst differentiation (Buikema and Haselkorn, 2001), and a functional HetR is required for symbiotic heterocyst differentiation (Wong and Meeks, 2002). Thus, we speculate that a plant signal would enter the initiation cascade prior to induction of *hetR* transcription, either at or before activation of NtcA. Because NtcA is required for induced transcription of genes expressed late in heterocyst maturation (Herrero *et al.*, 2001), it must also be activated during symbiotic heterocyst formation. NtcA is currently thought to perceive cellular nitrogen status as the concentration of 2-oxogluratate (Muro-Pastor *et al.*, 2001; Vazquez-Bermudez *et al.*, 2002). It has not been excluded that the symbiotic-plant signal (either directly or indirectly) elevates the cellular 2-oxoglutarate pool and thereby activates the combined nitrogen-limitation cascade. Both the synthesis and activity of isocitrate dehydrogenase needs to be examined in the symbiotic growth state.

The free-living pattern of heterocyst spacing, with single heterocysts separated by 10 to 20 vegetative cells, is altered in the symbiotic growth state. In free-living populations, growth in the presence of combined nitrogen inhibits heterocyst differentiation. The spacing pattern is established as the combined nitrogen is depleted. The vegetative cells subsequently grow and divide using fixed nitrogen provided by heterocysts, thereby lengthening the vegetative cell interval between heterocysts; the spacing pattern is maintained when a vegetative cell in the middle of the interval initiates differentiation. It is not known whether the same mechanism is involved in establishment and maintenance of the spacing pattern in the symbiotic state. In symbiotic associations between *Nostoc* and cycads (Lindblad *et al.*, 1985), *Gunnera* spp. (Söderbäck *et al.*, 1990), and the water fern, *Azolla* (Peters and Mayne, 1974), a developmental gradient of low-to-high heterocyst frequency is evident from the meristematic tip to the base of the symbiotic structure. Nitrogen fixation activity parallels the developmental gradient up to a heterocyst frequency of *ca.* 35% of the cells. At that frequency, single heterocysts are mostly present at a site with a reduced vegetative-cell interval (Meeks and Elhai, 2002). Heterocyst frequencies, which are greater than 30-35%, correlate with clusters of 2-3, or more, heterocysts at a site and a decline in the rate of nitrogen fixation. These clusters may result from new heterocysts arising adjacent to older non-functional heterocysts (Meeks and Elhai, 2002). The presence of a developmental gradient implies that the high functional heterocyst frequency in symbiosis of *ca.* 30-35% does not appear synchronously as would be expected from a mechanism analogous to that involved in initial establishment of the free-living pattern. Rather, the gradient is consistent with a symbiotic alteration in the maintenance of the pattern in the free-living state. Less is known about maintenance than of establishment of pattern. One gene product, HetN, however, has been directly implicated in maintenance of the pattern (Callahan and Buikema, 2001). Neither the lack nor over-expression of HetN has been examined in the symbiotic growth state.

Consistent with the fundamental difference in physiological responses of symbiotically associated *Nostoc*, the emerging models of the exquisitely complex rhizobia-legume interactions that lead to a root nodule (Downie and Walker, 1999; Perret *et al.*, 2000) do not provide a basis upon which to predict protein function from gene sequence in *N. punctiforme*. Reports of heterologous DNA:DNA hybridization signals using rhizobial *nod* gene probes (Rasmussen *et al.*, 1996) have not lead to the identification of symbiotic genes in cyanobacteria. If, in the process of establishing a functional nitrogen-fixing association, certain plants have evolved mechanisms to manipulate genes that *Nostoc* can express in the free-living growth state (Meeks and Elhai, 2002), then there is little reason to assume that homologues will be found in distantly related Proteobacteria that, as a result of evolutionary selective pressure, display a distinctly different life style. Signaling strategies and sequences of interaction, however, could be similar. Functional genomic analyses will likely be necessary to identify genes responsible for the symbiotic competence of *N. punctiforme*.

3. OVERVIEW OF THE *N. PUNCTIFORME* GENOME

The most recent assembly of the shotgun sequence with 11-X coverage yields 9.02 Mb in 203 contigs, representing about 98% of the estimated 9.25 Mb genome (Table 1). This database, computationally annotated by Frank Larimer at the Oak Ridge National Laboratory, yielded 7,281 open reading frames (ORFs), of which 73% can be associated with a previously recognized ORF and 27% are unique to *N. punctiforme* when *Anabaena* 7120 is not included in the comparisons (see below). This genome is among the largest of the microbial genomes that have been sequenced.

3.1. Comparisons to the genomes of unicellular Synechocystis *6803 and heterocyst-forming* Anabaena *7120*

The size and complexity of the *N. punctiforme* genome can be appreciated by comparisons with the genomes of two other cyanobacteria. The unicellular, non-nitrogen-fixing *Synechocystis* sp. strain PCC 6803 (*Synechocystis* 6803) was among the first microbial genomes sequenced (Kaneko *et al.*, 1996). The 3.57-Mb genome of *Synechocystis* 6803 putatively encodes 3,215 proteins, 47% and 53% of which can be associated with known and hypothetical proteins, respectively (Table 2). Fifty-five percent of the *N. punctiforme* ORFs find similarity in the *Synechocystis*

6803 genome, whereas 80% of the *Synechocystis* 6803 ORFs are present in *N. punctiforme*.

Table 1. Numerical summary of the genome of Nostoc punctiforme.

Current genome size[a]	~9,250,000 bases (11.4 x sequencing coverage; May, 2001); 41.37 mol % GC
Size of annotated sequence	9,020,037 bases (*ca*. 95% of presumed genome)
Preliminary analyses	7,281 candidate protein-encoding gene models or ORFs
Relative to total database	5,314 of the ORFs (73% of the total) can be associated with a previously recognized ORF 3,328 of the recognized ORFs (46% of the total) encode proteins with known or probable known function 1,986 of the recognized ORFs (27% of the total) encode conserved hypothetical proteins with no known function 1,967 of the ORFs (27% of the total) cannot be associated with a previously recognized ORF, excluding ORFs in *Anabaena* 7120

[a]Meeks *et al*., 2001; http://www.jgi.doe.gov/

Table 2. Comparisons of the genomes of N. punctiforme and Synechocystis *sp. strain PCC 6803* [a].

Synechocystis 6803 genome[b]	3,573,471 bases, yielding 3,215 ORFs; 47.7 mol % GC
Relative to the total database	1,521 ORFs (47% of the total) encode proteins that can be associated with known or probable known function 1,694 ORFs (53%) encode conserved hypothetical or hypothetical proteins
N. punctiforme vs. Synechocystis 6803	3,965 of the *N. punctiforme* ORFs (55%) are present in *Synechocystis* 6803 2,547 of the *Synechocystis* 6803 ORFs (80%) are present in *N. punctiforme* 668 of the *Synechocystis* 6803 ORFs (20%) are unique to *Synechocystis* 6803

[a]The comparative results are based on reciprocal BLAST analyses.
[b]Kaneko et al., 1996

These comparative values indicate that the larger *N. punctiforme* genome is not an exact multiple of the smaller *Synechocystis* 6803 genome, which would require that nearly all of the *N. punctiforme* ORFs find significant similarity in the *Synechocystis* 6803 genome. If larger cyanobacterial genomes arose by gene duplication in the smaller genomes, there appears to have been sufficient sequence divergence since that event to obscure any direct lineage. However, the fact that 80% of the *Synechocystis* ORFs find similarity in the *N. punctiforme* genome is consistent with the prediction that the *N. punctiforme* genome contains gene families that multiply duplicate those present in *Synechocystis* 6803. Multigene families do appear to be extensively represented in the *N. punctiforme* genome, except for core metabolic processes that define cyanobacteria, such as genes encoding photosynthetic functions (Meeks *et al.*, 2001).

Table 3. Comparisons of the N. punctiforme and Anabaena sp. strain PCC 7120 genomes[a]

Anabaena 7120 genome[b]	7,211,789 bases, yielding 6,132 ORFs; 41.2 mol% GC
Relative to the total database	4,157 ORFs (68% of the total) encode proteins that can be associated with known or probable known function 1,975 ORFs (32% of the total) cannot be associated with a previously recognized ORF
N. punctiforme vs. Anabaena 7120	5,431 of the *N. punctiforme* ORFs (75%) are present in *Anabaena* 7120
	4,814 of the *Anabaena* 7120 ORFs (79%) are present in *N. punctiforme*
	1,489 of the *Anabaena* 7120 ORFs (24%) are unique to *Anabaena* 7120
	486 of the 1,975 unique *Anabaena* 7120 ORFs are present in *N. punctiforme*

[a]The comparative results are based on reciprocal BLAST analyses.
[b]Kaneko *et al.*, 2001

The close relationship between the heterocyst-forming cyanobacteria *N. punctiforme* and *Anabaena* 7120 is reflected in their relative genome similarity (Table 3). The *Anabaena* 7120 genome, at 7.21 Mb, is about 78% of the size of *N. punctiforme* and encodes 6,132 ORFs, 1,975 (32%) of which cannot be associated with a previously recognized ORF (Kaneko *et al.*, 2001). Approximately 79% of the *Anabaena* 7120 ORFs find similarity in the *N. punctiforme* genome, whereas 75% of the *N. punctiforme* ORFs are similar to those in the *Anabaena* 7120 genome. A significant fraction of the extra coding capacity in *N. punctiforme* is present as ORFs encoding hypothetical proteins. Reciprocal BLAST analyses indicate that 486 ORFs, which encode hypothetical proteins, are present in both *N. punctiforme* and *Anabaena* 7120 genomes and absent in the genomes of all other organisms surveyed as of August 2000. Consequently, the ORFs unique to *N. punctiforme* and to *Anabaena* 7120 reduce to 1,481 and 1,489, respectively. The number of ORFs encoding hypothetical proteins in *N. punctiforme* and *Anabaena* 7120 are similar to the total coding capacity of a small cyanobacterial genome, such as

Prochlorococcus marinus strain MED4 (Hess *et al.*, 2001). An interesting consequence of this analysis is identification of the 486 ORFs encoding hypothetical proteins uniquely shared by *N. punctiforme* and *Anabaena* 7120. We predicted that many of these gene products will contribute to expression of phenotypic traits common to both organisms, such as heterocyst differentiation (Meeks *et al.*, 2001). Subsequently, 3 genes, which encode shared hypothetical proteins involved in heterocyst differentiation and pattern formation (see section 4), and 1 gene each that encode proteins enriched in akinetes (AvaK; Zhou and Wolk, 2002) and hormogonia (HomA; Campbell and Meeks, unpublished results), have been identified by either classical genetic techniques or protein isolation and characterization. The presence in *Anabaena* 7120 of genes that encode proteins associated with hormogonia and akinetes is consistent with both evidence that *Anabaena* 7120 is genotypically a *Nostoc* species (Rippka and Herdman, 1992; Wilmotte and Herdman, 2001) and suggestions that it has diminished phenotypic traits due to spontaneous mutation during nonselective serial passage in culture over an extended time period (Meeks *et al.*, 2002).

3.2. Genome characteristics relevant to cyanobacteria, plants, and the N. punctiforme *life style*

A preliminary analysis of the *N. punctiforme* genome (Meeks *et al.*, 2001) documented the extensive multiple occurrences of tandem repeated heptameric (STRR; Mazel *et al.*, 1990; Holland and Wolk, 1990) and dispersed repeated octameric (HIP1; Robinson *et al.*, 2000) sequences, as well as insertion sequences linked to transposases. There is an apparent dearth of hexameric palindromic sequences that are associated with restriction enzymes encoded by *Anabaena* and *Nostoc* species. Moreover, sequences of only two type II restriction/modification systems (isoschizomers of *Bgl*II and *Acy*I) are present in the genome. These observations indicate that the *N. punctiforme* genome is highly plastic and in a state of flux (Meeks *et al.*, 2001). They are also consistent with widespread exchange of DNA among heterocyst-forming cyanobacteria and extending to other organisms.

The identity of genes encoding proteins in basic metabolic categories, such as photosynthetic and respiratory energy metabolism, auto- and hetero-trophic carbon metabolism, inorganic and organic nutrient transport, transcription, and monomer, polymer and secondary product synthesis, including proteins, the cell envelope, and chromosome, were also surveyed and, with some notable exceptions, the numbers were found to be generally similar to their individual occurrence in other cyanobacteria and bacteria (Meeks *et al.*, 2001). *N. punctiforme* appears to contain the breadth of genetic information that collectively reflects nearly all of the physiological repertoire of cyanobacteria. Although peripherally relevant to the nitrogen-fixation physiology of *N. punctiforme*, those data will not be presented here. Three categories of genes, however, warrant elaboration; the presence of genes that encode proteins related to plant biology and those implicated in environmental sense and response will be discussed below, whereas gene products concerned with nitrogen assimilation will be described in Section 4.

3.3. Genes related to plant functions

Some cyanobacteria are now well known to exhibit a circadian rhythm similar to the rhythms characteristic of plants and animals (Golden *et al.*, 1998), but *N. punctiforme* has not been shown to express such a rhythm. The *N. punctiforme* genome contains homologues of four genes encoding proteins, KaiA, KaiB, KaiC and CikA, which are essential for circadian rhythm expression in *Synechococcus elongatus* strain PCC 7942. The C-terminal domain of the KaiA protein is thought to modulate KaiC autophosphorylation, whereas the N-terminal domain functions as a pseudo-receiver of an entrainment signal from CikA in *S. elongatus* (Williams *et al.*, 2002). However, the N-terminal domain of the KaiA protein in *N. punctiforme* is truncated. The CikA protein is a sensor histidine kinase, with a GAF chromophore-binding domain and a C-terminal phosphoreceiver domain that is characteristic of response regulator proteins. The nearest sequence homologue of CikA from *S. elongatus* in both *N. punctiforme* (45% identify, 65% similarity) and *Anabaena* 7120, lacks the C-terminal phosphoreceiver domain. A gene is present in *N. punctiforme* that would encode a protein with an identical domain architecture to CikA, but the sequence shows only 31% identity and 50% similarity to CikA of *S. elongatus*. The questions have not yet been asked whether the C-terminal receiver domain is essential for function and can be replaced by a separate protein (if either of these two proteins might function as a primary sensor in a rhythm entrainment in *N. punctiforme*) or whether the KaiA and CikA differences imply a quite different entrainment pathway.

The *cikA*-like genes reflect a large cluster of genes encoding putative chromophore-binding, phytochrome-like proteins in the *N. punctiforme* genome. Montgomery and Lagarias (2002) identified two Cph1, four Cph2 cyanobacterial phytochrome family proteins and 15 phytochrome-related proteins encoded in the *N. punctiforme* genome. None have yet been subject to either biochemical or genetic analyses. Nobles *et al.* (2001) established experimentally that *N. punctiforme,* and other cyanobacteria, synthesize cellulose and identified three putative cellulose synthase-encoding genes in *N. punctiforme*. Phylogenetic analyses implied that the higher plant cellulose synthase, CesA, and cyanobacterial cellulose synthases share a common evolutionary branch that is distinct from the branch containing heterotrophic bacteria.

A recent bioinformatic analysis of the *Arabidopsis thaliana* genome has revealed that perhaps 18% of its nuclear genes were derived from the endosymbiotic cyanobacterial ancestor of its extant chloroplast (Martin *et al.*, 2002). A significant majority of those genes show highest similarity to ORFs in *N. punctiforme* rather than to unicellular cyanobacteria. It is most likely that genes encoding proteins involved in circadian rhythm and phytochrome responses, as well as cellulose synthesis, are included in that 18% subset of the *A. thaliana* nuclear genome. Identification of others and their physiological role in growth and development of *N. punctiforme* is of interest.

3.4. Genes related to environmental sensing and cellular response

The multiple phenotypic traits and developmental alternatives of *N. punctiforme* are expressed in response to specific environmental signals, including those from the symbiotic plant partner. Therefore, one would anticipate a sufficiently adequate environmental sensing, signal transduction, and response capacity. However, *N. punctiforme* has an astonishingly higher than anticipated number of genes encoding proteins associated with environmental sensing/signaling (>350 protein kinases, response regulators and cyclases), protein modification (>150 molecular chaperones, proteases and sulfhydryl group modifiers) and transcriptional regulatory functions (>120 DNA binding proteins, including 60 DNA-binding domains associated with response regulator receiver domains [see below], as well as 13 sigma subunits of RNA polymerase). The above numbers are based on the computational annotation accessible at the JGI site and are, of course, subject to revision, dependent not only on completion of the genome sequence, but also on manual annotation and ultimately experimental demonstration of function. Even though subject to minor change, the numbers of sensory proteins identified in the genomes of *N. punctiforme* and *Anabaena* 7120 (Ohmori *et al.*, 2001) markedly exceed those in other bacteria and the specific proteins are complex with variable domain architecture. Here, examples of the families of sensor histidine protein kinases, signal-transduction response regulators, and serine/threonine protein kinases will be discussed because they are present in multiple copies, their domain organization is extremely diverse, and some are likely to be involved in cellular differentiation, including heterocyst development.

3.4.1. Sensor histidine kinases

Sensor histidine kinases are sensory input proteins containing sensory modules and activation domains (for a thorough discussion see Parkinson and Kofoid, 1992; Galperin *et al.*, 2001). A minimal sensor histidine kinase protein contains activation domains, which consist of an ATPase, identifed as HATPase, that is required for autophosphorylation, and a conserved histidine phosphoacceptor, called HisKA. Most, but not all, sensor histidine kinase proteins also contain one or more sensory modules that may include a PAS/PAC domain involved in heme- and flavin-binding, a GAF domain that binds cGMP, other nucelotides or chromophores, a small ligand-binding cache domain, a HAMP-linker domain, or a phosphotransfer Hpt domain. Some of these domains may also contribute to dimerization of the sensor histidine kinase, which is required for activity. The 126 complete sensor histidine kinases of *N. punctiforme* can arbitrarily be organized into three categories (Figure 2): (i) 63 proteins containing the minimal histidine kinase HATPase and HisKA domains, without or with sensory modules; (ii) 51 proteins containing the histidine kinase domains, plus a response regulator receiver only domain (in one instance also a response regulator output domain) also without or with various sensory domains; and (iii) 12 proteins containing the C-terminal histidine kinase domains, plus an N-terminal serine/threonine protein kinase domain, all of which contain a GAF domain, some also contain PAS/PAC domains, and one of which also contains a response regulator receiver only domain. In addition, 25 genes are present encoding putative proteins with a HATPase domain and lacking the HisKA domain; this category consists of proteins with just the HATPase domain (17), or with various sensory modules (7), plus one associated with a response regulator receiver only domain. These kinases most likely participate in phosphorylation of another protein. There are four genes present encoding proteins with only a HisKA domain and two with only a HAMP domain. In addition, there are five genes encoding CheA-like proteins with HATPase, response regulator receiver only domain and Hpt modules, and genes encoding six proteins with methyl-accepting chemotaxis protein domains, three of which also contain 14 or 7 GAF modules most likely involved in photosensing and one with a N-terminal 560 amino acid region that could be involved in chemosensing. Thus, the sum of putative proteins with both (or one or the other) HATPase and HisKA domains and identified by COGs analyses as sensor histidine kinases is currently 163. There are a further eight genes present that encode proteins only containing one or more sensory modules that may be involved in modulation of the activity of the sensor histidine kinases. Clearly, the capacity of *N. punctiforme* to sense its chemical and, perhaps, physical environment and possibly modulate expression of that sensing is extraordinarily extensive and complex. However, there has been little or no association of specific environmental signals with a sensor protein and a physiological response in *N. punctiforme*.

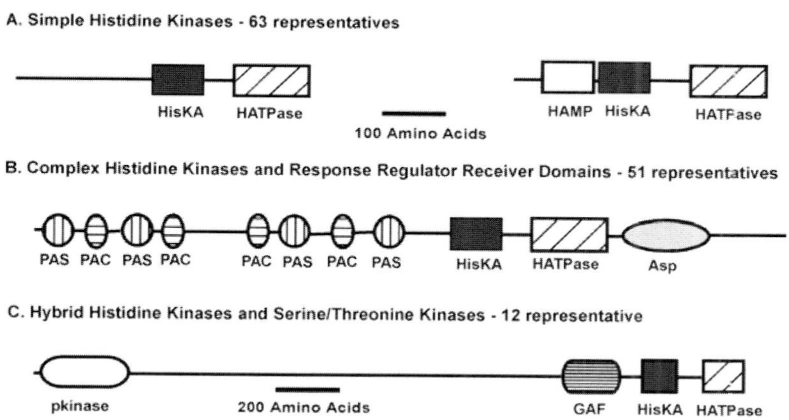

Figure 2. Protein architecture of representatives of the three classes of sensor histidine kinase proteins of N. punctiforme.

3.4.2. Response regulator proteins

In the current model, response regulator proteins interact with a cognate sensor histidine kinase to effect a response to the environmental signal (Parkinson and Kofoid, 1992; Hoch and Varushese, 2001). The response may target either protein activation, as in the chemotaxis signal-transduction cascade, or activation of gene expression, as in modulation of inorganic-nutrient acquisition such as phosphate or nitrogen (Stock *et al.*, 2000). Response regulators are, therefore, sensory output proteins that minimally contain a phosphoreceiver module, with a conserved aspartate residue that is phosphorylated, and most often an output domain predominately consisting of a helix-turn-helix DNA-binding module. The recently identified domains, GGDEF and EAL, associated with diguanylate cyclases and phosphodiesterases, and the metal-dependent phosphohydrolyase domain, HD-GYP, also have output functions linked to receiver modules (Galperin *et al.*, 2001). The 135 putative response regulator proteins encoded in the *N. punctiforme* genome can be arbitrarily organized into four groups (Figure 3): (i) 35 receiver domain only proteins; (ii) 53 receiver domain only proteins also linked to an input module (51 of these proteins were previously scored as sensor histidine kinases, one other is associated with a hybrid sensor histidine kinase and serine/threonine kinase protein and one with a HisKA domain protein); (iii) 8 receiver domain only proteins with associated either GGDEF (7) or HD (1) domains; and (iv) 60 proteins with a receiver domain and a helix-turn-helix DNA-binding output domain, only one of which is also associated with a histidine kinase domain.

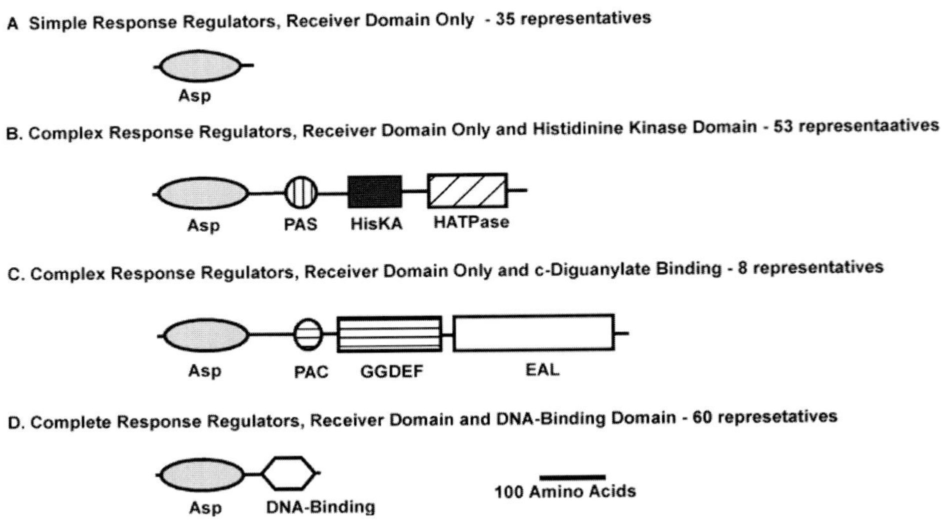

Figure. 3. Protein architecture of representatives of the four classes of response regulator proteins of N. punctiforme.

The response components of the environmental signal transduction pathways of *N. punctiforme* are obviously as extensive and complex as the sensory components. What is perhaps most striking is the number of receiver domain only proteins. Simple receiver domain only proteins are commonly thought to function by two mechanisms. One mechanism is in multicomponent phosphorelay signal transduction pathways, exemplified by SpoOF in the pathway initiating sporulation in *Bacillus* species (Hoch and Varushese, 2001). The other mechanism is by protein-protein interactions that need not involve phosphotransfer, as in CheY activation of the flagellar motor (Parkinson and Kofoid, 1992). The fact that receiver domain only proteins (with or without either kinase- or diguanylate-binding domains) outnumber those also with a helix-turn-helix output domain indicates extensive operation of integrative signal transduction phosphorelay pathways and/or activation of target proteins in *N. punctiforme*. A sensor histidine kinase (HepK) and receiver domain only response regulator (DevR) are essential for synthesis of the heterocyst envelope polysaccharide (Campbell *et al.*, 1996; Hagen and Meeks, 1999; Zhou and Wolk, 2003; Zhu *et al.*, 1998). Whether DevR-phosphate directly activates an effector target protein or serves as a phosphodonor to another protein in a relay pathway has not been determined.

3.4.3. Serine/threonine protein kinases

Serine/threonine (S/T) protein kinases were previously thought to be restricted to eukaryotic organisms. They are now recognized to also be distributed throughout the prokaryotic world (Shi *et al.*, 1998), including cyanobacteria (Zhang, *et al.*, 1998). The *N. punctiforme* genome contains 59 genes encoding proteins with S/T kinase domains. These protein kinases can be separated into three groups based on protein architecture (Figure 4): (i) 38 protein kinase only domain proteins; (ii) 12 hybrid histidine kinase domain and protein kinase domain proteins previously scored as sensor histidine kinases; and (iii) 9 protein kinase domain proteins associated with tetratrichopeptide (TPR) repeat domains (4) and 31-40 unit tryptophan-aspartate (WD) repeat domains (5). TPR (Lamb *et al.*, 1995) and WD (Smith *et al.*, 1999) repeat domains are associated with a variety of cellular functions, including signal transduction. Both domains influence peptide folding and the subsequent tertiary shape of the protein. There is some information on the physiological role of S/T protein kinases in cyanobacteria. Disruption of one S/T kinase in *Anabaena* 7120 led to the differentiation of heterocysts that were less active in nitrogen fixation in both the presence and absence of O_2 than the wild-type (Zhang, *et al.*, 1998). Mutation of a hybrid S/T kinase and histidine kinase protein in *Anabaena* 7120 also resulted in slower growth on N_2 than the parental strain (Phalip *et al.*, 2001). In neither case was the physiological defect precisely defined.

Lastly, the *N. punctiforme* genome contains seven genes encoding phosphoserine/threonine/tyrosine phosphatases and one gene encoding a phosphohistidine phosphatase. Although these observations establish a capacity to regulate the intensity and temporal period of a signal transduction phosphorelay pathway by dephosphorylation of the activated phosphoproteins, the specific capacity seems limited relative to the number and diversity of putative phosphoproteins. An unknown extent of autodephosphorylation or dephosphorylation by a cognate phosphorelay protein could be a primary mode of regulation. In any case, the wealth of putative sense-response proteins in the *N. punctiforme* genome exceeds the known environmental signals to which the organism physiologically responds. A challenge will be to identify the physiological process that may be regulated by the signal-transduction systems, determine whether the signals are of environmental or cellular origin, and the extent of interactions between signal-transduction elements and pathways.

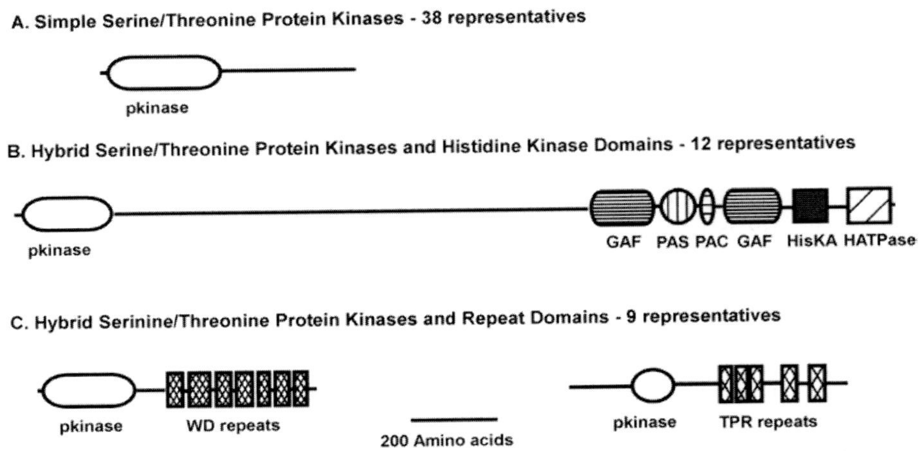

Figure 4. Protein architecture of representatives of the three classes of serine/threonine protein kinases of N. punctiforme.

4. *N. PUNCTIFORME* GENES INVOLVED IN HETEROCYST FORMATION, NITROGENASE EXPRESSION, AND AMMONIUM AND NITRATE ASSIMILATION.

Nitrogenase expression is dependent on heterocyst differentiation and maturation in essentially all heterocyst-forming cyanobacteria (see Volume 2 for a discussion of *A. variabilis*, the notable exception). Heterocyst differentiation in free-living species is repressed by the presence of either ammonium or nitrate. Therefore, understanding and potentially manipulating the regulation of nitrogen fixation in these cyanobacteria must be based on knowledge of the combined-nitrogen-sensitive formation of the microoxic mature heterocyst, a cell that can best be characterized as a sink of reductant from, and a source of fixed nitrogen for, the immediately adjacent vegetative cells in the filament. Nomenclature for classifying heterocyst- and nitrogen-fixation-defective mutants and genes was defined by Ernst *et al.* (1992). The intent in this section is not to discuss the mechanistic details of heterocyst differentiation and nitrogenase expression; these topics are covered in Volumes 5 and 2, respectively.

Genome sequence information is most valuable in predicting the enzymatic machinery required for expression of nitrogenase activity in the mature heterocysts; it is also valuable, but less so, for identifying proteins involved in synthesis of the unique components heterocyst envelope; and it is of little value in predicting the proteins of the regulatory cascade in initiation of differentiation and commitment to terminal differentiation.

4.1. Genes encoding products involved in heterocyst formation, classification based on phenotype.

Mutants that fail to form a morphologically distinct heterocyst are scored as defective in heterocyst differentiation, in contrast to mutants that form either structurally or functionally defective cells with the appearance of a heterocyst. Mutants defective in differentiation can be separated into those that do and those that do not initiate the process. Initiation of heterocyst differentiation can be screened microscopically by the presence of cells that have either little or no phycobiliprotein-induced fluorescence within 12 to 30 h following depletion for combined nitrogen; by that time, such cells are typically present in a spacing pattern within the filament that is similar to the ultimate heterocyst-spacing pattern. In the absence of nitrogen fixation, all nitrogen-starved cells will ultimately become deficient in phycobiliprotein-induced fluorescence. If a mutant fails to form weakly fluorescent cells in a spaced pattern within 30 h following combined-nitrogen deprivation, the mutation is linked to mechanisms involved in the initiation of differentiation. If a mutant produces weakly fluorescent cells in a spaced pattern that do not eventually develop into a cell with the appearance of a heterocyst, the mutant is considered competent to initiate differentiation, but blocked in commitment to terminal differentiation.

4.2. Genes encoding proteins required for initiation of differentiation and commitment to terminal differentiation.

The currently known gene products required for both initiation of heterocyst differentiation and commitment to terminal differentiation are listed in Table 4. All of the genes encoding these proteins have been identified by traditional genetic analyses of mutation, complementation, and phenotypic characterization; none of the genes could be associated with heterocyst differentiation by sequence analysis alone. For clarity, the gene products have been organized into five positive, five negative, and one uncertain element involved in the initiation of differentiation and two gene products associated with commitment to terminal differentiation. From 140 (Wolk, 2000) to over 1,000 (Lynn *et al.*, 1986) genes have been estimated to be involved in heterocyst differentiation and function. We have suggested (Meeks *et al.*, 2002), that the regulatory cascades involved in the initiation of heterocyst differentiation and commitment to terminal differentiation are as complex, if not more so, as that of *Bacillus subtilis* sporulation, in which 185 genes/gene products are projected to participate (Kunst *et al.*, 1997).

Table 4. Gene products essential for the initiation of heterocyst differentiation, pattern of heterocyst spacing, or commitment to terminal differentiation.

Gene Product	Properties	*N. punctiforme* vs. *Anabaena* 7120[a]		Citation
A. Positive:	Mutants differentiate no (or only terminal, *patA*) heterocysts	% identity	% similarity	
NtcA	Sequence specific DNA-binding protein, global nitrogen regulator in the Fnr family.	93	93	1
HanA (HU)	Sequence non-specific DNA-binding, DNA structure.	100	100 (82, 82, 84)	2
HetR	Primary activator, autoprotease, transcriptional autoinduction.	88	93	3, 4
HetF	Essential for HetR autoinduction, unique protein with a CHF protease domain.	65	74	5
PatA	Essential for differentiation of intercalary heterocysts, response regulator receiver domain lacking DNA-binding output domain.	67	80	6
B. Negative:	Mutants differentiate multiple heterocysts			
PatS	Small peptide (13 or 17 aa) repressor of differentiation, effective exogenously.	100	100	7
HetN	Ketoacyl reductase, maintenance of the spacing pattern.	49	65	8
PatB	C-terminal DNA-binding domain, N-terminal ferredoxin domain.	82	90	9
PatU	Unique protein, delayed and incomplete resolution of differentiating clusters.	67	84	10
PatN	Unique protein, multiple single heterocysts with reduced vegetative cell intervals.	53	60	10
C. Uncertain:	Null mutant same as wild type, over expression gives multiple heterocysts.			
HetL	Pentapeptide repeat, over expression phenotype insensitive to PatS.	28 *N. punctiforme* genes with BLAST values > 2 e-6; 12 highest with % similarity values of 48 to 55.		11

D. Commitment:	Mutants blocked prior to the transition to terminal differentiation,			
HetC	NtcA-dependent expression, autoregulated transcription, HlyB family of ABC protein exporters.	46	66 (67)	12
HetP	Unique protein, overexpression of *hetP* and 3' adjacent ORF yield multiple heterocysts.	51	70 (70, 66)	13

[a]Values based on BLAST results; numbers in parentheses are to extra copies of the gene product, relative to the primary copy. Citations: 1, Herrero *et al.*, 2001; 2, Khudyakov and Wolk, 1996; 3, Buikema and Haselkorn, 2001; 4, Zhou *et al.*, 1998; 5, Wong and Meeks, 2001; 6, Liang *et al.*, 1992; 7, Yoon and Golden, 1998; 8, Callahan and Buikema, 2001; 9, Liang *et al.*, 1993; 10, Meeks *et al.*, 2002; 11, Liu and Golden, 2002; 12, Khudyakov and Wolk, 1997; 13, Fernádez-Piñas *et al.*, 1994.

Because heterocyst differentiation occurs in response to a limitation in combined nitrogen, a basic question is how *N. punctiforme* senses the nitrogen status. Except for the *glnB*-encoded P_{II} protein (Arcondéguy *et al.*, 2001), the σ^{54} and other members of the Ntr signal-transduction system of Proteobacteria (including uridylylation of P_{II} and P_{II}-regulated adenylylation of glutamine synthetase) are lacking in all cyanobacterial genomes sequenced. In unicellular cyanobacteria, P_{II} is modified by serine phosphorylation in response to inorganic carbon and nitrogen status (Forchhammer and Tandeau de Marsac, 1994), but P_{II} modification was not detected in *N. punctiforme* and, because a null mutant could not be isolated, P_{II} is apparently essential under either nitrogen-limited or -replete conditions (Hanson *et al.*, 1998). Therefore, the fundamental Ntr model in Proteobacteria does not extrapolate to heterocyst differentiation or even general nitrogen control in cyanobacteria.

4.2.1. Genes encoding positive elements for initiation of differentiation

In cyanobacteria, the first gene product known to be activated following combined nitrogen deprivation is NtcA, a DNA-binding protein in the Fnr and Crp family of transcriptional regulators (Luque *et al.*, 1994). NtcA functions as a global regulator of nitrogen-responsive genes in all cyanobacteria examined (Herrero *et al.*, 2001). Analogous, in part, to P_{II}, NtcA appears to sense nitrogen status by the relative cellular concentration of 2-oxoglutarate (Muro-Pastor *et al.*, 2001; Vazquez-Bermudez *et al.*, 2002). Mutants defective in NtcA do not activate genes involved in the acquisition of alternative nitrogen sources to ammonium, such as nitrate or dinitrogen, including those of heterocyst differentiation (Herrero *et al.*, 2001).

Gene products are identified as having a positive developmental role if the mutant fails to initiate heterocyst differentiation, of which *ntcA* is an example. Mutants defect in the DNA-binding protein, HanA, fail to initiate heterocyst differentiation (Khudyakov and Wolk, 1996). However, *hanA* mutants of *Anabaena* 7120 are highly pleiotrophic, implying multiple functions for HanA, and the stable mutant that was analyzed appeared to have acquired secondary mutations that allowed for measurable growth in combined nitrogen. For those reasons, HanA is most often omitted from working models of the heterocyst-differentiation cascade. If HanA does have a differentiation role, it could be associated with activation of *hetR* transcription. HetR is modeled as the primary activator of heterocyst differentiation (Adams, 2000; Wolk; 2000; Meeks and Elhai, 2002). HetR is a wonderfully complex protein in terms of expression and function. Induced transcription of *hetR* is dependent on both NtcA (Herrero *et al.*, 2001; Wong and Meeks, 2001) and itself (Black *et al.*, 1993), and HetR appears to contribute to the induced transcription of *ntcA* (Muro-Pastor *et al.*, 2002). The protein may be modified, it has autoproteolytic activity (Zhou *et al.*, 1998), is essential for heterocyst differentiation, and is localized to differentiating cells and mature heterocysts (Buikema and Haselkorn, 2001; Wong and Meeks, 2001). There is no sequence similarity between HetR and characterized regulatory proteins that would allow one to predict these diverse responses and activities.

A functional HetF is also required for the initiation of heterocyst differentiation and the autoinduced transcription of *hetR* is dependent on HetF (Wong and Meeks, 2001). Overexpression in *trans* of either *hetR* or *hetF* yields clusters of heterocysts in the absence of combined nitrogen and excess HetR induces heterocyst formation in *Anabaena* 7120 in the presence of nitrate. *hetF* was originally among the 486 hypothetical genes shared by *N. punctiforme* and *Anabaena* 7120. However, recent sequence analyses have established that HetF is the first functionally identified member of a large family of the caspase-hemoglobinase-fold (CHF)-containing proteases (Aravind and Koonin, 2002). Most interestingly, this family of proteins is associated with signal transduction and programmed cell death, both of which are reflected in the developmental cascade of heterocysts and their physiological fate. These observations provide a link between terminal heterocyst differentiation and conventional apoptosis.

The phenotype of the PatA mutant differs from other positive regulators in that the mutant forms heterocysts, but only at the ends of the filaments (Liang *et al.*, 1992). Consequently, the heterocyst frequency is low and the pattern is disrupted. Epistatic analyses established that PatA operates downstream of HetR in the developmental program; clusters of heterocysts do not appear when HetR is overexpressed in a *patA* mutant (Liang *et al.*, 1992). PatA contains a C-terminal response regulator receiver only domain and may participate in modification of HetR (Buikema and Haselkorn, 2001).

4.2.2. Genes encoding negative elements for initiation of differentiation

If a mutant differentiates more heterocysts than the parental strain, the gene product is identified as having a negative role. An increase in heterocyst frequency must result in an alteration in the wild-type spacing pattern, therefore, such mutants are generally identified as pattern (Pat) strains. Two altered patterns have been observed, the most common of which is the presence of clusters of heterocysts at one site in the filaments. This pattern is referred to as multiple contiguous heterocysts (Mch) (Black *et al.*, 1993). One mutant has been isolated that forms a higher frequency of heterocysts that are located singly in the filaments with a correspondingly shortened vegetative-cell interval and is referred to as multiple singular heterocysts (Msh), in contrast to Mch (Meeks *et al.*, 2002).

Five such negative regulatory mutants have been reported; four, *patS*, *hetN*, *patB* and *patU*, result in Mch, whereas *patN* yields Msh. The *patS* gene encodes a small (13 or 17 amino acids) peptide that is considered to be the primary negative regulator in establishment of the pattern following nitrogen deprivation (Yoon and Golden, 1998). Exogenous supply of a C-terminal pentapeptide of PatS suppresses heterocyst differentiation. *patS* is primarily transcribed in differentiating heterocysts (Yoon and Golden, 2001). These results contribute to the idea that PatS is the long sought diffusible inhibitor of heterocyst differentiation that is produced by heterocysts. *patS*, at 39 or 51 bp, is too small to have been detected as a gene by any algorithm.

HetN contains a domain with similarity to β-ketoacyl reductase typically involved in either fatty acid or polyketide synthesis (Black and Wolk, 1994). The originally isolated and reconstructed mutants gave three different heterocyst phenotypes (wild-type, Mch, or no heterocysts), some of which may have resulted from secondary suppressor mutations. The conflicting results have been resolved by copper-dependent transcription, using the *petE* promoter, of *hetN* (Callahan and Buikema, 2001). Controlled overexpression of *hetN* suppressed heterocyst differentiation, whereas lack of expression resulted in Mch, thereby verifying a negative regulatory role that is similar to PatS. However, the Mch phenotyupe appeared only at 48 h after nitrogen deprivation and at 24 h after establishment of the typical pattern of single spaced heterocysts. This sequence of events is consistent with disruption of the mechanism that maintains rather than establishes the spacing pattern.

The *patB* gene encodes an unusual protein containing a C-terminal helix-turn-helix DNA-binding motif and a N-terminal bacterial ferredoxin domain (Liang *et al.*, 1993). Mutations in *patB* also gave variable phenotypes. Filaments of the original frame-shift mutant fragmented under nitrogen-limited incubation conditions, making analysis of heterocyst spacing patterns equivocal. In terms of intensity of response, the heterocyst phenotype appears to depend on the specific mutation (Jones *et al.*, 2003). A *patB* deletion mutant has little nitrogenase activity, but initially differentiates heterocysts in a wild-type spacing pattern, which is then followed by Mch. The original frame-shift mutation disrupted the C-terminal domain; this mutant had about 14% of the wild-type nitrogenase activity and a delayed Mch phenotype similar to the deletion mutant. Site-specific disruption of the ferredoxin domain yielded a less severe Mch pattern and higher nitrogenase activity. The phenotype is consistent with a negative regulatory role in heterocyst differentiation for PatB, but it may be associated with maintenance of pattern, similar to HetN.

The *patU* and *patN* mutants, which differentiate multiple heterocysts, have been preliminarily characterized in *N. punctiforme* (Meeks *et al.*, 2002). Both mutants grow slowly with N_2 as the nitrogen source and both genes encode hypothetical proteins whose sequences in the current database are shared only by *N. punctiforme* and *Anabaena* 7120. The *patU* phenotype is a delayed induction of clusters of differentiating cells following nitrogen deprivation that incompletely resolve to a variable number in the Mch clusters, as well as single heterocysts. The *patN* phenotype is the unique Msh pattern that forms within the same time frame as the normal spacing pattern. The *patN* Msh pattern is similar to the symbiotic spacing pattern in regions of highest nitrogenase activity (Meeks and Elhai, 2002).

Thus, PatS, PatN and, perhaps, PatU are negative regulatory elements of the mechanism(s) that establishes the heterocyst-spacing pattern, whereas HetN and PatB appear to function as negative elements in maintaining the pattern.

4.2.3. Genes encoding uncertain elements for initiation of differentiation

The *hetL* gene was identified in a survey of elements that suppress an overexpression of *patS* heterocyst inhibition phenotype in *Anabaena* 7120 (Liu and Golden, 2002). Such a suppressor effect indicates a positive regulatory role. However, a *hetL* insertion mutant had a wild-type heterocyst differentiation and dinitrogen-dependent growth phenotype. HetL almost entirely consists of pentapeptide repeats, with no obvious functional domains. Therefore, a specific role for HetL in heterocyst differentiation and/or pattern determination is difficult to predict at this time. The *N. punctiforme* genome contains multiple pentapeptide repeat proteins, none of which show strong similarity to HetL (Table 4).

4.2.4. Genes associated with commitment to terminal differentiation

The *hetC* (Kudyakov and Wolk, 1997) and *hetP* (Fernández-Piñas *et al.*, 1994) mutants of *Anabaena* 7120 initiate heterocyst differentiation as evidenced by the presence, in a spaced pattern in the filaments, of weakly fluorescent cells. However, these cells do not continue differentiation and do not fix N_2 in air, thus, they appear blocked prior to commitment to terminal differentiation. There have been no reports of regression of the partially differentiated *hetC* or *hetP* cells to the vegetative-cell state following re-addition of combined nitrogen, which would be consistent with such a categorization. The *hetC* mutant weakly fluorescent cells continue to divide in the absence of biomass increase (Xu and Wolk, 2001), similar to cells of hormogonium filaments (Meeks *et al.*, 2002). HetC has similarity to bacterial ABC protein exporters, whereas HetP has no sequence similarity. The genes are co-located in *Anabaena* 7120, but not in *N. punctiforme*. In addition, the *N. punctiforme* genome contains a second gene with similarity to *hetC* and two more with similarity to *hetP*. Because commitment to terminal differentiation is thought to require DNA replication (see Meeks and Elhai, 2002), it is reasonable to assume that more than these gene products are required for commitment.

Only three of the genes/gene products listed in Table 4 have been characterized genetically in both *N. punctiforme* and *Anabaena* 7120. Mutants with lesions in *ntcA*, *hetR* and *patS* yield identical phenotypes in the two organisms. The fact that these are among the gene products with highest sequence identity is probably not significant. Nevertheless, it would be beneficial to compare mutant phenotypes in two or more strains, especially in those gene products with less than 70% identity. Some of the gene products appear to be unique to *N. punctiforme* and *Anabaena* 7120, but many contain domains and motifs recognized by one or more algorithm and associated with either a biochemical or physiological function. However, in no case can the function be predicted as part of the regulation cascade of heterocyst differentiation in the absence of a mutant phenotype.

4.3. Genes encoding proteins required for maturation of the microoxic heterocyst and support of nitrogenase activity.

Mutants in this class differentiate cells with the overall appearance of a heterocyst, but the cultures are unable to fix N_2 in the presence, or in some cases absence, of O_2. The progression to terminal heterocyst differentiation results in structural and physiological changes that lead to the microoxic cytosol essential for nitrogenase expression. In addition to elimination of photosystem II-dependent O_2 production and enhanced respiratory O_2 consumption, an envelope bilayer is synthesized that is not present in vegetative cells and functions to retard diffusion of gases and the translocation of ions and other hydrophilic solutes (Wolk *et al.*, 1994). This envelope consists of an outer polysaccharide layer and an inner glycolipid layer, both of which are essential for O_2 protection (Murry and Wolk, 1989).

The structures of the polysaccharides have been chemically determined in three strains to consist of a 3:1 glucose-mannose backbone with various mono- and di-saccharide side chains (see Wolk, 2000). The glycolipids consist of very long chain (C_{26}–C_{32}) either keto- or hydroxyl-substituted alcohols, either ester- or ether-linked primarily to glucose (see Wolk, 2000). The length and modification of the hydrocarbon chain is more characteristic of a polyketide than a fatty acid. Utilizing the structural information, the basic biosynthetic pathways can be modeled, using similar types of pathways as precedence. Then, by applying genomic information to predict motifs and catalytic domains, the gene products that could be involved in synthesis of the monomers and their polymerization can be predicted; *e.g.*, what kind of enzyme is required and are any similar enzymes putatively encoded in the genome? This approach is in progress with the polysaccharides (Fan *et al.*, 2002).

The information currently available on synthesis and assembly of the unique heterocyst envelope is almost entirely derived from genes identified through functional genetic analyses (Table 5). Five genes have been identified as associated with the polysaccharide layer. They include regulation of synthesis by the two-component sensor histidine kinase, HepK (Zhu *et al.*, 1998), and receiver domain only response regulator, DevR (Campbell *et al.*, 1996; Hagen and Meeks, 1999). The genetic evidence is supported by biochemical evidence of predicted phosphotransfer from HepK to DevR (Zhou and Wolk, 2003). However, the target (either protein activation or protein synthesis) of the phosphorylated DevR has yet to be identified. Association of the remaining three gene products, which include two sugar transferases and a putative export component, with polysaccharide synthesis is supported by the respective mutant phenotype.

Table 5. Gene products essential for terminal differentiation of mature heterocysts functioning in an oxic environment.

Gene Product	Properties	*N. punctiforme* vs. *Anabaena* 7120[a]		Citation
A.Envelope1	Mutants lack the polysaccharide component	%identity	%similarity	
epK	Sensor histidine kinase with transmembrane domains.	68	78	1, 2
DevR	Response regulator lacking DNA-binding output domain.	93	98	3
HepA	ABC transporter, ATP-binding and transmembrane domain.	67	79	4
HepB	Glycosyltransferase.	74	86	5
HepC	Sugar transferase, gene adjacent to *hepA* in *N. punctiforme*.	53	68	5
B.Envelope2	Mutants lack the glycolipid component			
HglK	C-Terminal pentapeptide repeat, glycolipids synthesized but not assembled.	74	85	6
HglB (HetM)	Polyketide synthesis with N-terminal phosphopantetheine attachment site.	72	82	7
HglC	Beta-ketoacyl synthase with acyl transferase domain.	48	65 (76)	7
HglD	Beta-ketoacyl synthase domain only.	not in *N. punctiforme*		7
HglE	Beta-ketoacyl synthase with acyl transferase domain and N-terminal phosphopantetheine attachment site, chain elongation factor.	74	83	8
HetI	Phosphopantetheine-binding domain, no mutants analyzed.	66	81	7
DevA	ABC transporter, ATP-binding domain, glycolipids synthesized, but not assembled.	84	93	9
DevB	Membrane permease, glycolipds synthesized as DevA	69	81	9
DevC	Membrane permease, glycolipds synthesized as in DevA.	80	89	9

C. Cytosol	Proteins required for nitrogen fixation and nitrogen export.			
DevH	Regulatory protein with DNA binding domain.	99	99	10
Inv–2 copies	Neutral invertase, hydrolysis of sucrose, no mutants analyzed.	76; 86	85; 92	11
G6PD	Glucose-6-phosphate dehydrogenase.	94	96	12
6PGD	6-Phosphogluconate dehydrogenase, no mutants analyzed.	88	91	
ICD	Isocitrate dehydrogenase, no mutants analyzed.	91	96	13
HupS	Uptake hydrogenase, H_2 recycling.	88	93	14
HupL	Uptake hydrogenase, H_2 recycling.	88	93	14
Cox2	Aa_3-Cytochrome oxidase, A, B and C, heterocyst specific?	72-85	84-95	15
Cox3	Aa_3-Cytochrome oxidase, A, B and C, heterocyst specific?	75-88	86-94	15
Nif	See Table 6 and Figure 5.			
GlnA	Glutamine synthetase, ammonium assimilation.	91	96	16

[a]*Values based on BLAST results; multiple numbers and numbers in parentheses are to extra copies of the gene product relative to the primary copy.*

Citations: 1, Zhu et al., 1998; 2, Zhou and Wolk, 2003; 3, Hagen and Meeks, 1999; 4, Holland and Wolk, 1990; Wolk, 2000; 6, Black et al., 1995; 7, Bauer et al., 1997; 8, Campbell et al., 1997; Fiedler et al., 1998; 10, Hebbar and Curtis, 2000; 11, Curtti et al., 2002; 12, Summers et al., 1995; 13, Muro-Pastor et al., 1996; 14, Tamagnini et al., 2002; 15, Schmetter et al., 2002; 16, Turner et al., 1983.

Nine gene products are currently identified as being involved in either glycolipid synthesis or assembly, one (HetI) is included based on sequence and gene locus in lieu of a mutant phenotype. The identified presumptive enzymes are consistent with synthesis of a polyketide hydrocarbon chain rather than a fatty acid. However, this is not to imply that gene products, which are associated by motifs and protein architecture with enzymes of polyketide synthesis, will all be involved with glycolipid synthesis. All heterocyst-forming cyanobacteria most likely produce and export cyclic peptides (Christiansen *et al.*, 2001), which are non-ribosomally synthesized and contain side chains modified by enzymes analogous to polyketide-modifying enzymes. Such modifying and peptide synthetase-encoding genes are present in distinct clusters in the *N. punctiforme* genome (Meeks *et al.*, 2001). The functionally identified ABC transporter (DevABC) is required for assembly of the layer, but not for synthesis of the glycolipids (Fiedler *et al.*, 1998), presumably through translocation of the lipid across the cell and, perhaps, outer membrane. The

HglK protein is also not required for synthesis of a glycolipid, and is modeled as essential for assembly (Black *et al.*, 1995). Except for both HglC, of which there are two possible gene copies in the N. *punctiforme* genome, and HglD, the protein architecture of which is absent in the *N. punctiforme* genome, the remaining proteins have relatively high sequence similarity.

The metabolic potential of heterocysts has been modeled using a relatively large published biochemical and physiological database (Wolk *et al.*, 1994), but there has been little genetic confirmation of those experimental results. With two exceptions, the cytosolic gene products listed in Table 5 and identified primarily by genetic analyses, are involved in reductant generation, cycling, and consumption. The putative DNA-binding protein, DevH, was identified by a DNA hybridization transcriptional screen (Hebbar and Curtis, 2000). The mutant differentiated heterocysts, but the culture failed to reduce acetylene or to grow with N_2 as the nitrogen source, and the heterocysts lacked the characteristic polar bodies. These phenotypic properties imply an inability to induce transcription of one or more key factors necessary for continued physiological maturation. The signal-transduction pathway that could activate DevH and other genes whose transcription could be influenced by DevH are unknown.

Glutamine synthetase (GS, encoded by *glnA*), which incorporates ammoniun into glutamate to form glutamine in an ATP-dependent reaction, is about twice as concentrated, with twice the specific activity, in heterocyst as in vegetative cells (Wolk *et al.*, 1994). The increased concentration of GS in heterocysts is ascribed to transcription utilizing a promoter sequence with similarity to that of the *nifH* gene (Tumer *et al.*, 1983) and markedly different from the σ^{70}-like promoter sequence of vegetative-cell genes (Curtis and Martin, 1994). The extent of utilization of this *nifH*-like promoter by other genes whose transcription is enhanced in heterocysts is unknown. The loss of photosystem II-dependent O_2 evolution also restricts heterocysts to reduced carbon catabolic mechanisms for generation of the reductant necessary to fuel both nitrogenase and respiratory O_2 consumption.

Sucrose has long been considered as the reduced carbon compound translocated from vegetative cells to heterocysts. Bidirectional (synthesis and cleavage) sucrose synthase enzyme assays of wild-type and sucrose synthease mutants implied, by elimination, that only an invertase activity could hydrolyze sucrose in heterocysts (Curatti *et al.*, 2002). If so, an invertase mutant should be unable to fix N_2, but this has not been demonstrated. Genes in the *N. punctiforme* genome encode two putative higher plant-related neutral invertases. Enzyme assays have supported catabolism of glucose and fructose in heterocysts by the activity of the initial enzymes of oxidative pentose phosphate metabolism, glucose-6-phosphate dehydrogenase (G6PD) and 6-phosphogluconate dehydrogenase (6PGD) (Wolk *et al.*, 1994). Mutant analyses have verified that G6PD is essential for nitrogen fixation in *N. punctiforme* (Summers *et al.*, 1995). 6PGD is also likely to be required for heterocyst-based nitrogen fixation, but verification has yet to be published. The *N. punctiforme* genome contains two additional genes, each encoding G6PD and 6PGD. Their physiological role is obscure because the alternative G6PDs could not complement mutation of the first identified gene (*zwf*), which apparently encodes the primary G6PD.

Muro-Pastor *et al.* (1996) established that isocitrate dehydrogenase (IDH) is present in high concentrations in heterocysts, but the structural gene could not be inactivated. IDH is essential for synthesis of 2-oxoglutarate, the substrate for glutamate synthase-mediated glutamate formation, which apparently takes place in vegetative cells (Meeks and Elhai, 2002). IDH, together with G6PD and 6PGD, could all contribute to the NADPH pool.

One or more hydrogenases (either uptake or bifunctional) are also present in heterocysts (Tamagnini *et al.*, 2002). The *N. punctiforme* genome contains genes encoding only an uptake hydrogenase. The role of the hydrogenases is to recycle reducing equivalents lost through nitrogenase action as H_2 gas. Consistent with such a role, a *N. punctiforme* HupL mutant grew with N_2 as the sole nitrogen source, but also evolved measurable amounts of H_2 gas into the environment, in contrast to the wild-type strain (Lindberg *et al.*, 2002).

Both genetic and biochemical analyses imply that two cytochrome oxidase complexes participate in O_2 consumption in heterocysts of *Anabaena* 7120 (Schmetter *et al.*, 2002). Respiratory O_2 consumption is required to maintain the microoxic cytosol and could also contribute to the ATP pool. However, the extent of coupling of the respiratory electron transport to ATP synthesis has not been determined. The *N. punctiforme* genome contains four clusters of genes encoding cytochrome c oxidases, two of which are highly similar (76 to 94%) to the putative heterocyst-localized cytochrome oxidases of *Anabaena* 7120. Genome sequences will provide additional information on the metabolic potential of heterocysts, but mutant analyses and functional assays will be essential in establishing physiological relevance.

4.4. The nif *genes of* N. punctiforme

None of the nitrogenase (*nif*) and *nif*-related genes of *N. punctiforme* have been subjected to mutant analyses, as have very few in *Anabaena* 7120. Essentially all that is known of *nif* genes in these cyanobacteria is derived from the pioneering work of Rice *et al.* (1982) in defining gene sequence, physical map location, and transcription pattern and size (for details, see Vol. II). There has been considerable focus in *Anabaena* 7120 on gene-rearrangement events that excise elements interrupting the structural genes (Golden, 1996).

The major *nif*-gene cluster of *N. punctiforme* is depicted in Figure 5A and the sequence relationship between *N. punctiforme* and *Anabaena* 7120 gene products is presented in Table 6. Except for multi-copy genes, such as *nifS* and *nifV*, in which the alternative copies may be involved in core biosynthetic pathways, the *nif* gene products are highly conserved between the two organisms. The overall organization is also nearly identical in the two organisms with a few notable exceptions.

First, present in the *N. punctiforme nif* cluster but absent in *Anabaena* 7120, is a gene encoding a globin (*glbN*), commonly called cyanoglobin. Cyanoglobin is widely, but inconsistently, distributed in cyanobacteria (Potts, 2000). The physiological role of cyanoglobin is obscure. The presence of cyanoglobin in both non-symbiotic and non-nitrogen-fixing strains (Potts, 2000) implies that it is not specifically associated with either nitrogen fixation or symbiotic competence.

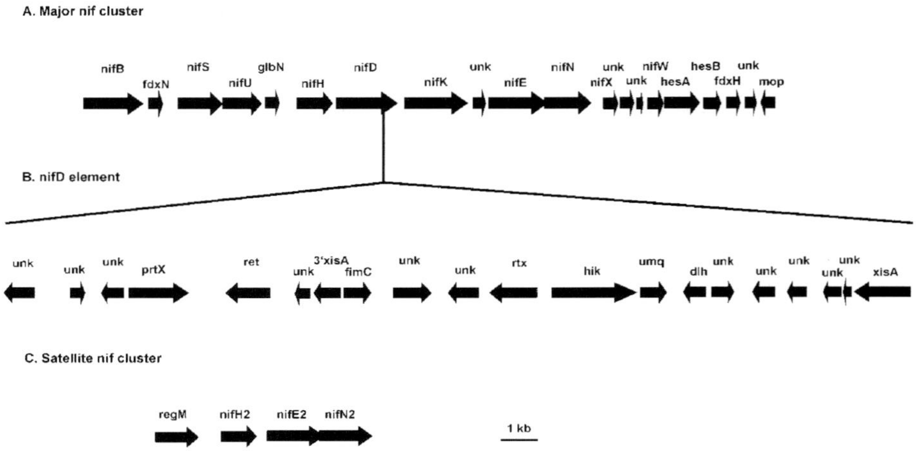

Figure 5. Physical organization of the major nif *cluster, the* nifD *element, and a satellite* nif *cluster in* N. punctiforme.

Second, although the *N. punctiforme nifD* gene contains an insertion element, there are no elements interrupting either the *fdxN* or the *hupL* (not shown in Figure 5) genes, however, both are present in *Anabaena* 7120.

Third, the *nifD* element is more than twice the size of the corresponding *nifD* element in *Anabaena* 7120. There have been two assemblies of the *N. punctiforme* shotgun sequence. Both assemblies yielded a 23.7-kb *nifD* element; however, computational annotation initially identified 20 ORFs and then later only 16 ORFs with three large gaps of unannotated sequence in which the missing ORFs previously resided. This discrepancy will be resolved when the genome sequence is completed. The ORF map of the *nifD* element in Figure 5B is based on the initial annotation. The two elements contain in common only the *xisA* gene, whose product catalyzes the site-specific recombination event that removes the element during the later stages of heterocyst differentiation and allows transcriptional read-through of the *nifHDK* operon (Golden, 1996), and an ORF identified as 3' from *xisA*. However, this ORF is quite some distance 3' of *xisA* in the *N. punctiforme nifD* element. The *N. punctiforme nifD* element contains genes encoding a bacterial reverse transcriptase (*ret*) and a sensor histidine kinase (*hik*) that have similarity to ORFs in *Anabaena* 7120, but they are not present within its 11-kb *nifD* element. Eleven of the putative ORFs encode conserved and unique hypothetical proteins, two of which are also found in *Synechococystis* 6803. The remaining ORFs encode proteins containing motifs that are associated with various bacterial proteins, including one each involved in S-layer (*rtx*) and fimbral (*fimC*) assembly, ubi- and mena-quinone (*umq*) synthesis, and dienelactone hydrolase (*dlh*), as well as one

Table 6. Nitrogenase synthetic, support, and structural gene products.

Gene product	Function	*N. punctiforme* vs. *Anabaena* 7120	
		%identify	%similarity
NifB	FeMo-cofactor biosynthesis, Fe and S donor	83	91
FdxN	Ferredoxin	76	85
NifS	Stabilization of the Fe protein	85	91
NifS2-3	Cysteine desulfurase?	37[x]	52-54[x]
NifU	Stabilizaton of the Fe protein	83	91
GlbN	Cyanoglobin, unknown function, inconsistent distribution	-	-
NifH	Fe protein – dinitrogenase reductase - FeMo-cofactor biosynthesis	87	91
NifH2	FeMo-cofactor biosynthesis?	84[x]	90[x]
NifH3	FeMo-cofactor biosynthesis?	83[x]	89[x]
NifD	MoFe protein, α subunit - dinitrogenase	81	91
NifK	MoFe protein, β subunit - dinitrogenase	77	87
NifE	FeMo-cofactor biosynthesis	85	89
NifE2	FeMo-cofactor biosynthesis?	60[x]	75[x]
NifN	FeMo-cofactor biosynthesis	84	92
NifN2	FeMo-cofactor biosynthesis?	71[x]	82
NifX	Determination of organic portion of FeMo-cofactor	73	89
NifW	Required for MoFe protein activity	71	86
FdxH	Heterocyst specific ferredoxin	85	94
NifZ	Required for MoFe protein activity	66	83
NifJ	Pyruvate-Flavodoxin oxidoreductase	68 (80)[y]	80 (87)[y]
NifV	Homocitrate synthase – FeMo-cofactor biosynthesis	89 (77)[y]	94 (88)[y]
NifV2-4	Homocitrate synthase – FeMo-cofactor biosynthesis or LeuA	26-33[x]	45-50[x]
NifT	Widely distributed protein of unknown function in nitrogen-fixing organisms	80	87
XisA	Site specific recombinase of the *nifD* insertion element	83	93
HesA	Mo processing	91	86
HesB	Widely distributed protein of unknown function in many organisms	73	86

[x]*Comparison to the primary gene product in N. punctiforme.*
[y]*Comparison of the N. punctiforme single gene product to the second gene product in Anabaena 7120.*

with high similarity to protein X (*prtX*) of *A. variabilis*. These putative proteins provide little insight into the selective advantage of the *nifD* element in *N. punctiforme*. The element is dispensable for laboratory growth in either the presence or absence of combined nitrogen (Brusca *et al.*, 1990; Meeks *et al.*, 1994). Nevertheless, a *nifD* element is nearly universally present in heterocyst-forming cyanobacteria. The dramatic differences in gene number, identity and organization within the *nifD* element in N. *punctiforme* and *Anabaena* 7120 imply that *nifD* elements in other cyanobacteria may be equally diverse. This diversity complicates models to explain the origins and retention of such elements interrupting cyanobacterial genes.

Additional *nif* genes are present. They are apart from the major *nif* cluster in *N. punctiforme*, but in the same chromosomal region. Sequences with similarity to *nifV* (homocitrate synthase), *nifZ*, *nifT* and *nifP* are about 9 kb 5' of *nifB*, whereas *hupLS* are about 10 kb 3' of *fdxH*. An additional copy of *nifV* is present in *Anabaena* 7120, but that copy is absent in *N. punctiforme*. Rather, *N. punctiforme* contains two copies of *nifV*-like genes whose gene products have markedly less sequence similarity with either of the two genes in *Anabaena* 7120 or their homologue in *N. punctiforme*. Most interesting is a cluster of four genes elsewhere in the genome (Figure 5C) that encode second copies of three proteins (NifH2, NifE2 and NifN2), the primary copies of which are likely to be involved in iron-molybdenum cofactor (FeMo-cofactor) biosynthesis for incorporation into the MoFe protein (dinitrogenase). NifH and NifEN appear to interact as a complex for the insertion of molybdenum into the FeMo-cofactor, with the NifB cofactor supplying the iron and sulfur and NifX specifying the homocitrate residue (Rangaraj and Ludden, 2002). The only identified copies of *nifB* and *nifX* in *N. punctiforme* are in the major *nif* cluster.

The protein encoded by the gene (*regM*) 5' of *nifH2* in this *N. punctiforme* cluster has a N-terminal helix-turn-helix DNA-binding domain and a C-terminal molybdate-binding domain. The detection of this cluster of *nif*-associated genes in *N. punctiforme* illustrates both the limitation and the power of genome sequence. The major limitation is that sequence information alone does not provide evidence of function. *N. punctiforme* synthesizes only a heterocyst-localized Mo-nitrogenase, therefore, one can only surmise what might be the roles of the proteins encoded by the *nifH2E2N2* cluster. If they are not for synthesis of the FeMo-cofactor, maybe they are for an alternative nitrogenase, which was either lost or is yet to be acquired by *N. punctiforme*. If they are not associated with such an alternative nitrogenase, it is difficult to resist speculating that, in response to the presence of molybdenum (sensed by the RegM protein), either dependent on or irrespective of the environmental nitrogen source, *N. punctiforme* induces the synthesis of enzymes for the biosynthesis of a different type of molybdenum-containing cofactor. This cofactor could lack both an organic moiety (and so be independent of NifX) and the presence (or source) of the Fe-S components (and so be independent of NifB) of the FeMo-cofactor, and may also differ from the molybdopterin-type of cofactor as found in nitrate reductases (Rubio *et al.*, 1999).

In addition to verification of the expected, as is illustrated in the basic organization of the major *nif* cluster in *N. punctiforme*, a power of genome sequencing is this discovery of the unexpected and unknown. Such genome sequence information leads to insightful and specific mutagenesis and expression assays (transcriptomics), as well as biochemical analyses (proteomics). From this approach, a new unanticipated line of research will emerge that contributes to understanding the evolution of cellular processes and competitive growth of an organism.

4.5. Ammonium and nitrate assimilation in N. punctiforme

By differentiating nitrogen-fixing heterocysts, *N. punctiforme* and its relatives have solved the problem of growth in the absence of combined nitrogen in their self-perpetuated oxic environment. One might then assume that *N. punctiforme* would have a relatively limited capacity to assimilate other forms of nitrogen. That appears not to be the case. Culture studies have established that, as predicted, *N. punctiforme* will grow using ammonium and nitrate/nitrite as the nitrogen sources and, when doing so, will repress heterocyst differentiation (Campbell and Meeks, 1992). Moreover, the growth rate on ammonium is about 1.6-fold faster than that supported by N_2 (Summers *et al.*, 1996). Present in the genome are two gene clusters, one containing five genes that encode a urea ABC transporter and another of four genes encoding urease and an associated maturation protein. The gene organization is similar to that in *Anabaena* 7120 (Valladares *et al.*, 2002) and the gene products have 81-97% similarity in the two organisms. In addition, there are 43 genes encoding putative amino-acid transporters (plus 13 periplasmic amino acid-binding proteins). Growth of *N. punctiforme* on urea and amino acids as sole nitrogen sources has not been systematically examined. Such a robust potential for amino-acid transport was not anticipated. Cyanobacteria are well known for their limited transcriptional regulation of amino acid-biosynthesis enzymes (Doolittle, 1979); why then retain a capacity to take up a metabolic end-product if it does not reduce the biosynthetic costs? Do these porters have another physiological role?

An unexpected observation is the potential robustness of both the ammonium- and nitrate-transport capacity. Two genes are present in the *N. punctiforme* genome whose gene products align with ammonium/methylammoniun transporters (*amt*) from a variety of prokaryotic and eukaryotic sources. *Synechocystis* 6803 and *Anabaena* 7120 each contain three genes encoding putative Amt proteins. The three genes are distributed at single sites in the *Synechocystis* 6803 genome, but are contiguous in the *Anabaena* 7120 chromosome in the order alr0990, alr0991 and alr0992. The alr0991 gene product shows the highest homology (58% identity) to sll0108 of *Synechocystis* 6803, which has been experimentally demonstrated to function as the primary ammonium transporter in that organism (Montesinos *et al.*, 1998). The two *N. punctiforme* putative Amt proteins are most homologous to alr0990 (84% identity) and alr0992 (86% identity), respectively. They are singly spaced in the *N. punctiforme* chromosome and show only 31% sequence identity (46% similarity) with each other. Neither of the *N. punctiforme* gene products have strong homology to sll0108 (30 and 37% identities) or alr0991 (34 and 41% identities).

Anabaena 71210 contains a fourth gene identified as encoding Amt1, although this identify is not supported by experimental evidence. *N. punctiforme* contains two genes encoding proteins with 86% similarity to this putative Amt1, but all three proteins are most similar in sequence and domain architecture to inositol monophosphatase, indicating their likely misidentification as an Amt. The Amt transporters could use as substrate either protonated or unprotonated NH_3. Because NH_3 can freely diffuse across the membrane, it has been assumed that NH_4^+ is the transport substrate at neutral and acidic external pH. However, recent experimental evidence supports the designation of these proteins as facilitators of the diffusion of a gas (Soupene, Lee and Kustu, 2002), in this case, unprotonated NH_3, at a rate consistent with the nitrogen metabolic demands for rapid growth. Also consistent with this designation is sequence analyses identifying the rhesus (Rh) blood group substance as a paralogue of the Amt proteins and the hypothesis that Rh proteins are CO_2 gas channels (Soupene, King, *et al.*, 2002). Functional analyses will be required to establish whether one or both Amt proteins transport low concentrations of ammonia in *N. punctiforme*.

N. punctiforme may synthesize two, or possibly three, nitrate/nitrite transport systems. In freshwater cyanobacteria, nitrate and nitrite uptake is through an ABC-type transport complex encoded by *nrtA*, *nrtB*, *nrtC* and *nrtD*. Genes encoding nitrate transporters in cyanobacteria are commonly flanked by one or both of the genes encoding nitrite reductase (*nirA*) and nitrate reductase (*narB*) and are expressed as an operon (Frías *et al.*, 1997). As an example, the gene organization in *Anabaena* 7120 is *nirA-nrtABCD-narB* (Frías *et al.*, 1997). In the marine filamentous cyanobacterium *Trichodesmium* sp. strain WH 9601, a permease encoded by *napA* transports nitrate/nitrite and the genes are organized as *nirA-napA-narB* (Wang *et al.*, 2000). *N. punctiforme* contains a *nirA-napA-narB* gene cluster with 73% NapA sequence similarity to the *Trichodesmium* 9601 NapA. This observation indicates that the distribution of cyanobacterial nitrate/nitrite transporters is not a sole consequence of a marine *versus* freshwater habitat. Moreover, three genes with high similarity to *nrtABC*, which encode components of the nitrate/nitrite ABC transporter, are present as the sole constituents of a contig; *ntrD* is likely to appear upon completion of the genome sequence. Although function must be established, *N. punctiforme* appears to be unique among cyanobacteria in that it can assimilate nitrate and nitrite either concurrently through both transporters, or through one or the other dependent upon environmental conditions. Genes with less similarity to *nrtC* and *nrtD* of *Anabaena* 7120 are contiguous in another chromosomal location, not at the end of a contig. Because it is difficult to distinguish between bicarbonate and nitrate/nitrite transporters by sequence, it is not clear what the actual substrate of these gene products might be in the absence of functional assays.

When microbial ecologists, including phycologists, examine field samples, the presence of heterocyst-forming, nitrogen-fixing cyanobacteria can be scored because of the immediate microscopic detection of the distinctive heterocysts. Often, but not always, these species appear in nutrient-poor habitats. Non-heterocyst-forming species tend to be enumerated from nutrient-rich habitats. Although recognizing that multiple environmental factors may limit growth of a physiological group of organisms, the above observations have led to the idea that heterocyst-forming species do not compete well in habitats enriched with high concentrations of combined nitrogen; this idea may be erroneous. Relatively low concentrations of either ammonium (3.2 to 7.6 μM) or nitrate (6.2 to 9.5 μM) can repress heterocyst differentiation (Meeks *et al.*, 1983). Therefore, heterocyst-forming cyanobacteria in which heterocyst differentiation has been repressed could be present, but undetected, in a sample that is analyzed by morphology alone. The probability of just such a scenario is strengthened by the inorganic nitrogen assimilatory capacity present in the genome of *N. punctiforme*. If expressed, and operative with efficient kinetic parameters, the redundant ammonia and nitrate/nitrite assimilatory capacity indicates *N. punctiforme* and its ilk could be highly competitive in both nitrogen-rich and nitrogen-poor habitats, but with their presence inconsistently scored. The robust nitrogen-assimilatory capacity, and either its retention or acquisition over evolutionary time in *N. punctiforme*, as well as the more rapid growth on ammonium relative to N_2 and repression of heterocyst differentiation by low concentrations of combined nitrogen, imply that heterocyst-localized nitrogen fixation is more of a survival response than a competitive way-of-life strategy. The biosynthetic costs of nitrogen fixation and the loss of a reproductive photosynthetic cell are apparently sufficiently high that the default growth mode may be to use sources of combined nitrogen.

5. SUMMARY AND CONCLUSIONS

A justification for basic scientific inquiry, beyond knowledge for the sake of knowledge, is to apply the concepts and/or end products generated to problems of the environment and human health and nutrition. Accrual of a considerable amount of information on heterocyst-localized nitrogen fixation remains before the process could be applied in, for example, an agronomic situation. Nevertheless, the organizationally simple cyanobacterial symbiotic associations may provide a more realistic paradigm for engineering new nitrogen-fixing plant symbioses than the more complex, but highly effective, rhizobia-legume associations. The genome and genetic tractability of *N. punctiforme* provide an experimental foundation and system to approach such applied goals. Only a minor fraction of the information will come from purely computational and comparative analysis of the genome sequence, although such analyses will be instrumental to the overall process.

I hope to have convinced the reader that the elements of the regulatory program that initiates and propagates terminal heterocyst differentiation in both free-living and symbiotic growth states, as well as initiating symbiotic interactions, have few (if any) homologues in other organisms. Therefore, identification of these elements will require functional genetic approaches; *i.e.*, the genes (and their gene products) need to be fished out of the genome. The rate of progress in fishing can be greatly accelerated by the application of high throughput global approaches of DNA microarray transcription assays and mass spectrometric protein analyses. The sequenced genome of *N. punctiforme* allows entry into such high throughput global approaches in experimental design, application, and analysis. The genome of *N.*

punctiforme contains a whole lot of gourmet fish, many of which are likely to be involved in free-living and symbiotic nitrogen fixation as well as the other numerous phenotypic traits expressed by this organism. They are primed to be caught, identified, characterized, and manipulated by any and all who want to fish.

ACKNOWLEDGEMENTS

I thank Elsie Campbell, Jeff Elhai, and Eric Kofoid for critical review of the manuscript. I thank John Ingraham for reminding us that fishing is fun (2001, Learning to fly fish, Annual Review of Microbiology 55, 1-19) and to whom this review is dedicated. Unpublished results cited in this chapter were generated with support of grants from the NSF (IBN 0080805) and DOE (DE-FG03-02ER63330).

REFERENCES

Adams, D. G. (2000). Heterocyst formation in cyanobacteria. *Curr. Opin. Microbiol.,* 3, 618-624.

Aravind, L., and Koonin, E. V. (2002). Classification of the caspase-hemoglobinase fold: Detection of new families and implications for the origin of eukaryotic separins. *Proteins, 46*, 355-367.

Arcondéguy, T., Jack, R., and Merrick, M. (2001). P_{II} signal transduction proteins, pivotal players in microbial nitrogen control. *Microbiol. Mol. Biol. Rev., 65*, 80-105.

Bauer, C. C., Ramaswamy, K. S., Endley, S., Scappino, L. A., Golden, J. W., and Haselkorn, R. (1997). Suppression of heterocyst differentiation in *Anabaena* PCC 7120 by a cosmid carrying wild-type genes encoding enzymes for fatty acid synthesis. *FEMS Microbiol. Lett., 151*, 23-30.

Black, K., Buikema, W. J., and Haselkorn, R. (1995). The *hglK* gene is required for localization of heterocyst-specific glycolipids in the cyanobacterium *Anabaena* sp. strain PCC 7120. *J. Bacteriol., 177*, 6440-6448.

Black, T. A., Cai, Y., and Wolk, C. P. (1993). Spatial expression and autoregulation of *hetR*, a gene involved in the control of heterocyst development in *Anabaena. Mol. Microbiol.,* 9, 77-84.

Black, T. A., and Wolk, C. P. (1994). Analysis of a Het⁻ mutation in *Anabaena* sp. strain PCC 7120 implicates a secondary metabolite in the regulation of heterocyst spacing. *J. Bacteriol., 176*, 2282-2292.

Brusca, J. S., Chastain, C. J. and Golden, J. W. (1990). Expression of the *Anabaena* sp. strain PCC 7120 *xisA* gene from a heterologous promoter results in excision of the *nifD* element. *J. Bacteriol., 172*, 3925-3931.

Buikema, W. J., and Haselkorn, R. (2001). Expression of the *Anabaena hetR* gene from a copper-regulated promoter leads to heterocyst differentiation under repressing conditions. *Proc. Natl. Acad. Sci. USA, 98*, 2729-2734.

Callahan, S. M. and Buikema, W. J. (2001). The role of HetN in maintenance of the heterocyst pattern in *Anabaena* sp. PCC 7120. *Mol. Microbiol., 40*, 941-950

Campbell, E. L, Brahamsha, B., and Meeks, J. C. (1998). Mutation of an alternative sigma factor in the cyanobacterium *Nostoc punctiforme* results in increased infection of its symbiotic plant partner, *Anthoceros punctatus. J. Bacteriol., 180*, 4938-4941.

Campbell, E. L., Cohen, M. F., and Meeks, J. C. (1997). A polyketide-synthase-like gene is involved in the synthesis of heterocyst glycolipids in *Nostoc punctiforme* strain ATCC 29133. *Arch. Microbiol., 167*, 251-258.

Campbell, E. L, Hagen, K. D., Cohen, M. F., Summers, M. L., and Meeks, J. C. (1996). The *devR* gene product is characteristic of receivers of two component regulatory systems and is essential for heterocyst development in the filamentous cyanobacterium *Nostoc* sp. strain ATCC 29133. *J. Bacteriol., 178*, 2037-2043.

Campbell, E. L. and Meeks, J. C. (1989). Characteristics of hormogonia formation by symbiotic *Nostoc* spp. in response to the presence of *Anthoceros punctatus* or its extracellular products. *Appl. Environ. Microbiol., 55*, 125-131.

Campbell, E. L. and Meeks, J. C. (1992). Evidence for plant-mediated regulation of nitrogenase expression in the *Anthoceros-Nostoc* symbiotic association. *J. Gen. Microbiol., 138*, 473-480.

Campbell, E. L., Wong, F. C. Y., and Meeks, J. C. (2003). DNA binding properties of the HrmR protein of *Nostoc punctiforme* responsible for transcriptional regulation of genes involved in hormogonium differentiation. *Mol. Microbiol., 47*, 573-582.

Castenholz, R. W. (2001). Phylum BX. Cyanobacteria oxygenic photosynthetic bacteria. In D. R. Boone and R. W. Castehnolz (Eds.), *Bergeys's manual of systematic bacteriology*, 2nd edition, Volume One, The Archaea and the deeply branching and phototrophic Bacteria (pp. 473-599). New York: Springer.

Christiansen, G., Dittmann, E., Ordorika, L. V., Rippka, R., Herdman, M., and Börner, T. (2001). Nonribosomal peptide synthetase genes occur in most cyanobacteria genera as evidenced by their distribution in axenic strains of the PCC. *Arch. Microbiol., 176*, 452-458.

Cohen, M. F., and Meeks, J. C. (1996). A hormogonium regulating locus, *hrmUA*, of the cyanobacterium *Nostoc punctiforme* strain ATCC 29133 and its response to an extract of a symbiotic plant partner *Anthoceros punctatus*. *Mol. Plant-Microbe Interact., 10*, 280-289.

Cohen, M. F., Wallis, J. G., Campbell, E. L., and Meeks, J. C. (1994). Transposon mutagenesis of *Nostoc* sp. strain ATCC 29133, a filamentous cyanobacterium with multiple cellular differentiation alternatives. *Microbiol., 140*, 3233-3240.

Cohen, M. F., and Yamasaki, H. (2000). Flavonoid-induced expression of a symbiosis-related gene in the cyanobacterium *Nostoc punctiforme*. *J. Bacteriol., 182*, 4644-4646.

Curatti, L., Flores, E., and Salerno, G. (2002). Sucrose is involved in the diazotrophic metabolism of the heterocyst-forming cyanobacterium *Anabaena* sp. *FEBS Lett., 513*, 175-178.

Curtis, S. E., and Martin, J. A. (1994). The transcription apparatus. In D. A. Bryant (Ed.), *The molecular biology of cyanobacteria* (pp. 613-639). The Netherlands: Kluwer Academic Publishers.

Damerval, T., Guglielmi, G., Houmard, J., and Tandeau de Marsac, N. (1991). Hormogonium differentiation in the cyanobacterium *Calothrix*: A photoregulated developmental process. *Plant Cell, 3*, 191-201.

Doherty, H. M., and Adams, D. G. (1999). The organization and control of cell division genes expressed during differentiation in cyanobacteria. In G. A. Peschek, W. Loffelhardt, and G. Schmetterer (Eds), *The phototrophic prokaryotes* (pp. 453-461). New York: Kluwer Academic/Plenum Publishers.

Doolittle, W. F. (1979). The cyabobacterial genome, its expression and control of that expression. *Adv. Microbial Physiol., 20,* 1-102.

Downie, J. A., and Walker, S. W. (1999). Plant responses to nodulation factors. *Curr. Opin. Plant Biol., 2*, 483-489.

Elhai, J., and Wolk, C. P. (1990). Developmental regulation and spatial pattern of expression of the structural genes for nitrogenase in the cyanobacterium *Anabaena*. *EMBO J., 9*, 3379-3388.

Enderlin, C. S., and Meeks, J. C. (1983) Pure culture and reconstitution of the *Anthoceros-Nostoc* symbiotic association. *Planta, 158*, 157-165.

Ernst, A., Black, T., Cai, Y., Panoff, J.-M., Tiwari, D. N., and Wolk, C. P. (1992). Synthesis of nitrogenase in mutants of the cyanobacterium *Anabaena* sp. strain PCC 7120 affected in heterocyst development or metabolism. *J. Bacteriol., 174*, 6025-6032.

Fan, Q., Li, Y., Wolk, C. P., Kaneko, T., and Tabata, S. (2002). Identification, by mutation and complementation, of developmental genes of *Anabaena* sp. strain PCC 7120. In L. A. Sherman and Y. Takahashi (Eds.). *Microbial and plant metabolism – function through genomics*. (pp. 13-14). Maui, Hawaii: NSF and JSPS

Fernández-Piñas, F., Leganés, F., and Wolk, C. P. (1994). A third genetic locus required for the formation of heterocysts in *Anabaena* sp. strain PCC 7120. *J. Bacteriol., 176*, 5277-5283.

Fiedler, G., Arnold, M., Hannus, S., and Maldener, I. (1998). The DevBCA exporter is essential for envelope formation in heterocysts of the cyanobacterium *Anabaena* sp. strain PCC 7120. *Mol. Microbiol., 27*, 1193-1202.

Fogg, G. E. (1949). Growth and heterocyst production in *Anabaena cylindrica* Lemm. II. In relation to carbon and nitrogen metabolism. *Ann. Bot., (NS), 13*, 241-259.

Forchhammer, K., and Tandeau de Marsac, N. (1994). The P_{II} potein in the cyanobacterium *Synechococcus* sp. strain PCC 7942 is modified by serine phosphorylation and signals the cellular N-status. *J. Bacteriol., 176*, 84-91.

Frías, J. E., Flores, E., and Herrero, A. (1997). Nitrate assimilation gene cluster from the heterocyst-forming cyanobacterium *Anabaena* sp. strain PCC 7120. *J. Bacteriol., 179*, 477-486.

Galperin, M. Y., Nikolskaya, A. N., and Koonin, E. V. (2001). Novel domains of the prokaryotic two-component signal transduction systems. *FEMS Microbiol. Lett., 203*,11-21.

Golden, J. (1998). Programmed DNA rearrangements in cyanobacteria. In F. J. de Bruijn, J. R. Lupski, and G. M. Weinstock (Eds.). *Bacterial genomes physical structure and analysis* (pp. 162-173). New York: Chapman and Hall.

Golden, J. W., and Yoon, H.-S. (1998). Heterocyst formation in *Anabaena. Curr. Opin. Microbiol., 1*, 623-629.

Golden, S. S., Johnson, C. H., and Kondo, T. (1998). The cyanobacterial circadian system: A clock apart. *Curr. Opin. Microbiol., 1*, 669-673.

Hagen, K. D. and Meeks, J. C. (1999). Biochemical and genetic evidence for the participation of DevR in a phosphorelay signal transduction pathway essential for heterocyst maturation in *Nostoc punctiforme* ATCC 29133. *J. Bacteriol., 181*, 4430-4434.

Hanson, T. E., Forchhammer, K., Tandeau de Marsac, N., and Meeks, J. C. (1998). Characterization of the *glnB* gene product of *Nostoc punctiforme* strain ATCC 29133: *glnB* or the P_{II} protein may be essential. *Microbiol., 144*, 1537-1547.

Haselkorn, R. (1998). How cyanobacteria count to 10. *Science, 282*, 891-892.

Hebbar, P. B., and Curtis, S. E. (2000). Characterization of *devH*, a gene encoding a putative DNA binding protein required for heterocyst function in *Anabaena* sp. strain PCC 7120. *J. Bacteriol., 182*, 3572-3581.

Herdman, M., Janvier, M., Rippka, R., and Stanier, R.Y. (1977). Genome size of cyanobacteria. *J. Gen. Microbiol., 111*, 73-85.

Herdman, M., and Rippka, R. (1988). Cellular differentiation: Hormogonia and baeocytes. *Methods Enzymol., 167*, 232-242.

Herrero, A., Muro-Pastor, A. M., and Flores, E. 2001. Nitrogen control in cyanobacteria. *J. Bacteriol., 183*, 411-425.

Hess, W. R., Rocap, G., Ting, C. S., Larimer, F., Stilwagen, S., Lamerdin, J., and Chisholm, S. W. (2001). The photosynthetic apparatus of *Prochlorococcus*: Insights through comparative genomics. *Photosynth. Res., 70*, 53-71.

Hoch, J. A., and Varughese, K. I. (2001). Keeping signals straight in phosphor-relay signal transduction. *J. Bacteriol., 183*, 4941-49494.

Holland, D., and Wolk, C. P. (1990). Identification and characterization of *hetA*, a gene that acts early in the process of morphological differentiation of heterocysts. *J. Bacteriol., 172*, 3131-3137.

Hunsucker, S. W., Tissue, B. M., Potts, M., and Helm, R. (2001). Screening protocol for the ultraviolet-photoprotective pigment scytonemin. *Anal. Biochem., 288*, 227-230.

Johansson, C., and Bergman, B. (1994). Reconstitution of the symbiosis of *Gunnera mannicata* Linden: cyanobacterial specificity. *New Phytol., 126*, 643-652.

Jones, K. M., Buikema, W. J., and Haselkorn, R. (2003). Heterocyst-specific expression of *patB*, a gene required for nitrogen fixation in *Anabaena* sp. strain PCC 7120. *J. Bacteriol., 185*, 2306-2314.

Joseph, N. A., and Adams, D. (2000). Use of transposon-lux mutagenesis to study the *Nostoc-Blasia* symbiosis. In R. Guerrero (ed.), *10th International symposium on phototrophic prokaryotes*, program and abstracts (p.125). Barcelona: ISPP

Kaneko, T., Nakamura, Y., Wolk, C. P., *et al.* (2001). Complete genome sequence of the filamentous nitrogen-fixing cyanobacterium *Anabaena* sp. strain PCC 7120. *DNA Res., 8*, 205-213

Kaneko, T., Sato, S., Kotani, H., *et al.* (1996). Sequence analysis of the genome of the unicellular cyanobacterium *Synechocystis* sp. strain 6803. II. Sequence determination of the entire genome and assignment of potential protein-coding regions. *DNA Res., 3*, 109-136.

Khudyakov, I., and Wolk, C. P. (1996). Evidence that the *hanA* gene encoding for HU protein is essential for heterocyst differentiation in, and cyanophage A-4(L) sensitivity of, *Anabaena* sp. strain PCC 7120. *J. Bacteriol., 178*, 3572-3577.

Khudyakov, I., and Wolk, C. P. (1997). *hetC*, a gene coding for a protein similar to bacterial ABC protein exporters, is involved in early regulation of heterocyst differentiation in *Anabaena* sp. strain PCC 7120. *J. Bacteriol., 179*, 6971-6978.

Knight, C. D. and Adams, D. G. (1996). A method for studying chemotaxis in nitrogen fixing cyanobacterium-plant symbioses. *Physiol. Mol. Plant Pathol., 49*, 73-77.

Koonin. E. V., and Aravind, L. (2002). Origin and evolution of eukaryotic apoptosis: The bacterial connection. *Cell Death Differ., 9*, 394-404.

Kunst, F., Ogasawara, N., Moszer, I., *et al.* (1997). The complete genome sequence of the Gram-positive bacterium *Bacillus subtilis*. *Nature, 390*, 249-256.

Lamb, J. R., Tugendreich, S. and Hieter, P. (1995). Tetratrico peptide repeat interactions: To TRP or not to TPR? *Trends Biochem. Sci., 20*, 257-259.

Leganés, F., Fernandéz-Piñas, F., and Wolk, C. P. (1994). Two mutations that block heterocyst differentiation have different effects on akinete differentiation in *Nostoc ellipsosporum*. *Mol. Microbiol., 12*, 679-684.

Levin, P. A., $ Losick, R. (2000). Asymmetric division and cell fate during sporulation in *Bacillus subtilis*. In Y. V. Brun and L. J. Skimkets (eds), *Prokaryotic development* (pp.167-189).Washington, D.C.: ASM Press.

Liang, J., Scappino, L., and Haselkorn, R. (1992). The *patA* gene product, which contains a region similar to CheY of *Escherichia coli*, controls heterocyst pattern formation in the cyanobacterium *Anabaena* 7120. *Proc. Natl. Acad. Sci. USA, 89*, 5655-5659.

Liang, J., Scappino, L., and Haselkorn, R. (1993). The *patB* gene product, required for growth of the cyanobacterium *Anabaena* sp. strain PCC 7120 under nitrogen-limiting conditions, contains ferredoxin and helix-turn-helix domains. *J. Bacteriol., 175*, 1697-1704.

Lindberg, P., Schültz, K., Happe, T., and Lindberg, P. (2002). A hydrogen-producing, hydrogenase-free mutant strain of *Nostoc punctiforme* ATCC 29133. *Intl. J. Hydrogen Energy, 27*, 1291-1296.

Lindblad, P., Hällbom, L., and Bergman, B. (1985). The cyanobacterium-*Zamia* symbiosis: C_2H_2-reduction and heterocyst frequency. *Symbiosis, 1*, 19-28.

Liu, D., and Golden, J. W. (2002). *hetL* overexpression stimulates heterocyst formation in *Anabaena* sp. strain PCC 7120. *J. Bacteriol., 184*, 6873-6881.

Luque, I., Flores, E., and Herrero, A. (1994). Molecular mechanisms of the operation of nitrogen control in cyanobacteria. *EMBO J., 13*, 2862-2869.

Lynn, M. E., Battle, J. A., and Ownby, J. W. (1986). Estimation of gene expression in heterocysts of *Anabaena variabilis* by using DNA-RNA hybridization. *J. Bacteriol., 136*, 1695-1699.

Martin, W., Rujan, T., Richly, E., Hansen, A., Cornelsen, S. Lins, T., Leister, D., Stoebe, B., Hasegawa, M., and Penny, D. (2003). Evolutionary analysis of *Arabidopsis*, cyanobacterial, and chloroplast genomes reveals plastid phylogeny and thousands of cyanobacterial genes in the nucleus. *Proc. Natl. Acad. Sci. USA, 99*, 12246-12251.

Mazel, D., Houmard, J., Castets, A. M., and Tandeau de Marsac, N. (1990). Highly repetitive DNA sequences in cyanobacterial genomes. *J. Bacteriol., 172*, 2755-2761.

Meeks, J. C. (1998). Symbiosis between nitrogen-fixing cyanobacteria and plants. *Bioscience, 48*, 266-276.

Meeks, J. C. (2003). Symbiotic interactions between *Nostoc punctiforme*, a multicellular cyanobacterium, and the hornwort *Anthoceros punctatus*. *Symbiosis, 35*, 55-71.

Meeks, J. C., Campbell, E. L., and Bisen, P. S. (1994). Elements interrupting nitrogen fixation genes in cyanobacteria: presence and absence of a *nifD* element in clones of *Nostoc* sp. strain Mac. *Microbiol., 140*, 3225-3232.

Meeks, J. C., Campbell, E. L., Summers, M. L., and Wong, F. C. (2002). Cellular differentiation in the cyanobacterium *Nostoc punctiforme*. *Arch. Microbiol., 178*, 395-403.

Meeks, J. C., and Elhai, J. (2002). Regulation of cellular differentiation in filamentous cyanobacteria in free-living and plant associated symbiotic growth states. *Microbiol. Mol. Biol. Rev., 65*, 94-121.

Meeks, J. C., Elhai, J. Thiel, T., Potts, M., Larimer, F., Lamerdin, J., Predki, P., and Atlas, R. (2001). An overview of the genome of *Nostoc punctiforme*, a multicellular, symbiotic cyanobacterium. *Photosynth. Res., 70*, 85-106.

Meeks, J. C., Wycoff, K. L., Chapman, J. S., and Enderlin, C. S. (1982). Regulation of expression of nitrate and dinitrogen assimilation by *Anabaena* species. *Appl. Environ. Microbiol., 45*, 1351-1359.

Mollenhauer, D., Mollenhauer, R., and Kluge, M. (1996). Studies on initiation and development of the partner association in *Geosiphon pyriforme* (Kutz) v. Wettstein, a unique encocytobiotic system of a fungus (Glomales) and cyanobacterium *Nostoc punctiforme* (Kutz) Hariot. *Protoplasma, 193*, 3-9

Montesinos, M. L., Muro-Pastor, A. M., Herrero, A., and Flores, E. (1998). Ammonium/ methylammonium permeases of a cyanobacterium, identification and analysis of three nitrogen-regulated *amt* genes in *Synechocystis* sp. PCC 6803. *J. Biol. Chem., 273*, 31463-31470.

Montgometry, B. L., and Lagarias, J. C. (2002). Phytochrome ancestry: Sensors of bilins and light. *Trends Plant Sci., 7*, 357-366.

Muro-Pastor, A. M., Valladares, A., Flores, E., and Herrero, A. (2002). Mutual dependence of the expression of the cell differentiation regulatory protein HetR and the global nitrogen regulator NtcA during heterocyst development. *Mol. Microbiol., 44*, 1377-1385.

Muro-Pastor, M. I., Reyes, J. C., and Florencio, F. J. (1996). The $NADP^+$-isocitrate dehydrogenase gene (*icd*) is nitrogen regulated in cyanobacteria. *J. Bacteriol., 187*, 4070-4076.

Muro-Pastor, M. I., Reyes, H. C., and Florencio, F. J. (2001). Cyanobacteria perceive nitrogen status by sensing intracellular 2-oxoglutarate levels. *J. Biol. Chem., 276*, 38320-38328.

Murry, M. A., and Wolk, C. P. (1989). Evidence that the barrier to the penetration of oxygen into heterocysts depends upon two layers of the cell envelope. *Arch. Microbiol., 151*, 469-474.

Nobles, D. R., Romanovicz, D. K., and Brown, R. M., Jr. (2001). Cellulose in cyanobacteria. Origin of vascular plant cellulose synthase? *Plant Physiol., 127*, 529-542.

Ohmori, M., Ikeuchi, M., Sato, N., Wolk, P., Kaneko, T., *et al.* (2001). Characterization of genes encoding multi-domain proteins in the genome of the filamentous nitrogen-fixing cyanobacterium *Anabaena* sp. strain PCC 7120. *DNA Res., 8*, 271-284.

Pate, J. S., Lindblad, P., and Atkins, C. A. (1988). Pathways of assimilation and transfer of fixed nitrogen in coralloid roots of cycad-*Nostoc* symbioses. *Planta, 176*, 461-471.

Perret, X., Staehelin, C., and Broughton, W. J. (2000). Molecular basis of symbiotic promiscuity. *Microbiol. Mol. Biol. Rev., 64*, 180-201.

Peters, G. A., and Mayne, B. C. (1974). The *Azolla-Anabaena azollae* relationship. I. Initial characterization of the association. *Plant Physiol., 53*, 813-819.

Phalip, V., Li, J. H., and Zhang, C. C. (2001). HstK, a cyanobacterial protein with both a serine/theorinine kinases domain and a histidine kinase domain: implication for the mechanism of signal transduction. *Biochem. J., 360*, 639-644.

Potts, M. (2000). Nostoc. In B. A. Whitton and M. Potts (Eds.), *The ecology of cyanobacteria, their diversity in time and space* (pp. 465-504). The Netherlands: Kluwer Academic Publishers.

Rangaraj, R. and Ludden, P. W. (2002). Accumulation of ^{99}Mo-containing iron-molybdenum cofactor precursors of nitrogenase on NifNE ad NifN, and NifX of *Azotobacter vinelandii. J. Biol. Chem., 277*, 40106-40111.

Rasmussen, U., Johansson, C., Renglin, A., Peterson, C., and Bergman, B. (1996). A molecular characterization of the *Gunnera-Nostoc* symbiosis; comparison with *Rhizobium*- and *Agrobacteirum*-plant interactions. *New Phytol., 133*, 391-398.

Rice, D., Mazur, B. J., and Haselkorn, R. (1982). Isolation and physical mapping of nitrogen fixation genes from the cyanobacterium *Anabaena* 7120. *J. Biol. Chem., 257*, 13157-13163.

Rippka, R., Deruelles, J., Waterbury, J. B., Herdman, M., and Stanier, R. Y. (1979). Generic assignments, strain histories and properties of pure cultures of cyanobacteria. *J. Gen. Microbiol., 111*, 1-61.

Rippka, R., and Herdman, M. (1992). *Pasteur Culture Collection of Cyanobacteria in Pure Culture* (pp. 4-57), Paris: Institut Pasteur.

Robinson, N. J., Rutherford, J. C., Pocock, M. R., and Cavet, J. S. (2000). Metal metabolism and toxicity: Repetitive DNA. In B. A.Whitton and M. Potts (Eds.), *The ecology of cyanobacteria their diversity in time and space* (pp. 443-463). The Netherlands: Kluwer Academic Publishers.

Rubio, L. M., Flores, E., and Herrero, A. (1999). Molybdopterin guanine dinucleotide cofactor in *Synechococcus* sp. nitrate reductase: Identification of *mobA* and isolation of a putative *moeB* gene. *FEBS Lett., 462*, 358-362.

Schmetterer, G., Pils, D., Ludwig, A., Wilken, C., Valladares, A., Herrero, A., and Flores, E. (2002). Oxygen protection in the heterocyst. In B. Bergman (Ed.), *Cyanofix final symposium: Cyanobacterial nitrogen fixation from molecules to ecological systems* (p. 62). Tomar, Portugal.

Shi, L., Potts, M., and Kennelly, P. J. (1998). The serine, threonine, and/or tyrosine-specific protein kinases and protein phosphatases of prokaryotic organisms: A family portrait. *FEMS Microbiol. Rev., 22*, 229-253.

Skimkets, L. J., and Brun, Y. V. (2000). Prokaryotic development: strategies to enhance survival. In Y. V. Brun and L. J. Skimkets (eds), *Prokaryotic development* (pp.1-7).Washington, D.C.: ASM Press.

Smith, T. F., Gaitazes, C., Saxena, K., and Neer, E. J. (1999). The WD repeat: A common architecture for diverse functions. *Trends Biochem. Sci., 24*, 181-185.

Söderbäck, E., Lindblad, P., and Bergman, B. (1990). Developmental patterns related to nitrogen fixation in the *Nostoc-Gunnera magellanica* Lam. symbiosis. *Planta, 182*, 355-362.

Soupene, E., King, N., Field, E., Liu, P., Niyogi, Huang, C.-H., and Kustu, S. (2002). Rhesus expression in a green alga is regulated by CO_2. *Proc. Natl. Acad. Sci. USA, 99*, 7769-7773.

Soupene, E., Lee, H., and Kustu, S. (2002). Ammonium/methylammonium transport (Amt) proteins facilitate diffusion of NH_3 bidirectionally. *Proc. Natl. Acad. Sci. USA, 99*, 3927-3931.

Steinberg, N. A., and Meeks, J. C. (1991). Physiological sources of reductant for nitrogen fixation activity in *Nostoc* sp. strain UCD 7801 in symbiotic association with *Anthoceros punctatus*. *J. Bacteriol., 173*, 7324-7329.

Stock, A. M., Robinson, V. L., and Gourdeau, P. N. (2000). Two-component signal transduction. *Annu. Rev. Biochem., 69*, 183-215.

Summers, M. L., Wallis, J. G., Campbell, E. L., and Meeks, J. C. (1995). Genetic evidence of a major role for glucose-6-phosphate dehydrogenase in nitrogen fixation and dark growth of the cyanobacterium *Nostoc* sp. strain ATCC 29133. *J. Bacteriol., 177*, 6184-6194.

Tamagnini, P., Axelsson, R., Lindberg, P., Oxelfelt, F., Wünschiers, R., and Lindblad, P., (2002). Hydrogenases and hydrogen metabolism of cyanobacteria. *Microbiol. Mol. Biol. Rev., 66*, 1-20.

Tandeau de Marsac, N. (1994). Differentiation of hormogonia and relationships with other biological processes. In D. A Bryant (Ed.) *The molecular biology of cyanobacteria* (pp. 825-842). The Netherlands:Kluwer Academic Publishers.

Tumer, N. E., Robinson, S. J., and Haselkorn, R. (1983). Different promoters for the *Anabaena* glutamine synthetase gene during growth using molecular or fixed nitrogen. *Nature, 306*, 337-342.

Valladares, A., Montesinos, M. L., Herrerso, A. and Flores, E. (2002). An ABC-type, high-affinity urea permease identified in cyanobacteria. *Mol. Microbiol., 43*, 703-715.

Vazquez-Bermudez, M. F., Herrero, A., and Flores, E. (2002). 2-oxoglutarate increases the binding affinity of the NtcA (nitrogen control) transcription factor for the *Synechococcus glnA* promoter. *FEBS Lett., 512*, 71-74.

Wang, Q., Li, H., and Post A. F. (2000). Nitrate assimilation genes of the marine diazotrophic, filamentous cyanobacterium *Trichodesmium* sp. strain WH9601. *J. Bacteriol., 182*, 1764-1767.

Williams, S. B., Vakonakis, I., Golden, S. S., and LiWang, A. C. (2002). Structure and function from the circadian clock protein KaiA of *Synechococcus elongatus*: a potential clock input mechanism. *Proc. Natl. Acad. Sci. USA, 99*, 15357-15362.

Wilmotte, A., and Herdman, M. (2001). Phylogenetic relationships among the cyanobacteria based on 16S rRNA sequences. In D. R. Boone and R. W. Castehnolz (Eds.), *Bergey's manual of systematic bacteriology*, 2nd edition, Volume One, The Archaea and the deeply branching and phototrophic Bacteria (pp. 487-493). New York: Springer.

Wolk, C. P. (2000). Heterocyst formation in *Anabaena*. In Y. V. Brun and L. J. Shimkets (Eds.), *Prokaryotic Development* (pp. 83-104). Washington, D.C.: American Society of Microbiology.

Wolk, C. P, Ernst, A., and Elhai, J. (1994). Heterocyst metabolism and development. In D. A. Bryant (Ed), *The molecular biology of cyanobacteria* (pp. 769-823). The Netherlands: Kluwer Academic Publishers.

Wong, F.C., and Meeks, J. C. (2001). The *hetF* gene product is essential to heterocyst differentiation and affects HetR function in the cyanobacterium *Nostoc punctiforme*. *J. Bacteriol., 183*. 26545-2661.

Wong, F. C., and Meeks, J. C. (2002). Establishment of a functional symbiosis between the cyanobacterium *Nostoc punctiforme* and the bryophyte hornwort *Anthoceros punctatus* requires genes involved in nitrogen control and initiation of heterocyst differentiation. *Microbiol., 148*, 315-323.

Xu, X., and Wolk, C. P. (2001). Role for *hetC* in the transition to a nondividing state during heterocyst differentiation in *Anabaena* sp. *J. Bacteriol., 141*, 183, 393-396.

Yoon, H.-S., and Golden, J. S. (1998). Heterocyst pattern formation controlled by a diffusible peptide. *Science, 282*, 935-938.

Yoon, H.-S., and Golden, J. S. (2001). PatS and products of nitrogen fixation control heterocyst pattern. *J. Bacteriol., 183*, 2605-2613.

Zhang, C.-C., Friry, A., and Peng, L. (1998). Molecular and genetic analysis of two closely linked genes that encode, respectively, a protein phosphatase 1/2A/2B homolog and a protein kinase homolog in the cyanobacterium *Anabaena* sp. strain PCC 7120. *J. Bacteriol., 180*, 2616-2622.

Zhang, C.-C., Gonzalez, L., and Phalip, V. (1998). Survey, analysis and genetic organization of genes encoding eukaryotic-like signaling proteins on a cyanobacterial genome. *Nucleic Acids Res., 26*, 3619-3625.

Zhou, R., Wei, X., Jiang, N., Li, H., Dong, Y., His, K-L., and Zhao, J. (1998). Evidence that HetR protein is an unusual serine-type protease. *Proc. Natl. Acad. Sci. USA, 95*, 4959-4963.

Zhou, R., and Wolk, C. P. (2002). Identification of an akinete marker gene in *Anabaena variabilis*. *J. Bacteriol., 184*, 2529-2532.

Zhou, R., and Wolk, C. P. (2003). A two-component system mediates developmentally regulated biosynthesis of a heterocyst polysaccharide. *J. Biol. Chem., 278*, 19939-19946.

Zhu, J., Kong, R., and Wolk, C. P. (1998). Regulation of *hepA* of *Anabaena* sp. strain PCC 7120 by elements 5' from the gene and by *hepK*. *J. Bacteriol., 180*, 4233-4242.

CHAPTER 5

THE *nif* GENES OF *RHODOBACTER CAPSULATUS, RHODOBACTER SPHAEROIDES* AND *RHODOPSEUDOMONAS PALUSTRIS*

R. HASELKORN[1] AND V. KAPATRAL[2]
[1]*Dept of Molecular Genetics and Cell Biology, University of Chicago, Chicago, IL 60637;* [2]*Integrated Genomics, 2201 West Campbell Park Drive, Chicago IL 60612, USA*

1. INTRODUCTION

The photosynthetic bacteria *Rhodobacter capsulatus* and its relatives were isolated by Howard Gest and shown by him to fix N_2 and to generate H_2 in the light. *R. capsulatus* became the preferred member of the family for genetic studies due to the identification of the generalized transducing agent (a bacteriophage) called GTA, which was shown to package and transfer DNA fragments from one strain to another with high efficiency (Marrs, 1974). *Rhodobacter sphaeroides* has been used extensively for biochemical and biophysical studies of the photosynthetic apparatus. One of the supreme mysteries of this field is the fact that photochemical reaction centers from *R. sphaeroides* can be purified and crystallized, whereas reaction centers from *R. capsulatus* cannot. Thus, there is a reasonable structure available for the *R. sphaeroides* reaction center but most of the molecular genetic studies, including amino-acid replacements, have been done using *R. capsulatus*.

Historically, genetic studies of nitrogen fixation in this group were carried out on *R. capsulatus*, starting with the isolation of nitrogen fixation (*nif*) mutants (Wall and Braddock, 1984) and their mapping using GTA. These mutants were used in turn as the starting point for the cloning and mapping of the *nif* genes (Avtges *et al.*, 1985). This work began with a library of chromosomal DNA fragments in a vector that could be transferred from *E. coli* to *R. capsulatus* by conjugation.

R. Palacios and W. E. Newton (eds.), Genomes and Genomics of Nitrogen-fixing Organisms, 71-82.

Complementation of the Wall *nif* mutants led to the isolation of DNA fragments that contained additional *nif* genes. These were identified initially by transposon insertion and transfer of the transposon to wild-type *R. capsulatus* and the creation thereby of new *nif* mutants. Eventually, the new *nif* genes were sequenced and *lac* fusions were constructed to study their regulation (Kranz and Haselkorn, 1985). Independently, Klipp and his collaborators cloned the *nif* genes of *R. capsulatus* and sequenced them (Klipp *et al.*, 1988). This last work led to the discovery of both multiple copies of the *nifA* gene, which explained why others had failed to isolate *nifA* mutants, and an alternative nitrogenase that lacks molybdenum and tungsten (Klipp *et al.*, 1988). Willison and Vignais also mapped *nif* genes in *R. capsulatus* using whole chromosome mobilization (Vignais *et al.*, 1985; Willison, 1993).

Global analysis of the *R. capsulatus* genome was begun by preparing a long-range physical map of the chromosome (Fonstein *et al.*, 1992), then constructing a cosmid library, and selecting an overlapping set of 192 cosmids that covered the chromosome and a 134-kb plasmid (Fonstein *et al.*, 1995). A fine-structure physical map was based on more than 500 mapped sites for six restriction enzymes. The sequencing of the cosmid set was carried out at the University of Chicago, The Institute for Genetics of the Academy of Sciences of the Czech Republic in Prague, and finally at Integrated Genomics, Inc. in Chicago (Vlcek *et al.*, 1997; Haselkorn *et al.*, 2001). Due to repeated elements, some of the overlaps in the original cosmid-based physical map were incorrect, leading to gaps in the final alignments. As of the spring of 2003, four gaps remained in the sequence. Two of these have been closed by lambda clones waiting to be sequenced. The last two gaps await closing by other methods, such as long-range PCR.

A current view of the genomes of the three species of photosynthetic bacteria is summarized in Table 1. The data for *R. capsulatus* are taken from the ERGO site at Integrated Genomics; the data for *R. sphaeroides* are from S. Kaplan; and data for *Rhodopseudomonas palustris* are from the Joint Genome Institute.

Table 1. Genome statistics

	R. capsulatus	*R. sphaeroides*	*R. palustris*
Genome size (Mb)	3.7	4.6	5.4
Number of contigs	9	14	2
Number of ORFs	3,616	4,488	5,208
Number of ORFs with function	2,594	3,096	3,539
Number of ORFs with no similarities	18	3	1
Ribosomal RNA operons	4	3	2

R. capsulatus has one chromosome and one 134-kb plasmid, so the number of contigs implies that there are still some unsequenced gaps in the data. *R. sphaeroides* has two chromosomes and five small plasmids, so seven gaps remain.

Finally, the *R. palustris* sequence is essentially complete. For each, the density of ORFs, about one ORF per kb, is typical of bacterial genomes. The frequency of functional assignments for the ORFs ranges from 72-67% for the three species and again this is typical for a good sequence analyzed using the tools in the ERGO suite.

The number of ribosomal RNA operons is not correlated with genome size at all. *R. palustris* has two such operons with the gene order 5S, 23S, tRNA-Ala, tRNA-Ile, 16S. One of the operons has, in addition, a gene encoding tRNA-Ser following the 16S gene. In *R. sphaeroides*, there are three complete operons: tRNA-Met, 5S, 23S, tRNA-Ala, tRNA-Ile, 16S. One of these operons has an additional tRNA-Ser gene at the 5' end. There is also a fragment of an operon consisting of a 16S RNA gene followed by tRNA-Val. In *R. capsulatus* there are four identical operons: tRNA-Met, 5S, 23S, tRNA-Ile, tRNA-Ala, 16S.

The best-studied system for nitrogen fixation remains that of *Klebsiella pneumoniae* and the system in *R. capsulatus* is similar in many details but with the following significant exceptions. First, nitrogenase activity can be modified, in response to either addition of ammonia or transfer of cells to darkness, by the covalent addition of an adenylyl group to the NifH protein. Second, as in *Azotobacter vinlandii*, there is an alternative nitrogenase that operates without molybdenum in its cofactor. Other minor differences will be noted below.

2. REGULATION OF THE NITROGEN-FIXATION SYSTEM

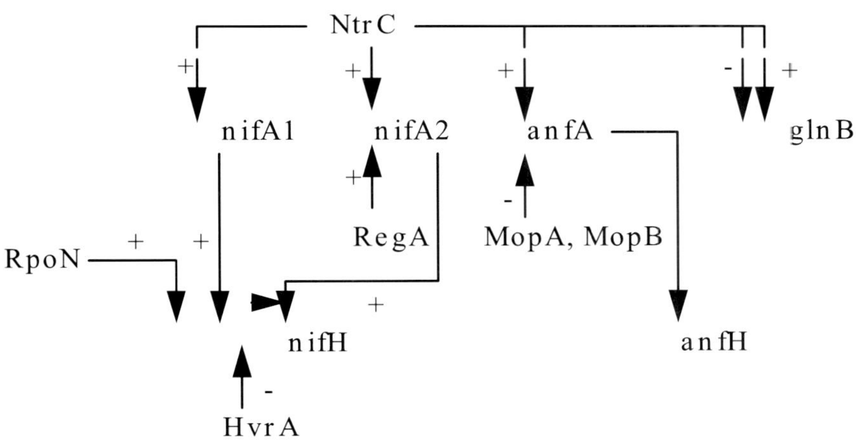

Figure 1. The regulatory circuits controlling expression of the nif *genes of* R. capsulatus. *See text for explanations.*

The regulatory circuits leading to transcription of the genes for nitrogenase and its co-factors are shown in Figure 1. The top row consists of NtrC alone. This protein was originally called NifR1 until its sequence similarity to the NtrC proteins of both *E. coli* and *K. pneumoniae* was noticed (Jones and Haselkorn, 1989). NtrC is a response regulator that binds to DNA and activates transcription by the major cellular RNA polymerase. NtrC is itself made active by phosphorylation by the histidine kinase NtrB, which responds to the nitrogen status of the cell. NtrC-P activates transcription of many genes, among them *nifA1, nifA2*, *anfA*, and *glnB* (P_{II}). NifA1 and NifA2 are essentially identical proteins (see below), either of which suffices to transcribe the structural *nif* operons, such as *nifHDK*. AnfA is a paralogue of NifA1 whose function is to activate transcription of the *anf* operons, such as *anfHDGK* (Masepohl *et al.*, 2002). All three factors (NifA1, NifA2, and AnfA) are inactivated by ammonia or O_2, making it impossible to express the *nif* operons under those conditions.

Expression of NifA2, but not NifA1, is enhanced by RegA, another response regulator that also functions in the expression of some genes for components of the photosynthetic apparatus (Elsen *et al.*, 2000). Transcription of *anfA* is repressed by MopA and MopB, which respond to the level of molybdenum in the cell (Masepohl *et al.*, 2002). Thus, under normal conditions, the alternative nitrogenase is not made, even in the absence of ammonia and O_2, unless the cells are also starved for molybdenum (Schuddekopf *et al.*, 1993).

Although the *nifA* genes are transcribed by NtrC-P/RNAP, the structural *nif* operons require a different RNA polymerase, one in which σ^{70} is replaced by σ^{54}. The latter factor is encoded by the *rpoN* gene, originally called *nifR4*. Expression of the *rpoN* gene is moderately complex (Preker *et al.*, 1992; Cullen *et al.*, 1994; Foster-Hartnett and Kranz, 1992). The gene is downstream of the *nifU2* gene in a small operon. One promoter, just upstream of *rpoN*, is transcribed constitutively by σ^{70}-RNAP to yield the RpoN needed by many operons. Under nitrogen-fixing conditions, a second promoter upstream of nifU2 is transcribed by σ^{54}-RNAP activated by NifA, yielding an enhanced level of RpoN.

Expression of the *nif* structural operons can be inhibited by low levels of drugs that interfere with DNA supercoiling (Kranz and Haselkorn, 1986). These results likely mean that special DNA structures are required to form transcription complexes at the major *nif* promoters. The small basic nucleoid-associated protein, H-NS, has a homologue in *R. capsulatus* called HvrA (Raabe *et al.*, 2002). The latter protein binds specifically to linear forms of the major *nif* promoters, *i.e.*, those of *nifH* and *nifB*. It seems to mediate the repression by ammonia of transcription from these promoters. Thus, productive transcription from, say, the *nifH* promoter requires σ^{54}-RNAP, either NifA1 or NifA2, IHF (to permit DNA bending for the interaction of NifA with σ^{54}-RNAP), and the absence of HvrA. Other proteins may enhance this transcription without being essential.

There is another set of controls over the entire system. The *glnB* gene encodes a small regulator protein, called P_{II}, which reports, through the adenylylation cascade, the nitrogen status of the cell. P_{II} is itself transcribed using NtrC, the *glnB* gene having both positive and negative binding sites for NtrC (Foster-Hartnett and Kranz, 1994). In the presence of ammonium, P_{II} prevents the transcription of the *nifA* and *anfA* genes, thereby shutting down the whole system. There is a second gene, called *glnK*, that encodes a paralogue of P_{II}, but this protein does not participate in the Ntr signal-transduction mechanism. Instead, it plays a role in a post-translational control of NifA activity. It also plays a role in the modification of NifH by the DraT/DraG system in response to ammonium. The significance of the two P_{II} proteins is manifest by the fact that cells can make highly active nitrogenase in the presence of ammonia if both *glnB* and *glnK* genes are mutated (Masepohl *et al.*, 2002).

As in *Rhodospirillum rubrum* and *Azospirillum brasilense*, *R. capsulatus* responds to either the addition of ammonium or transfer to darkness by covalent modification of the NifH component (the Fe protein) of nitrogenase. This modification, catalyzed by the DraT protein, is the addition of an ADP-ribosyl group to an arginine residue (Arg-101), resulting in total loss of activity. This modification can be reversed by a glycohydrolase, which is the product of the *draG* gene. The *draT* and *draG* genes form a small operon, about 1 Mb distant from *nifHDK*. It is curious that the DraTG system is missing from *R. sphaeroides*. Transfer of the *draTG* genes from *R. capsulatus* into *R. sphaeroides* endows the latter with the ability to modify nitrogenase in response to ammonium and darkness (Yakunin *et al.*, 2001). The DraTG system is regulated by P_{II} and, as mentioned above, a double *glnB/glnK* mutant makes active nitrogenase in the presence of ammonium.

3. OPERON STRUCTURE AND GENE ORGANIZATION

In the following figures, we show "pinned regions", which correspond to the neighborhoods of selected genes, taken from the ERGO suite at Integrated Genomics. In each figure, a single ORF is aligned in two or three genomes and then the neighboring genes can be compared. Details are given in the figure legends. Here, we comment on unexpected features compared to the situation in *K. pneumoniae*. Figure 2 reviews the *nif*-gene region of *K. pneumoniae*. Note the operon structure, indicated by the transcripts, and the placement of the *nifL* gene, which is missing from all the photosynthetic bacteria. NifL inactivates NifA in response to ammonia or O_2, whereas the NifA proteins of the photosynthetic bacteria take care of that by themselves.

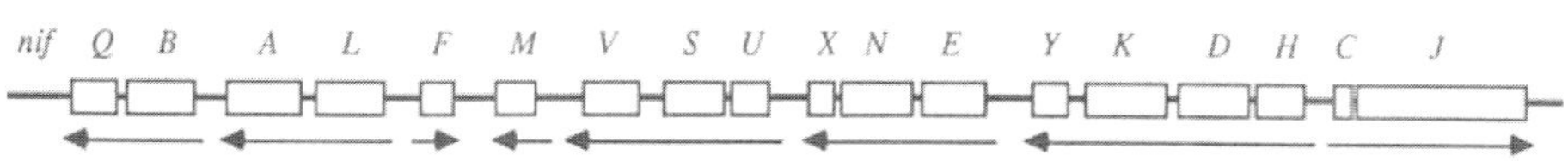

Figure 2. The nif *operons of* Klebsiella pneumoniae. *Arrows indicate transcription units.*

The operons that encode the structural proteins of the major Mo-containing nitrogenase of the photosynthetic bacteria are shown in Figure 3. This region contains the *nifHDK* operon, followed by the *nifUrpoN* operon (described above), a *nifAB* operon, and then a four-gene operon for Mo transport, followed by their repressor gene. The latter genes are slightly rearranged between *R. capsulatus* and *R. sphaeroides*. The *nifA* genes are duplicated in *R. capsulatus*, which made it initially impossible to determine the phenotype of a *nifA* mutant. Now, it is known that NifA is required for transcription of the other *nif* operons, as for *Klebsiella*. In fact, the genome sequence shows four paralogues of the *nifA* gene (Figure 4).

Figure 3. Region containing the structural genes for the major nitrogenase in photosynthetic bacteria.
Numbered ORFs are as follows: 1, NifH; 2, NifD; 3, NifK; 4, NifU2; 5, RpoN; 6, NifA; 7, NifB; 8, ModD (transporter); 9, ModC (ATP-binding); 10, ModB (permease); 11, Mo-binding protein; 12, MopB (Mo-pterin binding); 13, ModA (repressor).

```
NIFA1          MTDQQSRPASPRRRSTQSIADRLALDALYEIAKTFAAAPDPVAEVPQIFNVLSSFLD
NIFA2       MIDIRDRLVPQPQARHRSARATADRLALDALYEIAKTFAAAPDPVAEVPQIFNVLSSFLD
ANFA  MFGDDQVEALELGQASPEDEFGQCFTGECRVNLLPTLYRLNAVISQSPDPGESLGMILKTMRSEMR
NTRX                             MSDILIVDDEKDIRDLIAD--------------------
                                  .    :     :   ::

NIFA1 LRHGVLALLAEPGEGAGVNPYVIAATAFQRSPEAPAADVLPDAVARIVFRSGVPFVSFDLAAEFGA
NIFA2 LRHGVLALLAEPGEGAGVNPYVIAATAFQRSPEAPAADVLPDAVARIVFRSGVPFVSFDLAAEFGA
ANFA  MERGTVSLMDRRRG----QVFIHQSFGLTAEEESRGVYAMGEGITGKVAETGRVIIAPRLHESPDF
NTRX  --------------------------ILKDEGYATRTAANSDACMEAINTEAPALMILDIWLKDSR
                                 :  .  :  . .  :.    :   .  ::   :  . .

NIFA1 EAVPKRLRDAGQTLIAVPLRDPERSHFVLGVLAAYRSHDHNRSGFSDADVRVLTMVASLLEQALRF
NIFA2 EAVPKRLRDAGQTLIAVPLRDPERSHFVLGVLAAYRSHDHNRSGFSDADVRVLTMVASLLEQALRF
ANFA  LDRTGAHAGESRRRVAFVCVPIMLGKRVLGTIGGERSYLNQR--LLKQDAEFLAAIALMIAPTVEL
NTRX  MDGIDILKTVKRDNPDIPIIIISGHGNIEIAVAAIK----------QGAYDFISKPFNIDQLMVVI
                 :    .          :  .:.. :          .    .::    :    : :

NIFA1 RRRIARDRERALEDTRRMLQTVTEQRGPAAPVSLDGIVGSSPAIAEVVAQIKRVASTRMPVLLRGE
NIFA2 RRRIARDRERALEDTRRMLQTVTEQRGPAAPVSLDGIVGSSPAIAEVVAQIKRVASTRMPVLLRGE
ANFA  YLIANVEKLELERENRELRDALRERFKPGN-----IIGNSKAMMADVYDLIGKVSKTRATVLILGE
NTRX  SRAMETSRLRRENSSLRRRDLHSG-----------DMIGTSAAFRRLKDQLDKVTKSNGRVMLTGD
            .: .  ... .  :                :  .:.. :  :   : :*:.:.  *:: *:

NIFA1 SGTGKELFARAVHAQSPRAKGPFIRVNCAALSETLLESELFGHEKGAFTGATALKKGRFELADGGT
NIFA2 SGTGKELFARAVHAQSPRAKGPFIRVNCAALSETLLESELFGHEKGAFTGATALKKGRFELADGGT
ANFA  SGVGKELVASAIHYSSDRAAKPFVRFNCAAIPETLAESLLFGHEKGAFTGALASRKGLFEQADGGT
NTRX  PGSGKESAARYIHQHSTRAAAAFVTVNSATIAPERMEEVLFGRET----AERGIEKGLLEQAHGGV
      .* ***  *  :*  * **  .*: .*.*::.    *. ***:*.    .  . .** :* *.**.

NIFA1 LFLDEIGEISPAFQSKLLRVLQEGEFERVGGAKTIKVDTRIVAATNRDLEDAVARGQFRADLYFRI
NIFA2 LFLDEIGEISPAFQSKLLRVLQEGEFERVGGAKTIKVDTRIVAATNRDLEDAVARGQFRADLYFRI
ANFA  LFLDELGELSPSVQAKLLRVLQDRTLERVGGSTPVQVDVRVIAATNRELVRMVAEGRFREDLFYRL
NTRX  IYFDEVAEMPLGTQSKILRVLTEQQFSRVGGSDKVRVDLRVISSTTRNLTAEIAAGRFRQELYDRL
      :::**:.*:. . *:*:**** :  :.****:  ::** *::::*.*:*   :* *:** :*: *:

NIFA1 CVVPIVLPPLRNRKSDIKPLAQLFLDRFNKQNATNVK-FAADAFDQICRCQFPGNVRELENCVNRA
NIFA2 CVVPIVLPPLRNRKSDIKPLAQLFLDRFNKQNATNVK-FAADAFDQICRCQFPGNVRELENCVNRA
ANFA  NVVPITVPPLRERGSDIILLADHFVAKACKAMEKSVKRISTPALNMLMAYHWPGNVRELENVIERA
NTRX  NVVPIAVPSLAERREDVPLIAAHFIEVFNRTQGLALRPLSEEAVASLQTMDWPGNIRELRNVIERV
       ****.:*.* :* .*:  :*  *:    :     :: ::  *.  :   .:***:***.* ::*.

NIFA1 AALSDGA-IVLAEELACRQGACLSAELFRLQDGTSPIGGLAVGRVITPTVRVSAPPPEPAPAPEPA
NIFA2 AALSDGA-IVLAEELACRQGACLSAELFRLQDGTSPIGGLAVGRVITPTVRVSAPPPEPAPAPEPA
ANFA  VILSDDE-VIHAWNLP--------------------------------------------PSLQTA
NTRX  LILGDGTGPIEARELPGN---------------------------------------AALPEEGRI
        *.*.   : * :*.                                            *

NIFA1 PEAPPREEVPLRTKTAQLSREELLRALESAGWVQAKAARLLGMTPRQIAYALQKFEIELRKI
NIFA2 PEAPPREEVPLRTKTAQLSREELLRALESAGWVQAKAARLLGMTPRQIAYALQKFEIELRKI
ANFA  RESGTTLGLGLEEKVRLVESEMIVEALKTTQGNIGQAAELLQVSRRVLGLRMGRLGIDPHRYRAS.
NTRX  VLGGQLASLPLREARELFEREYLLTQINRFGGNISRTAAFVGMERSALHRKLKSLGVVTSAKSGR.
        .     : *.     .. * ::  ::      .::* :: :    :   :  : :
```

Figure 4. Amino-acid sequences of the four paralogues of NifA in the genome of Rhodobacter capsulatus.
** indicates identical residues in all the proteins; : indicates similarity.*

The two *nifA genes* are nearly identical except for the very N-terminal residues. The *anfA*-gene product is similar and required for transcription of the operon encoding the AnfHDGK proteins of the Mo-free nitrogenase. NtrX, which is part of a two-component signaling system that is clearly related to the NifA protein family (Ishida *et al.*, 2002), is shorter at the N-terminus but otherwise very similar to NifA through its central half and it has a good helix-turn-helix near the C-terminus.

Even though there is a single *rpoN* gene in *R. capsulatus*, there are four in the genome of *R. sphaeroides*. Their sequences are compared in Figure 5.

```
RCRPON                                                 MELAQ--TLSQRQTMQMAGQ
RSRPON1                                                MDMMQ--FQRQTTQLAMTQR
RSRPON2    MAAAAAGLASNRRHFTPVKLKFCPAGHFPPLCSYDRVGMTPGMQLYTAQ--SFAQRQSLVVTAQ
RSRPON3                                   MRHGRAGTRLDGTMKSRQRISIAQTQRLQLNLG
RSRPON4                                                   MQ---LRLGQRLAQRAV
                                                           *          :

RCRPON     MLHSLAILGMSSQDLSEHLTEQATSNPFLTYRAP------PAFIARG---GEDFDAVAAVAAHK
RSRPON1    MQESLRILQMSNADLADYLTAQALENPCLEVRVPEGASVAPALPSRGIQAGLDRDAFATVEGQP
RSRPON2    LQQAICLLQMPNAELSSFIESQSEENPFIELRLP---PAPVPSAFLGKTAPEDWDRVAGLAADP
RSRPON3    LTASIRVLNSDAEGLTRYLQEQAAENPHIQLEPATSTDWLPRWTSVLSRLAQGEGSAGGETVAA
RSRPON4    LVQRAEILEATGTDWAERIAAEAQRNPFLRVRQP-----------------ATAAPIPESAALP
           :     :*       :  :  ::  ** :  . .

RCRPON     --PSLMAHVVDQIEMAFTETPDRLLALRFAEALEPSGWLGQSLDSIALAAGVSLSRAESMLAVL
RSRPON1    --PSLLAHVEAQIDLAFFDPGDRRTALAFAEALEPSGWLGQPVSEVAAAAEVEEEEALVILERL
RSRPON2    G-PSLYVHVAAEIARLGFDAPQAAAAQVFLDALEPWGWLGRPMEELAFRAGLSLEAAEALLARL
RSRPON3    AGPSLMAHVMARIDTLYPRGPERRIAILLAEALEPTGWLGTGPDEIARQARVPSAEVEAVLAGL
RSRPON4    --PDLHAWLGSQIRMAMADPEDRALAFRLLEALEPSGWLGQPLSRLAPGA---EEQAARVLHRL
             *.* . :  .*        :   *  : :**** ****   . :*  *      .  :*  *

RCRPON     QGFEPTGLFARDLSDCLILQAREADILTWEVETLIRNIRLIAENRLSDLADLCDCLIGDIPEII
RSRPON1    QALEPAGLFARSLAECLALQLEDLGLLTWELRTMLDHLPLLAEGRIADLARRCDCEPEHIRENL
RSRPON2    QKIEPAGLFARTLAECLQLQAEEQGLLTPLFAAVLAHLPLLAAADLKGLCRACGCGMEDLKAVL
RSRPON3    QKIEPAGLFARTLAECLRLQAIEAERLDSTLSCLLDHLDLVAEGALGRLARLCNTDEAGVTARL
RSRPON4    QQMEPAGIFARDLRECLMLQARDRGQLDPAMAAVLDRLDCLASDGPAAVARAAGLEEQTVLRCL
           * :**:*:*** * :** **  :   *   .  :: .:  :*      :.  ..     :   :

RCRPON     KQIRHLNPKPGLAFDHQPTPVFPPDLIAVRGAEGWTVELNRATSPTITVREDRFADGTADAKAR
RSRPON1    ALIRSLSPKPGEAFAADRTPIQPPDVRVLRGPEGWEVELTRAQLPRIRVSE---AGDTGDRQAD
RSRPON2    RSLRGLNPKPGALFDAAPPPQRPPDLVVSRGAEGWRVDLNRSTLPSVVVRSD--AAEGFARTAA
RSRPON3    RLLRTFDPKPGAQFDPGAAPVREPDLIATKGEAGWEVSLNRSAMPTVQIRKP------DKRPTT
RSRPON4    ELIRRMDPKPGAAFAAEDAPLREPDLIARRTASGWSVELNRSLLPEVRVAP------LPDGSP
             :* :.****  *    .*   **: . :   ** *.*.*:  * : :             :

RCRPON     AERRKALAEARALAQALERRGDTLLRTAAVLVARQSAFLDKGPAHLVPLTLEDVASELGLHAST
RSRPON1    AWLARARSQARWLERAVERRQATLLRTAVCLVRHQADFLDQGPRALRPLSMEEVALELDLHPST
RSRPON2    PYVGERLSVAKWLARAVEHRNQTTLKIGAEVVRRQRGFLEEGLARMAPMTLREVADAVGVHEST
RSRPON3    PAARAAWTQAQAVGRMIENRNATLLRVAREILARQEAALDEGPSALVALTMTEVAEALGIHEST
RSRPON4    ADVRQMHQEALSLAKATGLRGRTLLAVGALVVERQRAFLDEGPAALVALGIGEAADALALHPST
           .        *  : :    *  * *  .  :: :*   *::*   : .: : :.*  : :* **

RCRPON     ISRAVSGRMIQTQTRALPLRAFFSRAVSTQGGGEAVSRDSALDFVQRTVGGEDPQNPLSDDAIV
RSRPON1    ISRATATRLIETPRGLIPLRAFFSRSVSSDGPEAPQSQDALMALVREIIAREDRTKPFSDDAIV
RSRPON2    VSRVSSGLMIATPQGTFPLKSFFTAALAAREGDTAGSAAAVRHRVRQLVQAESPDDPLSDDAIA
RSRPON3    VSRVVAGTCVDTPRGTWWLRRMFSGRLAEGGP----SAAAIRAAIARLVAQEDPAAPLSDGALV
RSRPON4    ISRTVTGLLMATPRGLVAVRDLFCARLAEGTG--ALSAPALQALIRETIAAESAHAPLEDGEIV
           :**. :   : *      :: :*   ::        *  :    : . :  *.   *:.*. :.

RCRPON     TLAERAGLRIARRTVAKYRSTLGLASSYERRRAAAAR
RSRPON1    KQAKLAGAVLARRTVTKYRETLGIPSSYDRKRAAAAA
RSRPON2    KIISDEGVTLARRTVAKYREQLNIPSSVQRRRQAIVTGAL
RSRPON3    EALAAEDMQLARRTVAKYREMLNIPPGHRRRRRPSRSA
RSRPON4    EALARRGIHVARRTVAKHRTLAGLPPAAVRRRAAPSGSEAVAFGRQPVPLGGRN
                 .  :*****:*:*   .:...  *:* .
```

Figure 5. Amino-acid sequences of the RpoN proteins of R. capsulatus *(top line) and* R. sphaeroides *(bottom four lines).*

At least two of these *rpoN* genes were known to be functional when it proved impossible to knock out a single *rpoN* gene with phenotypic consequences (Meijer and Tabita, 1992). The sequences are reasonably conserved through the central third of the molecule and all of the proteins have the conserved RpoN box (ARRTVAKYR) near the C-terminus.

The *ntrX* gene is adjacent to its cognate kinase gene, *ntrY*, in both *R. capsulatus* and *R. sphaeroides* (Figure 6). The same region contains the genes *nifR3, ntrB*, and *ntrC* (see Figure 1). The two chromosomal regions are very similar except for the insertion, in *R. sphaeroides*, of two genes encoding components of the potassium-uptake system, Trk, between *ntrX* and *hflX.* The small gene *nrfA* encoding a 77-amino acid protein is found in both genomes just upstream of *hflX.* NrfA is related to a nucleoid-binding protein of *E. coli*, Hfq, and, although it is not absolutely required for growth on N_2 as nitrogen source, it is required for maximum growth. It functions by regulating the level of transcription of both the *nifA* and *anfA* genes (Drepper *et al.*, 2002).

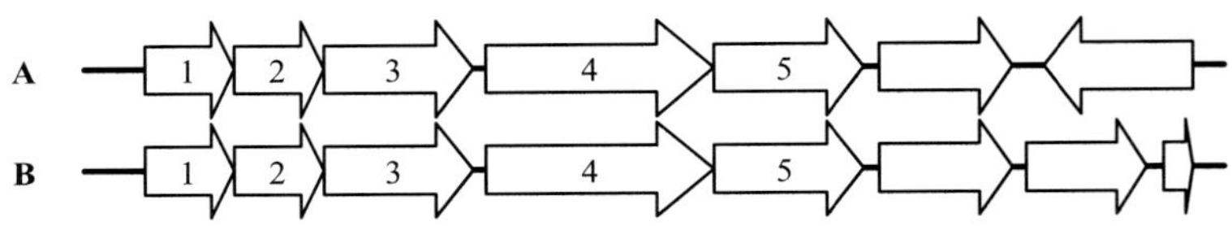

Figure 6. Pinned regions of the ntrB/ntrC *genes of* R. capsulatus (**A**) *and* R. sphaeroides (**B**). *Numbered ORFs are as follows: 1, NifR3; 2, NtrB (kinase); 3, NtrC (response regulator); 4, NtrY (kinase); 5, NtrX (response regulator).*

The genes encoding the structural components of the alternative nitrogenase in *R. capsulatus* are located about 100 kb away from the region shown in Figure 3. For comparison, we show also the pinned region from *Azotobacter vinelandii*, in which the alternative nitrogenases were discovered (Figure 7). The regions are similar in the two bacteria, except for the gene encoding an antibiotic-resistance protein, which is downstream of the *anf* operon in *R. capsulatus* but upstream and in reversed orientation in *A. vinelandii*. Note also that, between *anfD* and *anfK*, there is a small ORF for the δ-subunit, an extra component of the alternative nitrogenase. There is no alternative nitrogenase in *R. sphaeroides*.

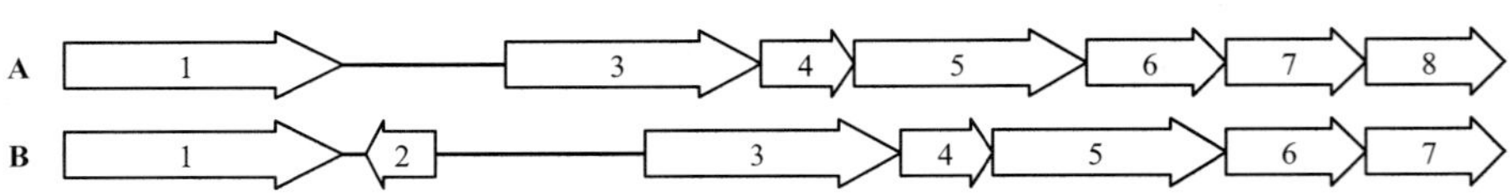

Figure 7. Pinned region of the anf operon in R. capsulatus (**A**) *and* A. vinelandii (**B**). *Numbered ORFs as follows: 1, AnfA; 2, AnfH; 3, AnfD; 4, AnfG (delta protein); 5, AnfK; 6, hypothetical; 7, hypothetical; 8, 5-nitromidazole resistance protein.*

The remaining *nif* genes of *R. capsulatus* are in a large cluster covering more than 13 kb, in which non-*nif* genes are interspersed. These are shown in Figure 8.

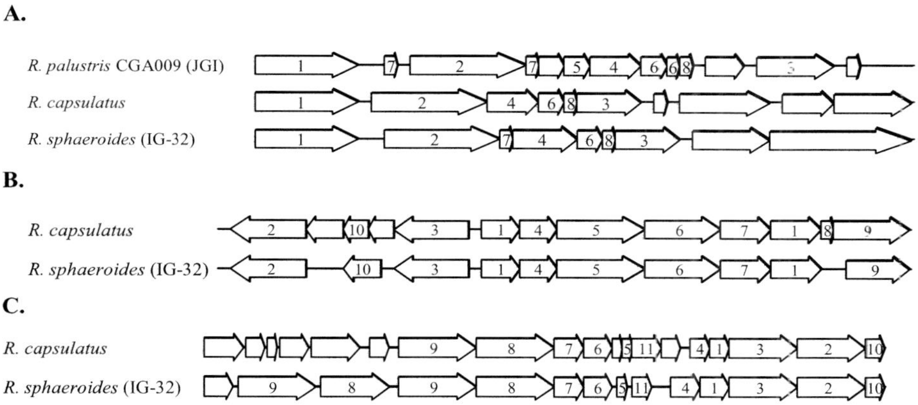

Figure 8. Pinned regions around the nifAB, nifUSVW, nifENX *and* rnfBCDGH *genes. Numbered ORFs are as follows:* ***A****. 1, NifA; 2, NifB; 3, hypothetical; 4, hypothetical;5, hypothetical; 6, NifZ, 7, ferredoxin; 8, FixU.* ***B****. 1, NifU; 2, NifV; 3, NifS; HesB family; 5, ferredoxin; 6, hypothetical, 7, NifX; 8, NifN; 9, NifE, 10, NifW.* ***C****. 1, RnfA; 4, RnfB; 5, RnfC; 6, RnfD; 7, RnfG; 1, RnfE; 8, RnfH; 9, flavoprotein precursor.*

Notable are a second *nifAB* operon (see Figure 3 for the NifA sequence) and a possible *nifUSVW* operon. This *nifU* gene, however, corresponds to just the C-terminal third of the real *nifU* gene. It contains a cys-X-X-cys sequence, so it could be involved in the synthesis of Fe-S clusters for nitrogenase (D.R. Dean, personal communication). The *nifV* gene encodes homocitrate synthase, which is needed for the FeMo-cofactor. The *nifS* gene is annotated as a cysteine desulfhydrase but this one is believed to encode the true NifS for nitrogenase cofactor synthesis. Following *nifQ*, ferredoxin III, and several hypothetical proteins, we come to the *nifENX* operon. Next, there are half a dozen ORFs for ferredoxins, a flavodoxin, a sigma factor, a subunit of a Na^+-translocating NADH-quinone reductase, and a thiamine-biosynthesis lipoprotein. Finally, there is a six-gene operon that encodes the Rnf electron-transport complex (*rnf BCDGEH*). This complex, which is membrane-bound, plays an important role in electron transfer to nitrogenase (Schmehl *et al.*, 1993).

The genome regions around several of the remaining regulatory protein genes (Figure 1) contain interesting features. The *glnB* gene, encoding P_{II}, precedes the *glnA* gene, encoding glutamine synthetase, in both *R. capsulatus* and *R. sphaeroides*. In both strains, the *glnK* gene, encoding a second copy of P_{II}, precedes an ammonium-transporter gene. Other genes in the neighborhood, not related to nitrogen fixation, are also similar in the two strains.

The last region to be considered is shown in Figure 9, which includes the regulatory pair *regA/regB* and the *hvrA* gene. RegB is a histidine kinase and RegA is the cognate response regulator (see Figure 1) that activates genes for photosynthesis as well as one of the *nifA* genes. An unusual feature is that these two genes are not co-transcribed; they are on opposite strands and separated by another ORF that encodes the regulatory protein PrrC. Downstream from the *regA* gene is another small ORF for the regulatory protein HvrA. These features are identical in *R. capsulatus* and *R. sphaeroides*.

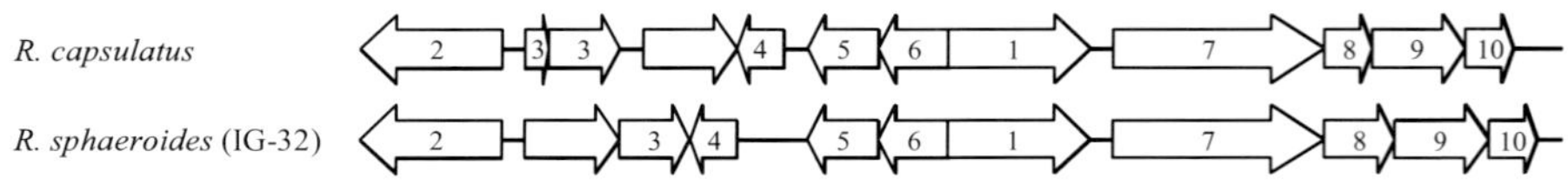

Figure 9. Gene organization at the RegB (1) in the R. capsulatus *and* R. sphaeroides *genomes.*
1. Histidine kinase RegB; 2. Adenosylhomocysteinase (EC 3.3.1.1); 3. Hypothetical cytosolic protein; (White ORF) AHCY transcriptional activator hvrB; 4. Trans-acting regulatory protein hvrA; 5. Photosynthetic response regulator PrrA; 6. Regulatory protein prrC; 7. Phosphate regulon sensor protein phoR (EC 2.7.3.-); 8. ATP/GTP hydrolase; 9. 7.5 kDa chlorosome protein; 10. Mannose-1-phosphate guanyltransferase (EC 2.7.7.13).

HvrA plays a strange role in nitrogen fixation, it is something of a mystery. It is a small basic protein related to the *E. coli* protein H-NS, which binds to bent DNA. In *R. capsulatus*, it activates transcription of genes for photosynthesis, but its role in nitrogen fixation appears to be negative. It binds specifically to the *nifH* promoter (Raabe *et al.*, 2002).

The identity between *R. capsulatus* and *R. sphaeroides* extends to another ORF in this region that is truly remarkable; one that encodes the chlorosome protein of *Chloroflexus*. The latter bacteria are considered to be very primitive phototrophs, most of which contain no conventional photochemical reaction centers but instead have football-shaped chlorosomes as light-harvesting antennae. These transfer energy to chlorophyll complexes attached to the cytoplasmic membrane. One species, *Chloroflexus auriantica*, actually has conventional reaction centers as well as chlorosomes. The fact that both *R. capsulatus* and *R. sphaeroides* have complete ORFs for the chlorosome protein suggests that the ORFs are functional and expressed. This suggestion is testable by mutation and expression experiments, which are eminently feasible.

ACKNOWLEDGEMENT

We are grateful for advice from William Buikema, Dennis Dean, and the entire genome sequencing team at Integrated Genomics, Inc. We regret the loss of Werner Klipp, who contributed so much to this field.

REFERENCES

Avtges, P., Kranz, R. G., and Haselkorn, R. (1985). Isolation and organization of genes for nitrogen fixation in *Rhodopseudomonas capsulata*. *Mol. Gen. Genet.*, *201*, 363-369.

Cullen, P. J., Foster-Hartnett, D., Gabbert, K. K., and Kranz, R. G. (1994). Structure and expression of the alternative sigma factor, RpoN, in *Rhodobacter capsulatus*; Physiological relevance of an autoactivated *nifU2-rpoN* superoperon. *Mol. Microbiol.*, *11*, 51-65.

Drepper, T., Raabe, K., Giaourakis, D., Gendrullis, M., Masepohl, B., and Klipp, W. (2002). The Hfq-like protein NrfA of the phototrophic purple bacterium *Rhodobacter capsulatus* controls nitrogen fixation via regulation of *nifA* and *anfA* expression. *FEMS Microbiol. Lett.*, *215*, 221-227.

Elsen, S., Dischert, W., Colbeau, A., and Bauer, C. E. (2000). Expression of uptake hydrogenase and molybdenum nitrogenase in *Rhodobacter capsulatus* is coregulated by the RegB-RegA two-component regulatory system. *J. Bacteriol.*, *182*, 2831-2837.

Fonstein, M., Koshy, E. G., Nikolskaya, T., Mourachov, P., and Haselkorn, R. (1995). Refinement of the high-resolution physical and genetic map of *Rhodobacter capsulatus* and genome surveys using blots of the cosmid encyclopedia. *EMBO. J.*, *14*, 1827-1841.

Fonstein, M., Zheng, S., and Haselkorn, R. (1992). Physical map of the genome of *Rhodobacter capsulatus* SB 1003. *J. Bacteriol.*, *174*, 4070-4077.

Foster-Hartnett, D., and Kranz, R. G. (1992). Analysis of the promoters and upstream sequences of nifA1 and nifA2 in *Rhodobacter capsulatus*; activation requires *ntrC* but not *rpoN*. *Mol. Microbiol.*, *6*, 1049-1060.

Foster-Hartnett, D., and Kranz, R. G. (1994). The *Rhodobacter capsulatus glnB* gene is regulated by NtrC at tandem *rpoN*-independent promoters. *J. Bacteriol.*, *176*, 5171-5176.

Haselkorn, R., Lapidus, L., Kogan, Y., Vlcek, C., Paces, J., Paces, V., *et al.* (2001). The *Rhodobacter capsulatus* genome. *Photosynthesis Res.*, *70*, 43-52.

Ishida, M. L., Assumpcao, M. C., Machado, H. B., Benelli, E. M., Souza, E. M., and Pedrosa, F. O. (2002). Identification and characterization of the two-component NtrY/NtrX regulatory system in *Azospirillum brasilense*. *Braz. J. Med. Biol. Res.*, *35*, 651-661.

Jones, R., and Haselkorn, R. (1989). The DNA sequence of the *Rhodobacter capsulatus ntrA, ntrB* and *ntrC* gene analogues required for nitrogen fixation. *Mol. Gen. Genet.*, *215,* 507-516.

Klipp, W., Masepohl, B., and Puhler, A. (1988). Identification and mapping of nitrogen fixation genes of *Rhodobacter capsulatus*: Duplication of a *nifA-nifB* region. *J. Bacteriol.*, *170,* 693-699.

Kranz, R. G., and Haselkorn, R. (1985). Characterization of *nif* regulatory genes in *Rhodopseudomonas capsulata* using *lac* gene fusions. *Gene*, *40*, 203-215.

Kranz, R. G., and Haselkorn, R. (1986). Anaerobic regulation of nitrogen-fixation genes in *Rhodopseudomonas capsulata*. *Proc. Natl. Acad. Sci. USA*, 83, 6805-6809.

Marrs, B. (1974). Genetic recombination in *Rhodopseudomonas capsulata*. *Proc. Natl. Acad. Sci. USA*, *71*, 971-973.

Masepohl, B., Drepper, T., Paschen, A., Gross, S., Pawlowski, A., Raabe, K., *et al.* (2002). Regulation of nitrogen fixation in the phototrophic purple bacterium *Rhodobacter capsulatus*. *J. Mol. Microbiol. Biotechnol.*, *4*, 243-248.

Meijer, W. G., and Tabita, F. R. (1992). Isolation and characterization of the *nifUSVW-rpoN* gene cluster from *Rhodobacter sphaeroides*. *J. Bacteriol.*, *174*, 3855-3866.

Preker, P., Hubner, P., Schmehl, M., Klipp, W., and Bickle, T. A. (1992). Mapping and characterization of the promoter elements of the regulatory *nif* genes *rpoN, nifA1* and *nifA2* in *Rhodobacter capsulatus*. *Mol. Microbiol.*, *6*, 1035-1047.

Raabe, K., Drepper, T., Riedel, K. U., Masepohl, B., and Klipp, W. (2002). The H-NS-like protein HvrA modulates expression of nitrogen fixation genes in the phototrophic purple bacterium *Rhodobacter capsulatus* by binding to selected nif promoters. *FEMS Microbiol. Lett.*, *216*, 151-158

Schmehl, M., Jahn, A., Meyer zu Vilsendorf, A., Hennecke, S., Masepohl, B., Schuppler, M., *et al.* (1993). Identification of a new class of nitrogen fixation genes in *Rhodobacter capsulatus*: A putative membrane complex involved in electron transport to nitrogenase. *Mol. Gen. Genet.*, *241*, 602-615.

Schuddekopf, K., Hennecke, S., Liese, U., Kutsche, M., and Klipp, W. (1993). Characterization of *anf* genes specific for the alternative nitrogenase and identification of *nif* genes required for both nitrogenases in *Rhodobacter capsulatus*. *Mol. Microbiol.*, *8*, 673-684.

Vignais, P. M., Colbeau, A., Willison, J. C., and Jouanneau, Y. (1985). Hydrogenase, nitrogenase, and hydrogen metabolism in the photosynthetic bacteria. *Adv. Microb. Physiol.*, *26*, 155-234.

Vlcek, C., Paces, V., Maltsev, N., Paces, J., Haselkorn, R., and Fonstein, M. (1997). Sequence of a 189-kb segment of the chromosome of *Rhodobacter capsulatus* SB1003. *Proc. Natl. Acad. Sci. USA*, *94*, 9384-9388.

Wall, J. D., and Braddock, K. (1984). Mapping of *Rhodopseudomonas capsulata nif* genes. *J. Bacteriol.*, *158,* 404-410.

Willison, J. C. (1993). Biochemical genetics revisited: the use of mutants to study carbon and nitrogen metabolism in the photosynthetic bacteria. *FEMS Microbiol. Rev.*, *10*, 1-38.

Yakunin, A. F., Fedorov, A. S., Laurinavichene, T. V., Glaser, V. M., Egorov, N. S., Tsygankov, A. A., *et al.* (2001). Regulation of nitrogenase in the photosynthetic bacterium *Rhodobacter sphaeroides* containing *draTG* and *nifHDK* genes from *Rhodobacter capsulatus*. *Can. J. Microbiol.*, *47*, 206-212.

CHAPTER 6

GENOMIC ARCHITECTURE OF THE MULTIPLE REPLICONS OF THE PROMISCUOUS *RHIZOBIUM* SPECIES NGR234

P. MAVINGUI[1], X. PERRET[2] AND W. J. BROUGHTON[2]

[1]*Laboratoire d'Ecologie Microbienne, UMR CNRS 5557, Université Claude Bernard Lyon 1, Villeurbanne, France, and* [2]*Laboratoire de Biologie Moléculaire des Plantes Supérieures, Université de Genève, Genève, Switzerland.*

1. INTRODUCTION

Rhizobium species NGR234 is a Gram-negative, α-proteobacterium that establishes nitrogen-fixing symbioses with more leguminous plants than any other micro-symbiont so far examined. What gives *Rhizobium* sp. NGR234 this broad host-range is not fully understood, however. Like many rhizobia, NGR234 has a composite genome. It consists of three replicons: a symbiotic plasmid pNGR234*a* of 536 kb, a megaplasmid pNGR234*b* (> 2,000 kb), and a chromosome possibly larger than that of *Sinorhizobium meliloti* (*ca.* 3,700 kb). Extensive genetic, metabolic, genome-dynamics, and sequencing-analyses research has been carried out on this bacterium. A landmark in understanding the genetics of nitrogen-fixing organisms was the complete nucleotide sequence of pNGR234*a*. This review focuses on the biology and genetics of the three replicons of NGR234 and intends to shed light on the possible mechanism responsible for broad host-range symbioses.

A large spectrum of Gram-negative bacteria, including *Burkholderia, Methylobacterium*, and *Ralstonia* are known to form nitrogen-fixing symbioses with legumes (Moulin *et al*., 2001a, b; Chen *et al*., 2001; Sy *et al*., 2001). Six genera of rhizobia are also recognized; these are *Allorhizobium, Azorhizobium, Bradyrhizobium, Mesorhizobium, Rhizobium,* and *Sinorhizobium* (van Berkum *et al*., 2003; Broughton, 2003). In compatible interactions, these soil bacteria penetrate plant roots *via* infection threads and new structures called nodules are formed (Broughton *et al*., 2000). Nodule organogenesis begins with the de-differentiation of few root cortical cells near the site of bacterial infection and then continues with the formation of a nodule primordium (Verma and Hong, 1996; Oke

R. Palacios and W. E. Newton (eds.), Genomes and Genomics of Nitrogen-fixing Organisms, 83-98.

and Long, 1999). Nodule formation not only involves a large number of plant genes, it also requires signals from the microsymbionts (Fisher and Long, 1992; Schultze *et al.*, 1994; Long, 1996; Broughton and Perret, 1999; Perret *et al.*, 2000a).

Symbiotic nitrogen-fixing associations show varying degrees of specificity (Martinez-Romero and Caballero, 1996). For instance, *Sinorhizobium meliloti* nodulates plants of the three genera, namely *Medicago*, *Melilotus* and *Trigonella* (Dénarié *et al.*, 1992), whereas *Azorhizobium caulinodans* is restricted to *Sesbania* species (M. Holsters, personal comunication; Lewin *et al.*, 1987). In stark contrast to these narrow host-range rhizobia, *Rhizobium* species NGR234 nodulates more than 112 genera of legumes and fixes nitrogen in association with plants that form either determinate or indeterminate nodules (Pueppke and Broughton, 1999). This extremely broad host-range makes NGR234 a convenient model to analyse the bacterial determinants involved in symbiotic promiscuity and to test the possible effects of alternative genome architectures on the nodulation capacities of this strain. In this chapter, we will summarise general information on symbioses with NGR234, especially focusing on genomic aspects that could help to understand the complexity of such a "universal symbiont".

2. PROMISCUITY OF NGR234

NGR234 was isolated in 1965 from nodules of the tropical legume *Lablab purpureus* in Papua New Guinea (Trinick, 1980). Soon after, it was shown that NGR234 was able to form nitrogen-fixing nodules with other legumes as well as the non-legume *Parasponia andersonii* (Broughton and Dilworth, 1971; Trinick, 1980). Since then, extensive studies have shown that the host-range of NGR234 includes more than 112 genera in all three subfamilies of *Leguminosae* (Pueppke and Broughton, 1999). Depending on the host-plant, both determinate and indeterminate nodules can be formed. This exceptional host-range was a further challenge to the concept of specificity in *Rhizobium*-legume symbioses and drove studies aimed at understanding the molecular basis for such promiscuity.

Initially, the control of specificity in legume-*Rhizobium* associations is mediated by chemical cross-talk between the symbionts. Flavonoids released by legume roots interact with rhizobial transcriptional regulators of the LysR family (*e.g.*, NodD) and induce the expression of nodulation genes (*nod*, *noe* and *nol*). In response, rhizobia secrete a family of lipo-chito-oligosaccharides (LCOs), called Nod factors, which induce nodule formation (Lerouge *et al.*, 1990; van Brussell *et al.*, 1992; Relić *et al.*, 1993). NodD1 of NGR234 is able to interact with a large spectrum of flavonoids and, partly for this reason, has an extended host-range (Fellay *et al.*, 1995). In addition, NGR234 produces a large variety of Nod factors that differ in the number and degree of their substitutions (Price *et al.*, 1992). Classical genetic methods combined with biochemical analyses led to the identification of all loci involved in the synthesis of NodNGR factors. These include the operon *nodSU*, which is involved in the carbamoylation and *N*-methylation of Nod-factors, and *nodZ*, which is required for their *O*-fucosylation (Jabbouri *et al.*, 1995; Quesada-Vincens *et al.*, 1997). Genes, such as *noeI*, *nolL* and *nolO*, encode enzymes that participate in *O*-methylation, *O*-carbamoylation, and

O-acetylation of Nod-factors, respectively (Relić *et al.*, 1994; Freiberg *et al.*, 1997; Jabbouri *et al.*, 1998). Another locus, *noeE* is involved in sulphation of NodNGR factors, thus, extending the host-range of NGR234 to include *Calopogonium caeruleum* (Hanin *et al.*, 1997). Although, many homologous loci are found individually in the genomes of other rhizobia, where they play a role in host specificity of nodulation (*hsn*), they coexist in NRG234. Taken together, these characteristics partially explain the symbiotic promiscuity of NGR234.

3. STRUCTURAL ORGANISATION OF THE NGR234 GENOME

3.1. Replicon number, size and geometry

Rhizobial genomes are complex. They are composed of a large chromosome plus none to many large plasmids. The complete set of plasmids of a given strain may represent up to 50% of the total genome (Honeycut *et al.*, 1993). Research on the structure of NGR234 genome began with the analysis of its plasmid profiles, which were shown initially to contain one, then later two, replicons in addition to the chromosome (Pankhurst *et al.*, 1983; Morrison *et al.*, 1983; 1984). After contradictory results and controversial debates (Freiberg *et al.*, 1997; Downie, 1997; Perret and Broughton, 1997a), it is now well established that NGR234 contains three replicons: one chromosome and two plasmids (pNGR234*a* and pNGR234*b*, Flores *et al.*, 1998). Pulse-field gel electrophoresis (PFGE) with restricted and intact genomic DNA of NGR234 confirmed that the three replicons are circular (Mavingui *et al.*, 2002; Perret and Mavingui, unpublished). The 536,165 bp of the symbiotic plasmid pNGR234*a* encodes most loci involved in nodulation and nitrogen fixation (Freiberg *et al.*, 1997). Electrophoretic analyses showed that the megaplasmid pNGR234*b* is larger than pSymB of *S. meliloti* (1,683 kb) and the NGR234 chromosome appeared larger than that of *S. meliloti* (3,654 kb) (Flores *et al.*, 1998; Mavingui *et al.*, 2002). Combined data from Echkardt gels, PFGE, and physical mapping give an estimate of the size of pNGR234*b* at more than 2,000 kb, making it probably the largest plasmid known in rhizobia (Table 1).

*3.2. The symbiotic plasmid pNGR234*a

Symbiotic loci are usually carried by large plasmids in *Rhizobium* strains and are therefore called "symbiotic plasmids" (Banfalvi *et al.*, 1981; Rosenberg *et al.*, 1981). In *B. japonicum* (Kündig *et al.*, 1993) and *M. loti* (Sullivan and Ronson, 1998), symbiotic loci are clustered in "symbiotic islands" on the chromosome. Curing and mobilisation experiments showed that NGR234 harbours a replicon that confers on bacterial transconjugants the ability to nodulate non-host plants or extend its host range (Morrison *et al.*, 1984; Broughton *et al.*, 1984; 1986). Hybridisation experiments confirmed the presence of nodulation and nitrogen-fixation (*nif, fix*) genes on this replicon (Pankhurst *et al.*, 1983; Broughton *et al.*, 1984). Later, a physical map of pNGR234*a* was constructed (Perret *et al.*, 1991) and its complete sequence of 536,165 nucleotides established (Freiberg *et al.*, 1997).

Table 1. Architecture and genome size of some members of rhizobia.

Strain	Replicon Number	Replicon size (kb)	Genome size (kb)	References
Agrobacterium tumefaciens C58	4	2841[a], 2075[a,e], 542, 214[b]	5672	Wood *et al.* 2001; Goodner *et al.* 2001
Bradyrhizobium japonicum USDA 110	1	9106[a,c,d]	9106	Kündig *et al.* 1993 ; Kaneko *et al.* 2000
Mesorhizobium loti MAFF303099	3	7036[a,c,d], 351, 208	7597	Kaneko *et al.* 2000.
Rhizobium sp. NGR234	3	3700[a], 2000[d], 536[c]	6236	Flores *et al.* 1998 Perret *et al.* 2000b Mavingui *et al.* 2002
Rhizobium etli CFN42	7	5000[a], 700, 500, 370[c], 270, 200[d], 150	7190	Bustos *et al.* 2001 González *et al.* 2003
Sinorhizobium meliloti 1021	3	3654[a], 1683[c], 1354[d]	6691	Galibert *et al.* 2001

[a,b,c,d] Indicate replicons containing: [a], *rRNA*; [b], Ti; [c], *nod-nif*; [d], *exo* genes.
[e] Most replicons of rhizobia are circular, except this linear chromosome of *A. tumefaciens*.

Both genetic characterisation and sequence annotation confirmed the presence of most symbiotic genes on pNGR234*a*. Except for *nodPQ* (Perret *et al.*, 1991), these include all the loci involved in Nod-factor biosynthesis as well as the transcriptional regulators, *nodD1*, *nodD2, syrM1, and syrM2.* In contrast to the *hsnI, hsnII,* and *hsnIII* loci, which are dispersed over the entire replicon, the *nif* and *fix* genes are clustered in a single large region that includes the regulator *nifA*, two nitrogenase structural *nifHDK* operons as well as genes for electron transport and ferredoxin synthesis (*fixABCXfdxBN*). A total of 19 *nod* boxes and 16 NifA-σ^{54} promoters were found. In addition, pNGR234*a* contains one copy of *dctA* and a 35-kb *locus*, which encodes a type three secretion system (TTSS) with homologues also present in *Rhizobium fredii* (de Lyra *et al.*, 2000), *M. loti* (Kaneko *et al.*, 2000), and *B. japonicum* (Göttfert *et al.*, 2001). First characterised in plant and animal pathogens where they play essential roles in virulence, TTSS modulates symbioses between rhizobia and legumes (Viprey *et al.*, 1998; Marie *et al.*, 2001). Finally, unlike pSymB of *S. meliloti*, which contains an Arg-tRNA gene (Finan *et al.*, 2001), no loci essential to transcription, translation, or primary metabolism were found on pNGR234*a*.

Strikingly, many duplicated sequences are present on pNGR234*a*, including the two *nifHDK* operons as well as both insertion (IS) and mosaic (MS) elements. Together, repeated sequences represent 18% of the whole replicon (Freiberg *et al.*,

1997). As IS/MS elements separate clusters of functionally related genes that have distinct G+C values, horizontal gene transfers and movement of IS/MS elements have shaped the mosaic structure of this plasmid. Several IS sequences, which are repeated on pNGR234*a*, are also found in multiple and identical copies on the other two replicons, further showing that they are mobile (Perret *et al.*, 1997; 2000b). In addition, many of the repeated sequences serve as points for homologous recombination, leading to genome rearrangements (see Palacios and Flores, this volume). Other interesting classes of reiterated sequences are short repeated DNA motifs (van Belkum, 1998) that are known to play a role in microbial pathogenesis and evolution (van Belkum, 1999). pNGR234*a* contains five mini-satellites (short tandem repeat units) with repeat units of more than 10 bp and a total length greater than 100 bp, that are randomly distributed throughout the replicon (Le Flèche *et al.*, 2001; http://minisatellites.u-psud.fr). One copy of RIME1 (*Rhizobium*-specific Intergenic Mosaic Elements) was identified between ORFs y4fQ and y4fR (Freiberg *et al.*, 1997).

Like many rhizobial plasmids, pNGR234*a* belongs to *repABC* replicon-type. Similarly replicating plasmids are pTiB63S and pRiA4b of *Agrobacterium* and pRL8JI of *Rhizobium* (Freiberg *et al.*, 1997). Also, a number of homologues of conjugal transfer loci of *Agrobacterium* (Moriguchi *et al.*, 2001) were found in pNGR234*a*, including the 12-bp *oriT*.

*3.3. The megabase-size replicons, pNGR234*b *and the chromosome*

Morrison *et al.* (1984) were the first to observe pNGR234*b* on Eckhardt gels (Eckhardt, 1978). A modified version of the in-gel cell lysis "Eckhardt" technique, however, allowed its consistent detection in NGR234 (Flores *et al.*, 1998). In addition, PFGE allowed the detection of pNGR234*b* together with the *ca.* 3.7-Mb chromosomal replicon (Figure 1; Mavingui *et al.*, 2002).

Although the sequencing of these two replicons is continuing, hybridisation experiments have provided much additional information and some genes involved in symbiosis have been found in this way on both replicons (Chen *et al.*, 1985; Perret, 1992; Gray *et al.*, 1990; Flores *et al.*, 1998). For instance, homologues of *exoBDFKL* (Becker *et al.*, 1993; Glucksmann *et al.*, 1993), which are involved in the synthesis of exo-polysaccharides, are located on pNGR234*b*. Two copies of *nodPQ* genes, which encode enzymes involved in the synthesis of PAPS, a precursor of Nod-factor sulphation (Schwedock and Long, 1990), are present; one on pNGR234*b* and the other on the chromosome. Homologues of *nodGE, fixLJ* (David *et al.*, 1988), and *fixK* (Batut *et al.*, 1989), whose products participate in the regulation of nitrogen fixation in addition to *fixNOPQ* (Preisig *et al.*, 1993), are also present on the chromosome.

Other chromosomal loci include: (i) both the micro-aerobically induced cytochrome oxidase complex and *fixGHIS* (Khan *et al.*, 1989), which encodes a membrane-bound complex that includes a cation pump involved in nitrogen fixation; (ii) the sigma factor *rpoN* (van Slooten *et al.*, 1990); (iii) the gene *pckA* for phosphoenolpyruvate carboxykinase (Østeras *et al.*, 1991); and (iv) the *hemA* gene for α-aminolaevulinic acid synthase (Stanley *et al.*, 1988). Moreover, as in many

Rhizobium species, the chromosome of NGR234 contains three 5S, 16S, and 23S rRNA gene clusters (Perret, 1992).

Further information on genes carried by both pNGR234*b* and the chromosome was obtained by randomly sequencing DNA of ANU265 (Viprey *et al.*, 2000), a derivative strain of NGR234 cured of pNGR234*a* (Morrison *et al.*, 1984). The position of several of the sequenced loci was mapped onto the partially ordered cosmid library constructed for each replicon. In total, 273 overlapping cosmids were grouped in contigs of 50 kb to 600 kb that covered almost 95% of both replicons (Perret, 1992). Blast analyses of 2,275 random "shot-gun sequences" from the ANU265 library showed that more than one thousand (1,130) matched putative proteins grouped by functional classes (see Table 2).

As expected, many sequences that were homologous to housekeeping genes were found, including homologues of genes involved in the biosynthesis of vitamins, such as biotin and thiamine, which are essential for growth of rhizobia (West and Wilson, 1939; Streit *et al.*, 1996). Various homologues of the chaperones genes, *groEL*, *groES*, *dnaJ*, and others which encode small heat-shock proteins (sHsps) that are present in many rhizobia (Münchbach *et al.*, 1999), were also identified. Surprisingly, homologues of *mocABC* and *mosA* were also found. These genes encode enzymes involved in the synthesis (*mos*) and catabolism (*moc*) of rhizopines that were thought to be specific to strains of *S. meliloti* (Rossbach *et al.*, 1994, 1995) where they provide a competitive advantage (Murphy and Saint, 1992).

With respect to replication, recombination, repair, and DNA transfer, a number of bacterial homologues were identified. For example, *dnaEQXZ, parC*, *gyrAB, recA*, and *uvrAC* were located on the chromosome, whereas *repB* and *traA* are found on pNGR234*b* (Viprey *et al.*, 2000; Perret *et al.*, unpublished data). IS/MS elements are also present on both pNGR234*b* and the chromosome, albeit to a lesser extent (around 2%) than on pNGR234*a*. Some of these IS/MS elements exist in multiple copies and are identical to those found on pNGR234*a* (Perret *et al.*, 2000b). In contrast to pNGR234*a* and pNGR234*b*, a large number of BIMEs (bacterial interspersed mosaic elements) were found in the chromosome and showed homology to those of the *S. meliloti* genome (Perret *et al.*, 2001).

3.4. Alternative architectures to the usual tripartite genome

As discussed above, the normal partitioning of NGR234 is into three replicons. Identical tripartite partitioning was also observed in freeze-dried cultures of the strain that were made in 1965 and 1971 (Flores *et al.*, 1998), suggesting that this genomic architecture is relatively stable. Nevertheless, as mentioned above, the NGR234 genome contains a number of repeated sequences, including IS/MS elements, some of which are shared between all three replicons (Perret *et al.*, 1997). Five copies of NGRIS4 (3,316 bp) exist, two on pNGR234*a*, one on pNGR234*b*, and two on the chromosome (Perret *et al.*, 2000b). It would, therefore, be surprising if recombination did not occur between these exact repeats. Recently, using a novel system for IS entrapment in Gram-negative bacteria (Schneider *et al.*, 2000), we were able to demonstrate IS movement in NGR234.

Table 2. Functional classes of predicted proteins encoded by pNGR234b and the chromosome (from Viprey et al., 2000; numbers in brackets are percentages of each group).

	Functional categories	Number
Cell envelope and cellular processes	Cell wall	17 (1.9)
	Transport/binding proteins and lipoproteins	184 (20.0)
	Sensors (signal transduction)	21 (2.3
	Membrane bioenergetics (electron transport and ATP synthase)	49 (5.3)
	Surface polysaccharide biosynthesis and export	25 (2.7)
	Sporulation	1 (0.1)
	Mobility and chemotaxis	26 (2.8)
	Cell division	5 (0.5)
	Protein secretion	13 (1.4)
	Chaperones/heat-shock proteins	12 (1.3)
	Cell killing	8 (0.9)
Metabolism	Carbohydrates and related molecules	69 (7.5)
	Amino acids and related molecules	91 (9.9)
	Nucleotides and nucleic acids	11 (1.2)
	Lipids	19 (2.1)
	Cofactors/prosthetic groups	37 (4.0)
	Phosphate	3 (0.3)
	Opine-like compounds	8 (0.9)
	Sulphur	2 (0.2)
Information pathways	DNA replication, restriction, modification and repair	26 (2.8)
	DNA segregation, recombination and transfer	10 (1.1)
	RNA synthesis and modification	19 (2.1)
	Protein synthesis and modification	63 (6.8)
	Regulatory functions	68 (7.4)
Other categories	Adaptation to atypical conditions and protection	27 (2.9)
	Transposon-related functions	51 (5.5)
	Phage-related functions	5 (0.5)
	Miscellaneous	52 (5.6)
Total		922 (100)

Among the IS trapped, hybridisation experiments indicated that two IS elements originated from pNGR234*a*, one from pNGR234*b*, and one from the chromosome (Mavingui *et al.*, unpublished data). Recombination between homologous repeats promotes intra-genomic rearrangements, such as deletions, amplifications, and inversions, of DNA regions throughout the NGR234 genome (Flores *et al.*, 2000; Mavingui *et al.*, unpublished data).

Furthermore, the genome of NGR234 is also prone to large-scale DNA rearrangements as shown by the ability of the original tripartite NGR234 genome to generate four alternative genomic architectures (Figure 1; Mavingui *et al.*, 2002). One structure consisted of a single large DNA molecule resulting from the co-integration of all three original replicons to give the form, chromosone-pNGR234*a*-pNGR234*b*. The other architectures corresponded to genomes with only two replicons. In these cases, the co-integrates were formed from only two of the three original replicons, either the chromosome and pNGR234*a* or the chromosome and pNGR234*b* or pNGR234*a* and pNGR234*b*. In other words, the genome of NGR234, like that of many bacterial species (Roth *et al.*, 1996; Arber, 2000, 2002), including rhizobia (Romero and Palacios, 1997; Guo *et al.*, 2003), is dynamic and prone to genetic rearrangements.

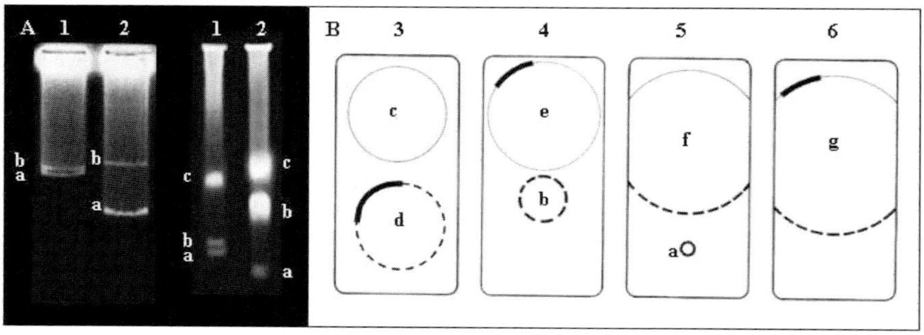

Figure 1. Multiple genomic architectures of Rhizobium *sp. NGR234. A, Plasmid and PFGE patterns. B, schematic representation of NGR234 alternative structures generated by cointegration between replicons.*

1 Sinorhizobium meliloti *containing: a, pSymA; b, pSymB; and c, chromosome. 2* Rhizobium *sp NGR234 harboring: a, pNGR234*a*; b, pNGR234*b*; c, chromosome. 3-6 NGR234 derivatives with cointegrates: d, pNGR234*b*-pNGR234*a*; e, chromosome-pNGR234*a*; f, chromosome-pNGR234*b*; g, chromosome-pNGR234*a*-pNGR234*b.

Amplifications of specific genes are often associated with both bacterial adaptation (Sonti *et al.*, 1999; Neuberger and Hartley, 1981) and virulence (Mekalanos *et al.*, 1983; Corn *et al.*, 1993). Amplification of either a particular DNA region or specific loci in rhizobia results in increased symbiotic capacity (Mavingui *et al.*, 1997; 1998; Castillo *et al.*, 1999). Whether genome plasticity in NGR234 plays a role in its promiscuous behaviour needs to be tested, however, fusions of the different replicons seem to have no effect on the nodulation properties of NGR234 on various hosts (Mavingui *et al.*, 2002).

4. CODING CAPACITY OF THE NGR234 GENOME

4.1. Mutational analysis

Usually, the construction of defined mutants by either transposon mutagenesis or allelic replacement is used for assigning gene function and for dissecting rhizobial determinants involved in symbiosis. In this way, genes that play a role in both the early and late stages of symbiosis have been identified in diverse bacteria (Spaink *et al.*, 1998; Triplett, 2000). Many functional genes in NGR234 were first characterised by mutagenesis, and the resulting null mutants were often either symbiotically inefficient or failed to establish symbioses with its hosts (Perret *et al.*, 2000a). As an example, mutation of the TTSS locus affects the ability of NGR234 to nodulate a variety of legumes (Viprey *et al.*, 1998). Nevertheless, either general or site-specific mutagenesis (producing "gene knockouts"), coupled with screening for varying phenotypes on an array of plants, fails to identify loci with subtle phenotypes or those that are refractory to mutagenesis. Alternative approaches are, therefore, required to identify and characterise less essential genes, and one of these approaches is to map transcripts produced during symbiosis.

4.2. Transcriptional analyses.

Availability of the complete DNA sequence of pNGR234*a* opened up the possibility of analysing transcription of the symbiotic plasmid on a global basis. Furthermore, "transcriptomics" permit analysis under both free-living and symbiotic conditions. PCR fragments representing all 416 predicted ORFs and their intergenic regions were amplified by PCR and transferred onto filter membranes. Then, hybridisations were performed with RNAs extracted either from cells of NGR234 grown in liquid cultures, both with and without induction by flavonoids, or from bacteroids isolated from nodules housing NGR234 (Frieberg *et al.*, 1997; Perret *et al.*, 1999). Most genes encoded by pNGR234*a* (60%) are specifically expressed under symbiotic conditions and many of them are controlled by either *nod* boxes or NifA-σ^{54} type promoters. Genes involved in Nod-factor synthesis are induced rapidly following flavonoid addition (including most *hsn* loci). In contrast, genes, such as those that form the TTSS locus and which are thought to be active only within the host plant, respond later to flavonoid induction (Viprey *et al.*, 1998). Levels of transcription of *nifA* and two transcriptional regulators (*y4qH* and *y4wC*) are much higher in bacteroids. More diverse transcripts were found in bacteroids from determinate as opposed to indeterminate nodules. Surprisingly, many IS sequences are strongly expressed both in free-living and bacteroid cells.

Similar, though less exhaustive, experiments were also performed to gain insights into the transcription patterns of both pNGR234*b* and the chromosome. A total of 921 selected clones (out of the 1,130 matches discussed in Section 3 above) were arrayed on filters that were subsequently probed with radioactively-labelled RNAs prepared from NGR234 cells grown under various free-living conditions (Perret *et al.*, 2000b). Positive hybridisation signals were obtained for many loci, including those for the subunits of RNA polymerase, both 50S and 30S ribosomal

RNA genes, and both *groEL* and *groES* homologues, as well as sequences that code for enzymes involved in intermediary metabolism. Interestingly, several sequences predicted to code for homologues of proteins of unknown function in other prokaryotes were expressed, suggesting that they also play a role in NGR234.

In contrast to the situation with pNGR234*a*, addition of flavonoids had little effect on the expression patterns of pNGR234*b* and the chromosome. Moreover, although 19% of the sequences (179 of 921) hybridised with RNAs from liquid cultures, only 3% gave positive signals with RNAs isolated from bacteroids, suggesting that differentiation into nitrogen-fixing symbionts induces repression of many housekeeping functions. Among the nodule-specific transcripts were homologues of glycosyl hydrolase (FixN; Preisig *et al.*, 1993) and GlnII, a glutamine synthetase involved in ammonium export (Carlson *et al.*, 1987; Martin *et al.*, 1988).

5. CONCLUSIONS AND PERSPECTIVES

The genome of NGR234 resembles that of many rhizobia. It is partitioned into a chromosome and two large plasmids. Random sequencing indicates that, as in *M. loti* (Kaneko *et al.*, 2000) and *S. meliloti* (Capela *et al.*, 2001), housekeeping loci are mostly located on the chromosome. The complete sequence of the symbiotic plasmid, pNGR234*a*, confirmed that most genes involved in symbiosis are present on this replicon. Comparative analyses revealed that pNGR234*a* shares many features with either symbiotic plasmids or symbiotic islands of other members of the *Rhizobiaceae*. For instance, the cytochrome P450 operon is almost identical to that of *B. japonicum* (Tully *et al.*, 1993; Göttfert *et al.*, 2001), whereas many nodulation and nitrogen-fixation loci are highly similar to those of *S. meliloti* (Galibert *et al.*, 2001). In addition, there were homologues of TTSS genes found in *B. japonicum*, *M. loti,* and *R. fredii*, (see Marie *et al.*, 2001). Moreover, the origins of replication and transfer, as well as the *tra* and *repABC* genes of pNGR234*a*, are highly similar to those of *Agrobacterium* plasmids (Freiberg *et al.*, 1997; Moriguchi *et al.*, 2001).

These findings, together with the absence of co-linearity between replicons, suggest that lateral transfer of genetic information has shaped the structure of the NGR234 genome. This assumption is also supported by the mosaic architecture of pNGR234*a*, where IS/MS elements divide clusters of functional genes. The dynamic nature of the NGR234 genome may also be a result of recombination between homologous repeats that generate DNA amplifications, deletions, inversions, and co-integrations (Flores *et al.*, 2000; Mavingui *et al.*, 2002). The biological consequences of this genomic plasticity remain to be assessed.

Comparison of different genomes has highlighted the many similarities among rhizobia. Nevertheless, important differences in the symbiotic capacity of *Rhizobium* species exist. As mentioned in the Introduction, NGR234 is at one end of the host-range spectrum and *Azorhizobium caulinodans* at the other. Despite the similarities between different members of the *Rhizobiaceae*, important genetic differences also exist. For example, in contrast to the majority of rhizobia that possesses some of the NGR234 *hsn* loci but lack others, NGR234 contains many diverse symbiotic genes. Two copies of the transcriptional regulators, *nodD* and *syrM*, exist and the central nodulation transcriptional protein, NodD1, can interact with diverse families of flavonoids. As a consequence, the expression of genes involved in the synthesis and decoration of Nod factors gives NGR234 many of the codes needed to initiate perfect dialogues with different plants. Furthermore, after establishing contact with a plant, NGR234 has other elements, such as the TTSS, that allow this dialogue to continue.

Although combined data from sequence and transcriptional analyses have provided a wealth of detailed information on the symbiotic properties of NGR234, much remains to be discovered about its genome. One important step towards a full understanding will be the completion of the whole genome sequence. Only then can full advantage be taken of important developments in post-genomic methods, including transcriptomics, proteomics and bioinformatics.

REFERENCES

Arber, W. (2000). Genetic variation: Molecular mechanisms and impact on microbial evolution. *FEMS Microbiol. Rev., 24,* 1-7.

Arber, W. (2002). Evolution of prokaryotic genomes. *Curr. Top. Microbiol. Immunol., 264,* 1-14.

Banfalvi, Z., Sakanyan, V., Koncz, G., Kiss, A., Dusha, I., and Kondorosi, A. (1981). Location of nodulation and nitrogen fixation genes on a high molecular weight plasmid of *Rhizobium meliloti. Mol. Gen. Genet., 184,* 318-325.

Batut, J., Daveran, M. L., David, M., Jacobs, J., Garnerone, A. M., and Kahn, D. (1989). *fixK*, a gene homologous with *fnr* and *crp* from *Escherichia coli*, regulates nitrogen fixation genes both positively and negatively in *Rhizobium meliloti. EMBO J., 8,* 1279-1286.

Becker, A., Kleickmann, A., Keller, M., Arnold, W., and Pühler, A. (1993). Analysis of the *Rhizobium meliloti* genes *exoU*, *exoV*, *exoW*, *exoT*, and *exoI* involved in exopolysaccharide biosynthesis and nodule invasion: *exoU* and *exoW* probably encode glucosyltransferases. *Mol. Plant-Microbe Interact., 6,* 735-744.

Broughton, W. J., and Dilworth, M. J. (1971). Control of leghaemoglobin synthesis in snake beans. *Biochem. J., 125,* 1075-1080.

Broughton, W. J., Heycke, N., Meyer, Z. A. H., and Pankhusrt, C. E. (1984). Plasmid link *nif* and *nod* genes in fast-growing rhizobia that nodulate *Glycine max*, *Psophocarpus tetragonolobus*, and *Vigna unguiculata. Proc. Natl. Acad. Sci USA, 82,* 3093-3097.

Broughton, W. J., Wong, C. H., Lewin, A., Samrey, U., Myint, H., Meyer, Z. A. H., *et al.* (19869. Identification of *Rhizobium* plasmid sequences involved in recognition of *Psophocarpus*, *Vigna*, and other legumes. *J. Cell Biol., 102,* 1173-1182.

Broughton, W. J., and Perret, X. (1999). Genealogy of legume-*Rhizobium* symbioses. *Curr. Opin. Plant Mol. Biol., 2,* 305-311.

Broughton, W. J, Jabbouri, S., and Perret, X. (2000). Keys to symbiotic harmony. *J. Bacteriol., 182,* 5641-5652.

Broughton, W. J. (2003). Roses by other name: Taxonomy of the *Rhizobiaceae. J. Bacteriol., 185,* 2975-2979.

Bustos, P., Cevallos, M. A., Collado-Vides, J., Gonzalez, V., Medrano, A., Moreno, G., *et al.* (2001). The symbiotic plasmid of *Rhizobium etli* CFN42. In T. M. Finan, M. R. O'Brian, D. B. Layzell, J. K. Vessey and W. E. Newton (Eds.), *Nitrogen fixation: global perspectives* (pp. 86-87). Wallingford, UK: CABI Publishing.

Capela, D., Barloy-Hubler, F., Gouzy, J., Bothe, G., Ampe, F., Batut, J., *et al.* (2001). Analysis of the chromosome sequence of the legume symbiont *Sinorhizobium meliloti* strain 1021. *Proc. Natl. Acad. Sci. USA, 98,* 9877-9882.

Carlson, T. A., Martin, G. B., and Chelm, B. K. (1987). Differential transcription of the two glutamine synthetase genes of *Bradyrhizobium japonicum. J. Bacteriol., 169,* 5861-5866.

Castillo, M., Flores, M., Mavingui, P., Martinez-Romero, E., Palacios R., and Hernandez, G. (1999). Increase in alfalfa nodulation, nitrogen fixation, and plant growth by specific DNA amplification in *Sinorhizobium meliloti*. *Appl. Environ. Microbiol., 65,* 2716-2722.

Chen, H., Batley, M., Redmond, J., and Rolfe, B. G. (1985). Alteration of the effective nodulation properties of a fast-growing broad host range *Rhizobium* due to changes in exopolysaccharide synthesis. *J. Plant Physiol., 120,* 331-349.

Chen, W. M., Laevens, S., Lee, T. M., Coenye, T., De Vos, P., Mergeay, M., *et al.* (2001). *Ralstonia taiwanensis* sp. nov., isolated from root nodules of *Mimosa* species and sputum of a cystic fibrosis patient. *Int. J. Syst. Evol. Microbiol., 51,* 1729-1735.

Corn, P. G., Anders, J., Takala, A. K., Käyhty, H., and Hoiseth, S. K. (1993). Genes involved in *Hemophilus influenzae* type b capsule expression are frequently amplified. *J. Infect. Dis., 167,* 356-364.

David, M., Daveran, M. L., Batut, J., Dedieu, A., Domergue, O., Ghai, J., *et al.* (1988). Cascade regulation of *nif* gene expression in *Rhizobium meliloti*. *Cell, 26,* 671-683.

de Lyra, M. C. P., Ollero, F. J., Madinabeitia, N., Espuny, M. R., Bellogin, R. A., Cubo, M., T., *et al.* (2000). Characterization of a *nolT* mutant of *Sinorhizobium fredii* HH103: Its role in type III secretion protein. In *Fourth European Nitrogen Fixation Conference Abstract Book* (p. 192). Sevilla, Spain: Viceconsejeria.

Dénarié, J., Debellé, F., and Rosenberg, G. (1992). Signaling and host range variation in nodulation. *Ann. Rev. Microbiol., 46,* 497-531.

Downie, A. (1997). Fixing a symbiotic circle. *Nature, 387,* 352-354.

Eckhardt, T. (1978). A rapid method for the identification of plasmid desoxyribonucleic acid in bacteria. *Plasmid, 1,* 584-588.

Fellay, R., Perret, X., Viprey, V., Broughton, W. J., and Brenner, S. (1995). Organization of host-inducible transcripts on the symbiotic plasmid of *Rhizobium* sp. NGR234. *Mol. Microbiol., 16,* 657-667.

Finan, T. M., Weidner, S., Wong, K., Buhrmester, J., Chain, P., Vorholter, F. J., *et al.* (2001). The complete sequence of the 1,683-kb pSymB megaplasmid from the N_2-fixing endosymbiont *Sinorhizobium meliloti*. *Proc. Natl. Acad. Sci. USA, 98,* 9889-9894.

Fisher, R. F., and Long, S. R. (1992). *Rhizobium*-plant signal exchange. *Nature, 357,* 655-660.

Flores, M., Mavingui, P., Girard, L., Peret, X., Broughton, W. J., Martínez-Romero, E., *et al.* (1998). Three replicons of *Rhizobium* sp. strain NGR234 harbor symbiotic gene sequences. *J. Bacteriol., 180,* 6052-6053.

Flores, M., Mavingui, P., Perret, X., Broughton, W. J., Romero, D., Hernández, G., *et al.* (2000). Prediction, identification, and artificial selection of DNA rearrangements in *Rhizobium*: Toward a natural genomic design. *Proc. Natl. Acad. Sci. USA, 97,* 9138-9143.

Freiberg, C., Fellay, R., Bairoch, A., Broughton, W. J., Rosenthal, A., and Perret, X. (1997). Molecular basis of symbiosis between *Rhizobium* and legumes. *Nature, 387,* 394-401.

Galibert, F., Finan, T. M., Long, S. R., Pühler, A., Abola, P., Ampe, F., *et al.* (2001). The composite genome of the legume symbiont *Sinorhizobium meliloti*. *Science, 293,* 668-672.

Glucksmann, M. A., Reuber, T. L., and Walker, G. C. (1993). Family of glycosyl transferases needed for the synthesis of succinoglycan by *Rhizobium meliloti*. *J. Bacteriol., 175,* 7033-7044.

González, V., Bustos, P., Ramírez-Romero, M. A., Medrano-Soto, A., Salgado, H., Hernández-González, I., *et al.* (2003). The mosaic structure of the symbiotic plasmid of *Rhizobium etli* CFN42 and its relation to other symbiotic genome compartments. *Genome Biol., 4,* R36.

Goodner, B., Hinkle, G., Gattung, S., Miller, N., Blanchard, M., Qurollo, B., *et al.* (2001). Genome sequence of the plant pathogen and biotechnology agent *Agrobacterium tumefaciens* C58. *Science, 294,* 2323-2328.

Göttfert, M., Rothlisberger, S., Kundig, C., Beck, C., Marty, R., and Hennecke, H. (2001). Potential symbiosis-specific genes uncovered by sequencing a 410-kilobase DNA region of the *Bradyrhizobium japonicum* chromosome. *J. Bacteriol., 183,* 1405-1412.

Gray, J. X., Djordjevic, M. A., and Rolfe, B. G. (1990). Two genes that regulate exopolysaccharide production in *Rhizobium* sp. strain NGR234: DNA sequences and resultant phenotypes. *J. Bacteriol., 172,* 193-203.

Guo, X., Flores, M., Mavingui, P., Fuentes, S. I., Hernández, G., Dávila, G., *et al.* (2003). Natural genomic design in *Sinorhizobium meliloti*: novel genomic architectures. *Genome Res., 13,* 1810-1817.

Hanin, M., Jabbouri, S., Quesada-Vincens, D., Freiberg, F., Perret, X., Prome, J. C., *et al.* (1997). Sulphation of *Rhizobium* sp. NGR234 Nod factors is dependent on *noeE*, a new host-specificity gene. *Mol. Microbiol., 24,* 1119-1129.

Honeycutt, R. J., McClelland, M., and Sobral, B. W. S. (1993). Physical map of the genome of *Rhizobium meliloti* 1021. *J.Bacteriol., 175,* 6945-6952.

Jabbouri, S., Felley, R., Talmont, F., Kamalaprija, P., Burger, U., Relić, B., *et al.* (1995). Involvement of *nodS* in N-methylation and *nodU* in 6-*O*-carbamoylation of *Rhizobium* sp. NGR234 Nod factors. *J. Biol. Chem., 270,* 22968-22973.

Jabbouri, S., Relić, B., Hanin, M., Prome, J. C., and Broughton, W. J., (1998). *nolO* and *noel* (HsnIII) of *Rhizobium* sp. NGR234 are involved in 3-*O*-carbamoylation and 2-*O*-methylation of Nod-factors. *J. Biol. Chem., 273,* 12047-12055.

Kahn, D., David, M., Domergue, O., Daveran, M., Ghai, J., Hirsch, P. R., *et al.* (1989). *Rhizobium meliloti fxGHI* sequence predicts involvement of specific cation pump in symbiotic nitrogen fixation. *J. Bacteriol., 171,* 929-939.

Kaneko, T., Nakamura, Y., Sato, S., Asamizu, E., Kato, T., Sasamoto, S., A., *et al.* (2000). Complete genome structure of the nitrogen-fixing symbiotic bacterium *Mesorhizobium loti*. *DNA Res., 7,* 331-338.

Kaneko, T., Nakamura, Y., Sato, S., Minamisawa, K., Uchiumi, T., Sasamoto, S., *et al.* (2002) Complete genomic sequence of nitrogen-fixing symbiotic bacterium *Bradyrhizobium japonicum* USDA110. *DNA Res., 9,* 189-197.

Kündig, C., Hennecke, H., and Göttfert, M. (1993). Correlated physical and genetic map of the *Bradyrhizobium japonicum* 110 genome. *J. Bacteriol., 175,* 613-622.

Le Flèche, P., Hauck, Y., Onteniente, L., Prieur, A., Denoeud, F., Ramisse, V., *et al.* (2001). A tandem repeats database for bacterial genomes: Application of the genotyping of *Yersinia pestis* and *Bacillus anthracis*. *BMC Microbiol., 1,* 2-8.

Lerouge P., Roche, P., Faucher, C., Maillet, F., Truchet, G., Prome, J. C., *et al.* (1990). Symbiotic host-specificity of *Rhizobium meliloti* is determined by a sulphated and acylated glucosamine oligosaccharide signal. *Nature, 344,* 781-784.

Long, S. (1996). *Rhizobium* symbiosis: Nod factors in perspective. *Plant Cell, 8,* 1885-1898.

Lewin, A., Rosenberg, C., Stanley, J., Dowling, D. N., Manen, J. F., Debelle, F., *et al.* (1987). Multiple host-specificity loci in the broad host-range *Rhizobium* NGR234. In D. P. S. Verma and N. Brisson (Eds.), *Molecular genetics of plant-microbe interactions* (pp. 232-2327). Dordrecht, The Netherlands: Martinus Nijhoff Publishers.

Marie, C., Broughton, W. J., and Deakin, W. J. (2001). *Rhizobium* type III secretion systems: Legume charmers or alarmers? *Curr. Opin. Plant Biol., 4,* 336-342.

Martin, G. B., Chapman, K.A., and Chelm, B. K. (1988). Role of the *Bradyrhizobium japonicum ntrC* gene product in differential regulation of the glutamine synthetase II gene (*glnII*). *J. Bacteriol., 170,* 5452-5459.

Martínez-Romero, E., and Caballero-Mellado, J. (1996). *Rhizobium* phylogenies and bacterial genetic diversity. *Critical Rev. Plant Sci., 15,* 113-140.

Mavingui, P., Flores, M., Romero, D., Martínez-Romero, E., Palacios, R. (1997). Generation of *Rhizobium* strains with improved symbiotic properties by random DNA amplification (RDA). *Nature Biotechnol., 15,* 564-569.

Mavingui, P., Laeremans, T., Flores, M., Romero, D., Martínez-Romero, E., and Palacios, R. (1998). Genes essential for Nod factor production and nodulation are located on a symbiotic amplicon (AMPRtrCFN299pc60) in *Rhizobium tropici*. *J. Bacteriol., 180,* 2866-2874.

Mavingui, P., Flores, M., Guo, X., Dávila, G., Perret, X., Broughton, W. J., *et al.* (2002). Dynamics of genome architecture in *Rhizobium* sp. strain NGR234. *J. Bacteriol., 184,* 171-176.

Mekalanos, J. J. (1983). Duplication and amplification of toxin gene in *Vibrio cholorae*. *Cell, 35,* 253-263.

Moriguchi, K., Maeda, Y., Satou, M., Hardayani, N. S. N., Kataoka, M., Tanaka N., *et al.* (2001). The complete nucleotide sequence of a plant root-inducing (Ri) plasmid indicates its chimeric structure and evolutionary relationship between tumor-inducing (Ti) and symbiotic (Sym) plasmids in *Rhizobiaceae*. *J. Mol. Biol., 307,* 771-784.

Morrison, N. A., Hau, C. Y., Trinick, M. J., Shine, J., and Rolfe, B. G. (1983). Heat curing of a Sym plasmid in a fast-growing *Rhizobium* sp. that is able to nodulate legumes and the nonlegume *Parasponia* sp. *J. Bacteriol., 153,* 527-531.

Morrison, N. A., Cen, Y. H., Chen, H .C., Plazinski, J., Ridge, R., and Rolfe, B. G. (1984). Mobilization of a sym plasmid from a fast-growing cowpea *Rhizobium* strain. *J. Bacteriol., 160,* 483-487.

Moulin, L., Munive, A., Dreyfus, B., and Boivin-Masson, C. (2001a). Nodulation of legumes by members of the beta-subclass of proteobacteria. *Nature, 411,* 948-950.

Moulin L., Chen, W. M., Béna, G., Dreyfus, B., and Boivin-Masson, C. (2001b). Rhizobia: the family is expanding. In T. M. Finan, M. R O'Brian, D. B. Layzell, J. K. Vessey and W. E. Newton (Eds.), *Nitrogen fixation: global perspectives* (pp. 61-65). Wallingford, UK: CABI Publishing.

Münchback, M., Nocker, A., and Narberhaus, F. (1999). Multiple small heat shock proteins in rhizobia. *J. Bacteriol., 181,* 83-90.

Murphy, P. J., and Saint, C. P. (1992). Rhizopines in the legume-*Rhizobium* symbiosis. In D. P. S. Verma (Eds.), *Molecular signals in plant-microbe communications* (pp. 377-390). Boca Raton, FL: CRC Press.

Neuberger, M. S., and Hartley, B. S. (1981). Structure of an experimentally evolved gene duplication encoding ribitol dehydrogenase in a mutant of *Klebsiella aerogenes. J. Gen. Microbiol., 122,* 181-191.

Oke, V., and Long, S. (1999). Bacteroid formation in the *Rhizobium*-legume symbiosis. *Curr. Opin. Microbiol., 32,* 837-849.

Østeras, M., Finan, T. M., and Stanley, J. (1991). Site-directed mutagenesis and DNA sequence of *pckA* of *Rhizobium* NGR234, encoding phosphoenolpyruvate carboxykinase: gluconeogenesis and host-dependent symbiotic phenotype. *Mol. Gen. Genet., 230,* 257-269.

Østeras, M., Stanley, J., and Finan, T. M. (1995). Identification of *Rhizobium* specific intergenic mosaic elements within an essential two-component regulatory system of *Rhizobium* species. *J.Bacteriol., 177,* 5485-5494.

Pankhurst, C. R., Broughton, W. J., Bachem, C., Kondorosi, E., and Kondorosi, A. (1983). Identification of nitrogen fixation and nodulation genes on a large plasmid from a broad host range *Rhizobium* sp. In A. Pühler (Ed.), *Molecular genetics of the bacteria-plant microbe interaction* (pp. 169-176). Berlin and Heidelberg, Germany: Springler-Verlag.

Perret, X., Broughton, W. J., and Brenner, S. (1991). Canonical ordered cosmid library of the symbiotic plasmid of *Rhizobium* species NGR234. *Proc. Natl. Acad. Sci. USA, 88*, 1923-1927.

Perret, X. (1992). Cartographie physique et génétique du génome de *Rhizobium* species NGR234. Ph.D. thesis #2489, University of Geneva, Switzerland.

Perret, X., and W.J. Broughton. (1997). How many replicons make a nodule? *Nature, 387,* 767.

Perret, X., Viprey, V., Freiberg, C., and Broughton, W. J. (1997). Structure and evolution of NGRRS-1, a complex, repeated element in the genome of *Rhizobium* sp. NGR234. *J. Bacteriol., 179,* 7488-7496.

Perret, X., Freiberg, C., Rosenthal, A., Broughton, W. J., and Fellay, R. (1999). High-resolution transcriptional analysis of the symbiotic plasmid of *Rhizobium* sp. NGR234. *Mol. Microbiol., 32,* 415-425.

Perret, X., Staehelin, C., and Broughton, W. J. (2000a). Molecular basis of symbiotic promiscuity. *Microbiol. Mol. Biol. Rev., 64,* 180-201.

Perret, X., Viprey, V., and Broughton, W. J. (2000b). Physical and genetic analysis of the broad-host range *Rhizobium* sp. NGR234. In E. W. Triplett (Ed.), *Prokaryotic nitrogen fixation. A model system for the analysis of a biological process* (pp. 679-692). Wymondham, UK: Horizon Scientific Press.

Perret, X., Parsons, J., Viprey, V., Reichwald, K., and Broughton, W. J. (2001). Séquences répétées des génomes de *Rhizobium* sp. NGR234 et *Sinorhizobium meliloti*: Une analyse comparative par séquençage aléatoire. *Can. J. Microbiol., 47,* 548-558.

Preisig O, Anthamatten, D., and Hennecke, H. (1993). Genes for a microaerobically induced oxidase complex in *Bradyrhizobium japonicum* are essential for a nitrogen-fixing endosymbiosis. *Proc. Natl. Acad. Sci. USA, 90*, 3309-3313.

Price, N. J. P., Relić, B., Talmont, F., Lewin, A,. Promé, D., Pueppke, S. G., *et al.* (1992). Broad-host range *Rhizobium* species NGR234 secretes a family of carmoylated and fucosylated, nodulation signals that are *0*-acetylated or sulphated. *Mol. Microbiol., 6,* 3575-3584.

Pueppke, S. G., and Broughton, W. J. (1999). *Rhizobium* sp. strain NGR234 and *R.fredii* USDA257 share exceptionally broad, nested host-ranges. *Mol. Plant-Microbe Interact., 12,* 293-318.

Quesada-Vincens, D., Fellay, R., Nassim, T., Viprey, V., Burger, U., Prome, J. C., *et al.* (1997). *Rhizobium* sp. NGR234 NodZ protein is a fucosyltransferase. *J. Bacteriol., 179,* 5087-5093.

Relić, B., Talmont, F., Kopcinska, J., Golinowsky, W., Promé, J. C., and Broughton, W. J. (1993). Biological activity of *Rhizobium* sp. NGR234 Nod-factors on *Macroptilium atropurpureum*. *Mol. Plant-Microbe Interact., 6,* 764-774.

Relić, B., Perret, X., Estrada-Garcia, M. T., Kpocinska, J., Golinowski, W., Krishnan, H. B., *et al.* (1994). Nod factors of *Rhizobium* are a key to the legume door. *Mol. Microbiol., 13,* 171-178.

Romero , D., and Palacios, R. (1997). Gene amplification and genomic plasticity in prokaryotes. *Ann. Rev. Genet., 31,* 91-111.

Rossbach, S, Kulpa, D. A., Rossbach, U., de Bruijn, F. J. (1994). Molecular and genetic characterization of the rhizopine catabolism (*mocABRC*) genes of *Rhizobium meliloti* L5-30. *Mol. Gen. Genet., 245,* 11-24.

Rossbach, S, Rasul G, Schneider, M, Eardly, B, de Bruijn, F. J. (1995). Structural and functional conservation of the rhizopine catabolism (*moc*) locus is limited to selected *Rhizobium meliloti* strains and unrelated to their geographical origin. *Mol. Plant Microbe Interact., 8,* 549-559.

Rosenberg, C., Boistard, P., Dénarié, J., and Casse-Delbart, F. (1981). Genes controlling early and late functions in symbiosis are located on a megaplasmid in *Rhizobium meliloti*. *Mol. Gen. Genet., 184,* 326-333.

Roth, J. R., Benson, N., Galitski, T., Haack, K., Lawrence, J. G., and Miesel, L. (1996). Rearrangements of the bacterial chromosome: Formation and applications. In F. C. Neidhardt, R. Curtiss III, J L. Ingraham, E. C. C. Lin, K. Low Jr, B. Magasanik *et al.* (Eds.), Escherichia coli *and* Salmonella typhimurium*: Cellular and molecular biology*, 2nd Ed. (pp. 2256-2276). Washington, DC: American Society for Microbiology Press.

Schultze, M., Kondorosi, E., Dénarié, J., Buiré, M., and Kondorosi, A. (1994). Cell and molecular biology of *Rhizobium*-plant interactions. *Int. Rev. Cytol., 156,* 1-74.

Schwedock, J., and Long, S. (1990). ATP sulphurylase activity of the *nodP* and *nodQ* gene products of *Rhizobium meliloti*. *Nature, 348,* 644-647.

Schneider, D., Fuare, D., Noirclerc-Savoye, M., Barrière, A. C., Coursange, E., and Blot, M. (2000). A broad-host-range plasmid for isolating mobile genetic elements in Gram-negative bacteria. *Plasmid, 4,* 201-207.

Spaink, H. P., Kondorosi, A., and Hooykass, P. J. J. (1998). *The* Rhizobiaceae. *Molecular biology of model plant-associated bacteria*. Dordrecht, The Netherlands: Kluwer Academic Publishers.

Stanley, J., Dowling, D. N., and Broughton, W. J. (1988). Cloning of *hemA* from *Rhizobium* NGR234 and symbiotic phenotype of a gene-directed mutant in diverse legume genera. *Mol. Gen. Genet., 215,* 32-37.

Sonti, R.V., and Roth, J. R. (1999). Role of gene duplications in the adaptaion of *Salmonella typhimurium* to growth on limiting carbon sources. *Genetics, 123,* 19-28.

Streit, W. R., Josep, C. M., and Phillips, D. A. (1996). Biotin and other water-soluble vitamins are key growth factors for alfalfa rhizosphere colonization by *Rhizobium meliloti* 1021. *Mol. Plant-Microbe Interact., 5,* 330-338.

Sullivan, J. T., Patrick, H. N., Lowther, W. L., Scott, D. B., and Ronson, C. W. (1995). Nodulating strains of *Rhizobium loti* arise through chromosomal symbiotic gene transfer in the environment. *Proc. Natl. Acad. Sci. USA, 92,* 8985-8989.

Sullivan, J. T., and Ronson, C. W. (1998). Evolution of rhizobia by acquisition of a 500 kb symbiosis island that integrates into a phe-tRNA gene. *Proc. Natl. Acad. Sci. USA, 95,* 5145-5149.

Sy, A., Giraud, E., Jourand, P., Garcia, N., Willems, A., de Lajudie, P., *et al.* (2001). Methylotrophic *Methylobacterium* bacteria nodulate and fix nitrogen in symbiosis with legumes. *J. Bacteriol., 183,* 214-220.

Trinick, M. J. (1980). Relationships among the fast-growing rhizobia of *Lablab purpureus*, *Leucaena leucocephala*, *Mimosa* sp., *Acacia farnesiana* and *Sesbania grandiflora* and their affinities with other rhizobia groups. *J. Appl. Bacteriol., 49,* 39-53.

Triplett, W. E. (2000). *Prokaryotic nitrogen fixation. A model system for the analysis of a biological process*. Wymondham, UK: Horizon Scientific Press.

Tully, R. E., and Keister, D. L. (1993). Cloning and mutagenesis of a cytochrome P-450 locus from *Bradyrhizobium japonicum* that is expressed anaerobically and symbiotically. *Appl. Environ. Microbiol., 59,* 4136-4142.

van Belkum, A, S. Scherer, L. van Alphen, and H. Verbrugh. (1998). Short-sequence DNA repeats in prokaryotic genomes. *Microbiol. Mol. Biol. Rev., 62,* 275-293.

van Belkum, A. (1999). Short sequence repeats in microbial pathogenesis and evolution. *Cell Mol. Life Sci., 56,* 729-734.

van Berkum, P., Terefework, Z., Paulin, L., Suomalainen, S., Lindström, K., and Eardly, B. D. (2003). Discordant phylogenies within the *rrn* loci of rhizobia. *J. Bacteriol., 185,* 2988-2998.

van Brussel, A. A. N., Bakhuizen, R., van Spronsen, P. C., Spaink, H. P., Tak, T., Lugtenberg, B. J. J., *et al.* (1992). Induction of preinfection thread structures in the leguminous host plant by mitogenic lipo-oligosaccharides of *Rhizobium*. *Science, 257,* 70-72.

van Slooten, J. C., Cervantes, E., Broughton, W. J., Wong, C. H., and Stanley, J. (1990). Sequence and analysis of the *rpoN* sigma factor gene of *Rhizobium* sp. strain NGR234, a primary coregulator of symbiosis. *J. Bacteriol., 172,* 5563-5574.

Verma, D. P. S., and Hong, Z. (1996). Biogenesis of the peribacteroid membrane in root nodules. *Trends Microbiol., 4,* 364-368.

Viprey, V., Del Geco, A., Golinowski, W., Broughton, W. J., and Perret, X. (1998). Symbiotic implications of type III protein secretion machinery in *Rhizobium*. *Mol. Microbiol., 28,* 1381-1389.

Viprey, V., Rosenthal, A., Broughton, W. J., and Perret, X. (2000). Genetic snapshots of the *Rhizobium* species NGR234 genome. *Genome Biol., 1,* 14.1-14.7.

West, P. M., and Wilson, P. W. (1939). Growth factor requirements of the root nodule bacteria. *J. Bacteriol., 37,* 161-185.

Wood, D. W., Setubal, J. C., Kaul, R., Monk, D. E., Kitajima, J. P., Okura, V. K., *et al.* (2001). The genome of the natural genetic engineer *Agrobacterium tumefaciens* C58. *Science, 294,* 2317-2323.

CHAPTER 7

FACETS OF THE *BRADYRHIZOBIUM JAPONICUM* 110 GENOME

M. GÖTTFERT[1], H. HENNECKE[2], AND S. TABATA[3]

[1]Technische Universität Dresden, Institut für Genetik, D-01062 Dresden, Germany; [2]Eidgenössische Technische Hochschule, Institut für Mikrobiologie, CH-8092 Zürich, Switzerland; [3]Kazusa DNA Research Institute, 2-6-7 Kazusa-kamatari, Kisarazu, Chiba 292-0812, Japan.

1. INTRODUCTION

Bradyrhizobium japonicum (Jordan, 1982), formerly known as *Rhizobium japonicum*, is the symbiont of cowpea, mungbean, siratro, and soybean. Strain USDA3I1b110, here referred to as strain 110, was isolated in 1959 from Florida soil (van Berkum and Fuhrmann, 2001). Besides propagating *in planta*, strain 110 is an aerobic heterotroph, which can also grow chemoautotrophically (Hanus *et al.*, 1979) and anaerobically with nitrate as the terminal electron acceptor (Daniel *et al.*, 1982).

Early studies on the genome of *B. japonicum* strains dealt with the G+C content (DeLey and Rassel, 1965; Elkan, 1969) and included DNA denaturation and renaturation experiments (Chakrabarti *et al.*, 1984). Hybridisations revealed that *B. japonicum* strains, unlike *Rhizobium* strains, harbour the *nif* genes on the chromosome (Masterson *et al.*, 1985; Prakash and Atherly, 1986). The development of pulsed-field gel electrophoresis was the basis of a more detailed characterization of the *B. japonicum* genome (Sobral *et al.*, 1991). Using this technology, the first correlated physical and genetic map of the strain 110 genome was created (Kündig *et al.*, 1993). This map proved that the genome consists of a single circular chromosome with the nodulation (*nod*) and nitrogen-fixation (*nif, fix*) genes clustered. Later, the nucleotide sequence of a 410-kb region (Göttfert *et al.*, 2001) gave a detailed insight into the coding capacity of the symbiotic gene region. The next step towards a whole genome characterization was the establishment of an ordered BAC library (Tomkins *et al.*, 2001). The publication of the complete genome sequence of strain 110 now reveals, in principle, its full coding potential

R. Palacios and W. E. Newton (eds.), Genomes and Genomics of Nitrogen-fixing Organisms, 99-111.

(Kaneko *et al.*, 2002a). The genome has a size of 9,105,828 bp. There are about 8317 protein-coding ORFs, 50 tRNA genes, an *rrn* operon, and a split tmRNA.

2. MATERIAL AND METHODS

Source of DNA and protein sequences was RhizoBase (http://www.kazusa.or.jp /rhizobase/). If not otherwise stated, sequences were analysed using the European Molecular Biology Open Software Suite (EMBOSS; Rice *et al.*, 2000). Intergenic regions were extracted with the 'extractseq' module, motifs were searched with either the 'fuzznuc' or 'fuzzpro' module. BLAST similarity searches were performed either at the National Center for Biotechnology Information website or at the RhizoBase website. Rho-independent terminators were identified using TransTerm (Ermolaeva *et al.*, 2000). G+C content and GC skew (calculated as G-C/G+C) were analysed with Quickie-Calc™ (Molecular Programming LLC). Genome comparisons were done using TaxPlot at the National Center for Biotechnology Information (www.ncbi.nlm.nih. gov/sutils/taxik2.cgi).

3. GENOME CHARACTERISTICS

3.1. Nucleotide composition

The overall G+C content is 64.1 % (Kaneko *et al.*, 2002a). Apart from a few small regions, there are two larger stretches with the coordinates 1681-2362 kb and 8975-0-75 kb that have a significantly lower G+C content of 59.4 % and 60.2 %, respectively (Figure 1). The asymmetric intrastrand distribution between G and C (GC skew) that is often observed in bacterial chromosomes indicates the location of the *oriC* and *ter* regions (Lobry, 1996; Lobry and Louarn, 2003; McLean *et al.*, 1998). In *B. japonicum*, two shifts are observed at 700 kb and 4890 kb. In addition, several genes likely involved in replication are located around the 700-kb position, *e.g.*, *parA* and *parB*, so the *oriC* should be located in the vicinity. The replication forks might meet around position 4890 kb. A *dif*-like sequence potentially involved in chromosome separation was identified nearby (Kaneko *et al.*, 2002a).

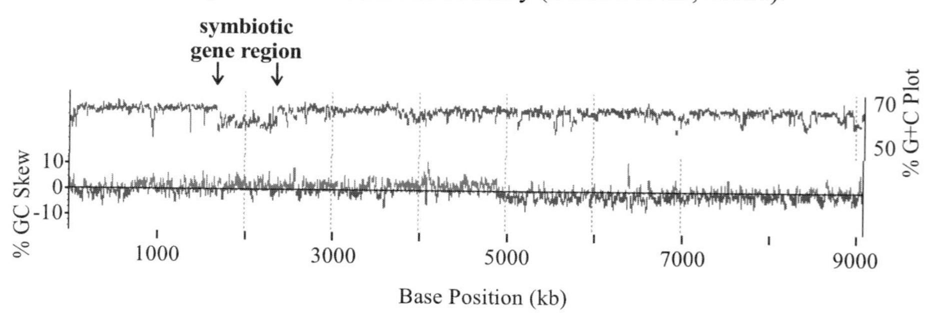

Figure 1. G+C content and GC skew of the strain 110 genome.
Window size 10000 nucleotides.

3.2. IS-related elements and genome stability

The genome harbours a total of 167 transposase-encoding/IS-related elements (Kaneko *et al.*, 2002b). 130 of them are located within the symbiotic region (1681-2362 kb) and the 8975-0-75 kb region. Out of the 20 groups reported, 9 have representatives in both regions and all have a copy in at least one of the two regions. The ten complete or partial copies of ISBj7 are exclusively located within the two regions. So far, there is no report on the activity of these elements, however, deletions that are obviously due to recombination between repeated elements have been described (Hahn *et al.*, 1984). Considering the fact that so many repetitive elements are closely spaced, it is surprising that the chromosome is comparatively stable. For most of the chromosome, there is no difference between the map published by Kündig *et al.* (1993) and the map based on the nucleotide sequence published by Kaneko *et al.* (2002a) despite the fact that the strains were maintained separately for several years. A significant deviation is around position 8975-0-75 kb, where a difference in the number and size of *Pac*I and *Pme*I fragments exists (data not shown). It is unclear if this difference is due to a rearrangement caused by a transposable element. At several other positions, there are indications of DNA insertions of varying sizes into tRNA genes (Kaneko *et al.* 2002a; 2002b). For some rhizobia, the phenomenon of genome plasticity is well established (Flores *et al.* 1988; Flores *et al.* 2000).

3.3. Basic replication, transcription and translation machinery

As expected, the basic replication, transcription, and translation machineries are very similar to those of other bacteria. Besides both DNA polymerase III and I, strain 110 encodes two proteins similar to DNA polymerase IV of *Escherichia coli*. Pol IV of *E. coli* is involved in error-prone repair (Napolitano *et al.*, 2000). Neither Pol II nor UmuD-like proteins were detected. Both MutS and MutL, which are required for mismatch repair (Yang, 2000), are present, however, there is no close homologue either of MutH, the sequence specific endonuclease, or of the DNA adenine methylase (Dam).

The Rho factor is encoded close to the putative *oriC* region. Rho-independent transcription terminator structures are present in 276 intergenic regions (98 % confidence level). The presence of 21 sigma factor-like proteins indicates the versatility with which strain 110 is able to respond to environmental changes. The protein encoded by blr2469 has weak similarity to poly(A) polymerase, however, there is no experimental evidence for poly(A) tailing in the *Rhizobiaceae*.

When genomes are compared, genes involved in translation are always the best conserved group. Strain 110 is no exception, nevertheless, there are two notable observations. First, there is no asparaginyl-tRNA synthetase. The charged tRNA is probably formed by transamidation of mischarged Asp-tRNA(Asn) as described for many prokaryotes (Tumbula *et al.*, 2000). Strain 110 has two sets of proteins, which are similar to the heterotrimeric glutamyl-tRNA amidotransferase (encoded by the *gatA*, *gatB*, and *gatC* genes), that could serve this function. Second, strain 110 has a two-piece tmRNA (Keiler *et al.*, 2000). The function of tmRNA is to rescue stalled ribosomes and to mark the unfinished proteins for degradation (Withey and Friedman, 2002). The SmpB protein (bll5070), which is known to be required for tmRNA activity (Karzai *et al.*, 1999), is also present. A mutation of the tmRNA gene in strain 110 leads to a deficiency in symbiosis (Ebeling *et al.*, 1991).

3.4. Restriction modification

The method of choice to transfer plasmids from *E. coli* into strain 110 is by conjugation. This method is a very time-consuming step. Therefore, electroporation protocols were elaborated, however, still with low yields of transformants if DNA propagated in *E. coli* was used (Guerinot *et al.*, 1990; Hattermann and Stacey, 1990). This failure is probably due to the presence of a restriction modification system of type I. Deleting this system should increase transformation efficiency.

3.5. The symbiotic island

The lower G+C content of the symbiotic-gene region suggests that it was transmitted by horizontal transfer from an unknown origin. Such a transmissible symbiotic element exists in *Mesorhizobium loti* (Sullivan and Ronson, 1998). There, the integration took place in a phe-tRNA gene and a phage integrase was found close to one end of the integrated element. Interestingly, in strain 110, there are proteins encoded within the symbiotic-gene region (encoded by bll1584/bll1585/bll1586) that are similar to the integrase/excisionase system of the *S. meliloti* phage 16-3. In *Sinorhizobium meliloti*, the phage integrates into proline tRNA (Papp *et al.*, 1993; Semsey *et al.*, 1999; Semsey *et al.*, 2002). No proline tRNA gene was detectable next to the integrase gene of strain 110; however, there is a val-tRNA gene at one end of the symbiotic region (Kaneko *et al.*, 2002a). No duplication was found at the other end. Instead, a 3'-terminal portion of a val-tRNA gene is located at a different position within the genome adjacent to a small region of lower G+C content (59.2 %; Kaneko *et al.*, 2002a). It has been suggested that the two parts originate from a rearrangement within the original symbiotic island.

In addition, the 8975-0-75 kb region might form a third part of the symbiotic island. As mentioned above, this region also has low G+C content and harbours many IS-related elements. Interestingly, the 8975-0-75 kb region contains a set of genes similar to the *trb* genes of the Ti-plasmid of *Agrobacterium tumefaciens*, which are required for conjugative transfer. Five of the genes reside as a copy also within the symbiotic gene region.

The Nod factor of strain 110 is a pentasaccharide of *N*-acetylglucosamine modified by a C18:1 fatty acyl chain and the sugar 2-O-methylfucose (Sanjuan *et al.*, 1992). The genes required for synthesis of the factor are located within two separate clusters. One contains the already known nodulation genes *nodABC*, which are essential for synthesis of the backbone, plus several other *nod* genes. The second cluster consists of *nodM*, which probably encodes a glucosamine synthetase (Baev *et al.*, 1991), *noeK* and *noeL*, which are likely to be involved in synthesis of the fucose residue, and *noeD*, which influences the acetylation degree of the Nod factor (Lohrke *et al.*, 1998).

The symbiotic-gene region also encodes all of the proteins known to be required for nitrogen-fixation activity (Göttfert *et al.*, 2001). However, it should be noted that other genes that support symbiotic nitrogen fixation are located outside the symbiotic gene region. Among these are the genes *fixNOQP*, which are essential for respiration under the micro-oxic conditions prevailing in the nodule (see also 3.8).

An unavoidable by-product of nitrogen fixation is the generation of H_2. Although strain 110 recycles this H_2 (Arp and Burris, 1979), only a relatively small cluster, encoding the small and large subunits of hydrogenase, and a few other genes, which were partly interrupted by frame-shifts, were identified within the symbiotic-gene region (Göttfert *et al.*, 2001). The genome sequence now reveals a large cluster (bll6924-bll6449) that encodes all components required for H_2 utilization. Three of the proteins (encoded by bll6926, bll6925 and bll6924) belong to the two-component regulatory family and two proteins are very similar to HoxX and HoxA, which are required for expression of *hup* structural genes in free-living *B. japonicum* (Durmowicz and Maier, 1997; Van Soom *et al.*, 1997).

The symbiotic region not only carries the *nod* and *nif* determinants, but also genes that encode both plant cell wall-degrading enzymes (Caldelari Baumberger *et al.*, 2003) and many proteins involved in transport processes (see also 3.7). One cluster encodes a type-III secretion system that influences symbiosis in a host-dependent manner (Krause *et al.*, 2002).

3.6. Regulatory capacity

In *B. japonicum*, there are several well-studied regulatory circuits and DNA elements that influence promoter activity. One example is the regulation of heat-shock genes. Two negatively regulating elements, CIRCE (controlling inverted repeat of chaperone expression; Minder *et al.*, 2000) and ROSE (repression of heat-shock gene expression; Nocker *et al.*, 2001), have been defined. Furthermore, three σ^{32}-like proteins indicate the flexibility of strain 110 and how it copes with stress situations.

Other well-characterized elements are the *nod* box (Wang and Stacey, 1991), which is the binding site of NodD, the FixK consensus sequence (Fischer, 1994), the –24/–12 promoter (Adams and Chelm, 1984; Thöny and Hennecke, 1989), and the putative housekeeping promoter that was identified upstream of the *rrn* operon (Beck *et al.*, 1997). In order to get an idea about the use of these promoters and regulatory nucleotide motifs, they were sought in intergenic regions (Table 1). In view of missing experimental evidence and to avoid too many false positives, the search was restricted to either perfect or close matches with respect to the consensus sequence. The *nod* box sequences were selected after an additional alignment (not shown). Five out of seven potential *nod* box elements were identified within the symbiotic-gene region, indicating that most, if not all, nodulation genes are located there. This result does not necessarily mean that other genes are of no relevance for nodulation. For example, *nwsB*, which is located outside the symbiotic region, is involved in the complex regulation of *nod* genes (Loh *et al.*, 2002b; Loh and Stacey, 2003). Furthermore, the novel signal molecule, bradyoxetin, which originates from an unknown biosynthetic pathway, mediates population density control of the nodulation genes (Loh *et al.*, 2002a).

Table 1. Conserved DNA elements in 5'-regions of genes.

σ^{54}	σ^{54}	FixK	*nod* box	σ^{70}
bll1044	blr3677	blr1201	*bll1718*	*rrn*
blr1719	bll4798	bsr2670	*bsr1863*	*bsl1986*
blr1755	blr5598	blr2767	*blr2024*	bll2905
blr1769	bll5679	bll2821	*bll1845*	bsl2907
blr1850	blr6145	bsr2822	*bll2016*	bll3735
bll1944	bll6702	bll2851	blr3139	blr5051
bll2063	bll6937	blr4641	bll5092	blr6795
bsl2575	bll7180	bll5772		blr6804
blr2725	blr7289	bll6060		
bll3193	bll8310	bsr7036		
		bll7787		

ORFs located within the symbiotic region are in italics.
σ^{54} consensus: TGGCACN(5)TTGC[TA], no mismatch allowed.
fixK consensus: TTGANCNNGATCAANG, no mismatch allowed.
nod box: ATCN(3,6)GATGN(4,6) ATCCAAACAATCGATTTTACCA, up to 9 mismatches allowed.
σ^{70} consensus: TTGACAN(16,18)TATAA[TC], 1 mismatch allowed.

14 out of 20 selected σ^{54}-dependent sequences are located outside the symbiotic region. Although they do not precede classical *nif* genes, some of the encoded proteins have interesting similarities to known proteins.

- bll1044 encodes a protein with an amidase/glutamyl-tRNA(Gln) amido-transferase motif and might be involved in nitrogen metabolism.
- blr6145 encodes one of the DctA copies, a protein belonging to the sodium:dicarboxylate symporter family.
- bll6937 is the first ORF in a long operon encoding components of the H_2-utilization complex.

The fact that *B. japonicum* has two σ^{54} genes (Kullik *et al.*, 1991), one inside and one outside the symbiotic region, suggests an important function of –24/–12 promoters in strain 110 apart from their contribution to symbiosis (Nienaber et al., 2000).

That FixK-dependent promoters are located outside the symbiotic-gene region is not surprising because this regulator is required for activation of transcription under microaerobic conditions (Anthamatten *et al.*, 1992; Nellen-Anthamatten *et al.*, 1998). Strain 110 contains two *fixK* genes, with $fixK_2$ being central for the regulation of O_2-responsive genes. $FixK_2$-regulated genes include *fixNOQP* and *fixGHIS*, which are required for formation of the high-affinity cbb_3-type cytochrome oxidase (Preisig *et al.*, 1996), and *napE* (bsr7036), which - together with genes probably in the same operon - encodes a periplasmic nitrate reductase. Because only a single nucleotide motif was searched, many genes that are known to be regulated by $FixK_2$ are not listed here, *e.g.*, *nirK*, encoding nitrite reductase, and *norCBQD*, encoding nitric oxide reductase (Mesa *et al.*, 2002; Velasco *et al.*, 2001).

There was an interesting low abundance of σ^{70}-like promoters. It was shown previously that the *rrn* promoter consists of the typical σ^{70}-dependent –35/–10 elements. Including the *rrn* promoter, there are only 8 operons that are preceded by –35/–10 consensus elements (1 mismatch allowed). One of the operons bll1985/bll1986 is located within the symbiotic-gene region and encodes proteins similar to HipA/HipB of *E. coli*, which are involved in the regulation of cell division (Hendricks *et al.*, 2000).

This lack of σ^{70} promoter-like sequences may indicate that most of the promoters require transcriptional activators. Most numerous are members of the two-component regulatory family (Table 2), followed by LysR- and TetR-type regulators. In addition, 21 sigma factor-related proteins indicate the complexity of the regulatory networks in strain 110. None of the putative sigma factors exhibited high similarity to σ^{38} (SigS). In *E. coli*, σ^{38} regulates the synthesis of many growth phase-related proteins (Hengge-Aronis, 2002). It will be interesting to see how the transition from exponential to stationary phase is regulated in strain 110.

Table 2. Predominant regulatory families

Family	Number of genes
Two-component	170
LysR	69
TetR	58
AraC	44
MarR	44
GntR	35
ArsR	17
LuxR	16
IclR	16
Crp	11
AsnC	9
Fis	7
LacI	7

3.7. Transport capacities

Based on gene quantity, strain 110 has an extraordinary broad transport capacity. There are as many as 511 genes that are likely to encode components of ABC transporters. These high-affinity uptake systems will guarantee the supply of strain 110 with nutrients even under very unfavourable conditions. Some of the systems may have export rather than import functions. For example, there are several proteins that have high similarity to HlyB and HlyD of *E. coli*. HlyB and HlyD are required for export of haemolysin (Koronakis *et al.*, 1992), a fatty acylated protein (the product of HlyA that is modified by HlyC activity) that forms pores in the membrane of the target cell. Strain 110 does not encode HlyA- or HlyC-like proteins and secreted proteins still have to be determined.

There are 7 proteins that have high similarity to Na^+/H^+-dicarboxylate symporters (DctA), which supply bacteria with corresponding carbon sources (Janausch *et al.*, 2002). It is unclear how similar the substrate spectra of these systems are. In this respect, it is worthwhile to mention that DctA of *S. meliloti* is also able to transport orotate, a monocarboxylic acid (Yurgel *et al.*, 2000).

Strain 110 appears very versatile with respect to protein export. Several proteins (encoded by bll0666, bll1439, bll1842, blr3500, and blr6022) have weak similarity to PulD and are predicted to be outer-membrane proteins. Therefore, they could be part of protein-secretion systems. One of these proteins is located within a cluster encoding a functional type-III secretion system (Krause *et al.*, 2002). Moreover, all rhizobial strains sequenced so far that do not contain a type-III secretion system (*S. meliloti*, *M. loti* R7A) encode proteins similar to the VirB system of *A. tumefaciens*. In contrast, no VirB-like transport system was found in strains (*B. japonicum* strain 110, *Rhizobium* sp. strain NGR234, *M. loti* strain MAFF303099) that harbour a type-III secretion system.

A twin-arginine translocation system seems also to be present (bll4749, bll4750, bll4751). Membrane-bound hydrogenases are secreted by the Tat pathway due to the presence of a N-terminal twin-arginine signal peptide within the small subunit (Wu *et al.*, 2000). Such a signal is also detectable in the corresponding protein of strain 110 (blr1720). In *Rhizobium leguminosarum* bv. *viciae*, the Tat system is also present (Meloni *et al.*, 2003). A *tatABC* mutant lost the ability for membrane targeting of hydrogenase and was impaired in symbiosis.

3.8. Respiratory chains

A total of five terminal oxidase complexes (or the encoding genes) were previously discovered in *B. japonicum*. The genome sequence has now revealed the presence of three additional complexes (Table 3). Thus, with eight terminal oxidases, *B. japonicum* has the most highly branched respiratory chain of all aerobic prokaryotes known to date. Four of these eight are cytochrome *c* oxidases, whereas the other four are quinol oxidases. A new member of the cytochrome *c* oxidase class is encoded by a putative bll4479-4483 operon. Some of the encoded proteins share similarity with PQQ-dependent dehydrogenases. The presence of biosynthesis genes (bsr6735-blr6739) for PQQ supports the notion that this unusual quinol serves as an electron donor to one of the *c*-type cytochromes of the new oxidase. The specific substrate oxidized by the PQQ-dependent oxidase remains to be determined.

The two newly detected quinol oxidases are homologs of previously described examples. The blr0149-0152 genes share high sequence similarity with *coxWXYZ* and might encode a cytochrome *o*-type quinol oxidase (hence the *cyo* nomenclature; Table 3). The blr3728-3729 genes are close homologs of a *cydAB*-encoded cytochrome *d*-type quinol oxidase. Why *B. japonicum* employs so many respiratory oxidases and under which growth condition each of them is predominantly active, are important questions to be answered in future research.

Table 3. B. japonicum *terminal oxidase genes*

Terminal oxidase	Genome ORF	Gene name	Subunit (SU), property, cofactor
1. aa_3-type heme-copper cytochrome *c* oxidase (Bott *et al.*, 1990)	blr1170	*coxB*	SU II, CuA center
	blr1171	*coxA*	SU I, heme A, heme A_3, CuB center
	blr1175	*coxC*	SU III
	blr1173	*coxF*	SU IV
2. Alternative heme-copper cytochrome *c* oxidase (Bott *et al.*, 1990)	bll3785	*coxM*	SU II, CuA center
	bll3784	*coxN*	SU I, diheme protein, CuB center
	bll3783	*coxO*	SU IIIA
	bll3782	*coxP*	SU IIIB
	bll3781	*coxQ*	SU IV
3. Novel type of heme-copper cytochrome oxidase (This report)	bll4481		SU II, CuA center
	bll4480		SU I, diheme protein, CuB center
	bll4479		diheme cytochrome *c*
	bll4483		diheme cytochrome *c*
	bll4482		triheme cytochrome *c*
4. cbb_3-type heme-copper cytochrome oxidase (Preisig *et al.*, 1993)	blr2763	*fixN*	SU I, heme B, heme B_3, CuB center
	blr2764	*fixO*	SU II, monoheme cytochrome *c*
	blr2765	*fixQ*	SU III
	blr2766	*fixP*	SU IV, diheme cytochrome *c*
5. bb_3-type heme-copper quinol oxidase (Surpin *et al.*, 1996)	blr2714	*coxW*	SU II
	blr2715	*coxX*	SU I, heme B, heme B_3, CuB center
	blr2716	*coxY*	SU III
	blr2717	*coxZ*	SU IV
6. New heme-copper quinol oxidase (This report)	blr0149	*cyoA*-like	SU II
	blr0150	*cyoB*-like	SU I, diheme protein, CuB center
	blr0151	*cyoC*-like	SU III
	blr0152	*cyoD*-like	SU IV
7. *bd*-type quinol oxidase (Arslan 2001)	bll0282	*cydB*	SU II
	bll0283	*cydA*	SU I, heme B, heme B-D binuclear center
8. New *bd*-type quinol oxidase (This report)	blr3729	*cydB*-like	SU II
	blr3728	*cydA*-like	SU I, heme B, heme B-D binuclear center

3.9. Genome comparison

It is known from 16S rDNA analysis that strain 110 is closer to *Rhodopseudomonas palustris* than to fast-growing rhizobia. In order to check if this is true for the complete genome, the whole protein set of strain 110 was compared to the sequenced genomes. As can be deduced from Figure 2, strain 110 is slightly more similar to *M. loti* than to *S. meliloti*. However, most hits were obtained with *R. palustris*. This close relation of strain 110 and *R. palustris* is also reflected by the fact that strain 110 shares more similarity with the intergenic regions of *R. palustris* than with those of the two rhizobial strains (data not shown).

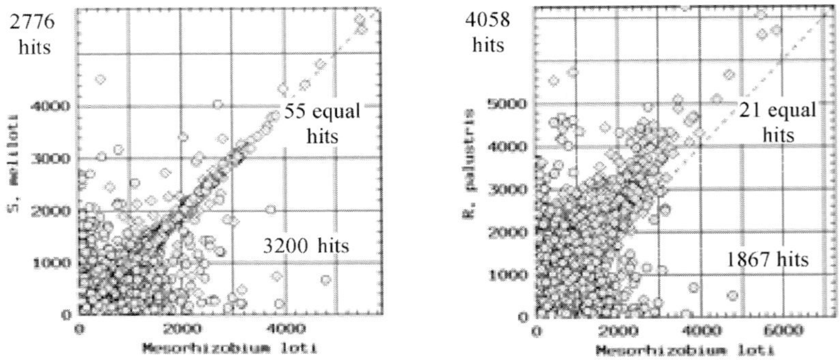

Figure 2. Whole genome comparisons of B. japonicum *strain 110 with* S. meliloti, M. loti *and* R. palustris.

The 8317 proteins of strain 110 were chosen as query. Each circle represents a single query genome protein, plotted by its BLAST scores to the highest scoring protein from each of the selected organisms. The preset cut-off was used.

4. PERSPECTIVES

The nucleotide sequence still leaves many questions to be answered. The functions of numerous transporters and transcriptional regulators are unclear. Several functions seem to be encoded by multiple gene copies. For example, there are about 27 proteins that have similarity to the glutathione S-transferase family and none of them has been studied so far. Furthermore, there are over 1400 genes that have no similarity to published sequences and 2500 genes have similarity to other hypothetical genes. Technical progress in transcript and protein analysis, *i.e.*, transcriptomics and proteomics, will soon yield a rough picture of gene expression under varying environmental conditions including symbiosis. It will be more difficult to follow changes in the bacterial cell envelope and to shed light on the signal exchange with the plant and within the bacterial community.

REFERENCES

Adams, T. H., and Chelm, B. K. (1984). The *nifH* and *nifDK* promoter regions from *Rhizobium japonicum* share structural homologies with each other and with nitrogen-regulated promoters from other organisms. *J. Mol. Appl. Genet., 2*, 392-405.

Anthamatten, D., Scherb, B., and Hennecke, H. (1992). Characterization of a *fixLJ*-regulated *Bradyrhizobium japonicum* gene sharing similarity with the *Escherichia coli fnr* and *Rhizobium meliloti fixK* genes. *J. Bacteriol., 174*, 2111-2120.

Arp, D. J., and Burris, R. H. (1979). Purification and properties of the particulate hydrogenase from the bacteroids of soybean root nodules. *Biochim. Biophys. Acta, 570*, 221-230.

Arslan, E. (2001). The *cbb*$_3$- and *bd*-type oxidases of the soybean symbiont *Bradyrhizobium japonicum*. Ph.D. thesis ETH Nr. 14188, Zurich, Switzerland.

Baev, N., Endre, G., Petrovics, G., Banfalvi, Z., and Kondorosi, A. (1991). Six nodulation genes of *nod* box locus 4 in *Rhizobium meliloti* are involved in nodulation signal production: *nodM* codes for D-glucosamine synthetase. *Mol. Gen. Genet., 228*, 113-124.

Beck, C., Marty, R., Kläusli, S., Hennecke, H., and Göttfert, M. (1997). Dissection of the transcription machinery for housekeeping genes of *Bradyrhizobium japonicum*. *J. Bacteriol., 179*, 364-369.

Bott, M., Bolliger, M., and Hennecke, H. (1990). Genetic analysis of the cytochrome *c-aa*$_3$ branch of the *Bradyrhizobium japonicum* respiratory chain. *Mol. Microbiol., 4*, 2147-2157.

Bott, M., Preisig, O., and Hennecke, H. (1992). Genes for a second terminal oxidase in *Bradyrhizobium japonicum*. *Arch. Microbiol., 158*, 335-343.

Caldelari Baumberger, I., Fraefel, N., Göttfert, M., and Hennecke, H. (2003). New NodW- or NifA-regulated *Bradyrhizobium japonicum* genes. *Mol. Plant-Microbe Interact., 16*, 342-351.

Chakrabarti, S. K., Mishra, A. K., and Chakrabarty, P. K. (1984). Genome size variation of rhizobia. *Experientia, 40*, 1290-1291.

Daniel, R. M., Limmer, A. W., Steele, K. W., and Smith, I. M. (1982). Anaerobic growth, nitrate reduction and denitrification in 46 *Rhizobium* strains. *J. Gen. Microbiol., 12*, 1811-1815.

DeLey, J., and Rassel, A. (1965). DNA base composition, flagellation and taxonomy of the genus *Rhizobium*. *J. Gen. Microbiol., 41*, 85-91.

Durmowicz, M. C., and Maier, R. J. (1997) Roles of HoxX and HoxA in biosynthesis of hydrogenase in *Bradyrhizobium japonicum*. *J. Bacteriol., 179*, 3676-3682.

Ebeling, S., Kündig, C., and Hennecke, H. (1991). Discovery of a rhizobial RNA that is essential for symbiotic root nodule development. *J. Bacteriol., 173*, 6373-6382.

Elkan, G. H. (1969). Deoxyribonucleic acid base composition of isolates of *Rhizobium japonicum*. *Can. J. Microbiol., 15*, 490-493.

Ermolaeva, M. D., Khalak, H. G., White, O., Smith, H. O., and Salzberg, S. L. (2000). Prediction of transcription terminators in bacterial genomes. *J. Mol. Biol., 301*, 27-33.

Fischer, H.-M. (1994). Genetic regulation of nitrogen fixation in rhizobia. Microbiol Rev 58, 352-386.

Flores, M., Gonzalez, M. A., Pardo, A., Leija, E., Martinez, D., Romero, D., *et al.* (1988) Genomic instability in *Rhizobium phaseoli*. *J. Bacteriol., 170*, 1191-1196.

Flores, M., Mavingui, P., Perret, X., Broughton, W. J., Romero, D., Hernandez, G., *et al.* (2000). Prediction, identification, and artificial selection of DNA rearrangements in *Rhizobium*: Toward a natural genomic design. *Proc. Natl. Acad. Sci. USA, 97*, 9138-9143.

Göttfert, M., Röthlisberger, S., Kündig, C., Beck, C., Marty, R., and Hennecke, H. (2001). Potential symbiosis-specific genes uncovered by sequencing a 410-kilobase DNA region of the *Bradyrhizobium japonicum* chromosome. *J. Bacteriol., 183*, 1405-1412.

Guerinot, M. L., Morisseau, B. A., and Klapatch, T. (1990). Electroporation of *Bradyrhizobium japonicum*. *Mol. Gen. Genet., 221*, 287-290.

Hahn, M., Meyer, L., Studer, D., Regensburger, B., and Hennecke, H. (1984). Insertion and deletion mutations within the *nif* region of *Rhizobium japonicum*. *Plant Mol. Biol., 3*, 159-168.

Hanus, F. J., Maier, R. J., and Evans, H. J. (1979). Autotrophic growth of H_2-uptake-positive strains of *Rhizobium japonicum* in an atmosphere supplied with hydrogen gas. *Proc. Natl. Acad. Sci. USA, 76*, 1788-1792.

Hattermann, D. R., and Stacey, G. (1990). Efficient DNA transformation of *Bradyrhizobium japonicum* by electroporation. *Appl. Environ. Microbiol., 56*, 833-836.

Hendricks, E. C., Szerlong, H., Hill, T., and Kuempel, P. (2000). Cell division, guillotining of dimer chromosomes and SOS induction in resolution mutants (*dif*, *xerC* and *xerD*) of *Escherichia coli*. *Mol. Microbiol., 36*, 973-981.

Hengge-Aronis, R. (2002). Stationary phase gene regulation: what makes an *Escherichia coli* promoter σ^S-selective? *Curr. Opin. Microbiol., 5*, 591-595.

Janausch, I. G., Zientz, E., Tran, Q. H., Kröger, A., and Unden, G. (2002). C_4-dicarboxylate carriers and sensors in bacteria. *Biochim. Biophys. Acta, 1553*, 39-56.

Jordan, D. C. (1982). Transfer of *Rhizobium japonicum* Buchanan 1980 to *Bradyrhizobium* gen. nov., a genus of slow-growing, root nodule bacteria from leguminous plants. *Int. J. Syst. Bacteriol., 32*, 136-139.

Kaneko, T., Nakamura, Y., Sato, S., Minamisawa, K., Uchiumi, T., Sasamoto, S., *et al.* (2002a). Complete genomic sequence of nitrogen-fixing symbiotic bacterium *Bradyrhizobium japonicum* USDA110. *DNA Res., 9*, 189-197.

Kaneko, T., Nakamura, Y., Sato, S., Minamisawa, K., Uchiumi, T., Sasamoto, S., *et al.* (2002b). Complete genomic sequence of nitrogen-fixing symbiotic bacterium *Bradyrhizobium japonicum* USDA110 (supplement). *DNA Res., 9*, 225-256.

Karzai, A. W., Susskind, M. M., and Sauer, R. T. (1999). SmpB, a unique RNA-binding protein essential for the peptide-tagging activity of SsrA (tmRNA) *EMBO J., 18*, 3793-3799.

Keiler, K. C., Shapiro, L., and Williams, K. P. (2000). tmRNAs that encode proteolysis-inducing tags are found in all known bacterial genomes: A two-piece tmRNA functions in *Caulobacter*. *Proc. Natl. Acad. Sci. USA, 97*, 7778-7783.

Koronakis, V., Stanley, P., Koronakis, E., and Hughes, C. (1992). The HlyB/HlyD-dependent secretion of toxins by gram-negative bacteria. *FEMS Microbiol. Immunol., 5*, 45-53.

Krause, A., Doerfel, A., and Göttfert, M. (2002). Mutational and transcriptional analysis of the type III secretion system of *Bradyrhizobium japonicum*. *Mol. Plant Microbe Interact., 15*, 1228-1235.

Kullik, I., Fritsche, S., Knobel, H., Sanjuan, J., Hennecke, H., and Fischer, H.-M. (1991). *Bradyrhizobium japonicum* has two differentially regulated, functional homologs of the σ^{54} gene (*rpoN*). *J. Bacteriol., 173*, 1125-1138.

Kündig, C., Hennecke, H., and Göttfert, M. (1993). Correlated physical and genetic map of the *Bradyrhizobium japonicum* 110 genome. *J. Bacteriol., 175*, 613-622.

Lobry, J. R. (1996). Asymmetric substitution patterns in the two DNA strands of bacteria. *Mol. Biol. Evol., 13*, 660-665.

Lobry, J. R., and Louarn, J. M. (2003). Polarisation of prokaryotic chromosomes. *Curr. Opin. Microbiol., 6*, 101-108.

Loh, J., and Stacey, G. (2003). Nodulation gene regulation in *Bradyrhizobium japonicum*: a unique integration of global regulatory circuits. *Appl. Environ. Microbiol., 69*, 10-17.

Loh, J., Carlson, R. W., York, W. S., and Stacey, G. (2002a). Bradyoxetin, a unique chemical signal involved in symbiotic gene regulation. *Proc. Natl. Acad. Sci. USA, 99*, 14446-14451.

Loh, J., Lohar, D. P., Andersen, B., and Stacey, G. (2002b). A two-component regulator mediates population-density-dependent expression of the *Bradyrhizobium japonicum* nodulation genes. *J. Bacteriol., 184*, 1759-1766.

Lohrke, S. M., Day, B., Kolli, V. S., Hancock, R., Yuen, J. P., de Souza, M. L., Stacey, G., Carlson, R., Tong, Z., Hur, H. G., Orf, J. H., and Sadowsky, M. J. (1998). The *Bradyrhizobium japonicum noeD* gene: A negatively acting, genotype- specific nodulation gene for soybean. *Mol. Plant-Microbe Interact., 11*, 476-488.

Masterson, R. V., Prakash, R. K., and Atherly, A. G. (1985). Conservation of symbiotic nitrogen fixation gene sequences in *Rhizobium japonicum* and *Bradyrhizobium japonicum*. *J. Bacteriol., 163*, 21-26.

McLean, M. J., Wolfe, K. H., and Devine, K. M. (1998). Base composition skews, replication orientation, and gene orientation in 12 prokaryote genomes. *J. Mol. Evol., 47*, 691-696.

Meloni, S., Rey, L., Sidler, S., Imperial, J., Ruiz-Argueso, T., and Palacios, J. M. (2003). The twin-arginine translocation (Tat) system is essential for *Rhizobium*-legume symbiosis. *Mol. Microbiol., 48*, 1195-1207.

Mesa, S., Velasco, L., Manzanera, M. E., Delgado, M. J., and Bedmar, E. J. (2002). Characterization of the *norCBQD* genes, encoding nitric oxide reductase, in the nitrogen fixing bacterium *Bradyrhizobium japonicum*. *Microbiology, 148*, 3553-3560.

Minder, A. C., Fischer, H.-M., Hennecke, H., and Narberhaus, F. (2000). Role of HrcA and CIRCE in the heat shock regulatory network of *Bradyrhizobium japonicum*. *J. Bacteriol., 182*, 14-22.

Napolitano, R., Janel-Bintz, R., Wagner, J., and Fuchs, R. P. (2000). All three SOS-inducible DNA polymerases (Pol II, Pol IV and Pol V) are involved in induced mutagenesis. *EMBO J., 19*, 6259-6265.

Nellen-Anthamatten, D., Rossi, P., Preisig, O., Kullik, I., Babst, M., Fischer, H. M., and Hennecke, H. (1998). *Bradyrhizobium japonicum* $FixK_2$, a crucial distributor in the FixLJ- dependent regulatory cascade for control of genes inducible by low oxygen levels. *J. Bacteriol., 180*, 5251-5255.

Nienaber, A., Huber, A., Göttfert, M., Hennecke, H., and Fischer, H. M. (2000). Three new NifA-regulated genes in the *Bradyrhizobium japonicum* symbiotic gene region discovered by competitive DNA-RNA hybridization. *J. Bacteriol., 182*, 1472-1480.

Nocker, A., Krstulovic, N. P., Perret, X., and Narberhaus, F. (2001). ROSE elements occur in disparate rhizobia and are functionally interchangeable between species. *Arch. Microbiol., 176*, 44-51.

Papp, I., Dorgai, L., Papp, P., Jonas, E., Olasz, F., and Orosz, L. (1993). The bacterial attachment site of the temperate *Rhizobium* phage 16-3 overlaps the 3' end of a putative proline tRNA gene. *Mol. Gen. Genet., 240*, 258-264.

Prakash, R. K., and Atherly, A. G. (1986). Plasmids of *Rhizobium* and their role in symbiotic nitrogen fixation. *Int. Rev. Cytol., 104*, 1-25.

Preisig, O., Anthamatten, D., and Hennecke, H. (1993). Genes for a microaerobically induced oxidase complex in *Bradyrhizobium japonicum* are essential for a nitrogen fixing symbiosis. *Proc. Natl. Acad. Sci. USA, 90*, 3309-3313.

Preisig, O., Zufferey, R., and Hennecke, H. (1996). The *Bradyrhizobium japonicum fixGHIS* genes are required for the formation of the high-affinity cbb_3-type cytochrome oxidase. *Arch. Microbiol., 165*, 297-305.

Rice, P., Longden, I., and Bleasby, A. (2000). EMBOSS: the European Molecular Biology Open Software Suite. *Trends Genet., 16*, 276-277.

Sanjuan, J., Carlson, R. W., Spaink, H. P., Bhat, U. R., Barbour, W. M., Glushka, J., and Stacey, G. (1992). A 2-O-methylfucose moiety is present in the lipo-oligosaccharide nodulation signal of *Bradyrhizobium japonicum. Proc. Natl. Acad. Sci. USA, 89*, 8789-8793.

Semsey, S., Blaha, B., Koles, K., Orosz, L., and Papp, P. P. (2002). Site-specific integrative elements of rhizobiophage 16-3 can integrate into proline tRNA (CGG) genes in different bacterial genera. *J. Bacteriol., 184*, 177-182.

Semsey, S., Papp, I., Buzas, Z., Patthy, A., Orosz, L., and Papp, P. P. (1999). Identification of site-specific recombination genes *int* and *xis* of the *Rhizobium* temperate phage 16-3 *J. Bacteriol., 181*, 4185-4192.

Sobral, B. W., Honeycutt, R. J., and Atherly, A. G. (1991). The genomes of the family Rhizobiaceae: size, stability, and rarely cutting restriction endonucleases. *J. Bacteriol., 173*, 704-709.

Sullivan, J. T., and Ronson, C. W. (1998). Evolution of rhizobia by acquisition of a 500-kb symbiosis island that integrates into a phe-tRNA gene [published erratum appears in (1998) *Proc. Natl. Acad. Sci. USA, 95*, 9059]. *Proc. Natl. Acad. Sci. USA, 95*, 5145-5149.

Surpin, M. A., Lübben, M., and Maier, R. J. (1996). The *Bradyrhizobium japonicum coxWXYZ* gene cluster encodes a bb_3-type ubiquinol oxidase. *Gene, 183*, 201-206.

Thöny, B., and Hennecke, H. (1989). The –24/–12 promoter comes of age. *FEMS Microbiol. Rev., 5*, 341-357.

Tomkins, J. P., Wood, T. C., Stacey, M. G., Loh, J. T., Judd, A., Goicoechea, J. L., Stacey, G., Sadowsky, M. J., and Wing, R. A. (2001). A marker-dense physical map of the *Bradyrhizobium japonicum* genome. *Genome Res., 11*, 1434-1440.

Tumbula, D. L., Becker, H. D., Chang, W. Z., and Soll, D. (2000). Domain-specific recruitment of amide amino acids for protein synthesis. *Nature, 407*, 106-110.

van Berkum, P., and Fuhrmann, J. J. (2001). Characterization of soybean bradyrhizobia for which serogroup affinities have not been identified. *Can. J. Microbiol., 47*, 519-525.

Van Soom, C., de Wilde, P., and Vanderleyden, J. (1997). HoxA is a transcriptional regulator for expression of the *hup* structural genes in free-living *Bradyrhizobium japonicum. Mol. Microbiol., 23*, 967-977.

Velasco, L., Mesa, S., Delgado, M. J., and Bedmar, E. J. (2001). Characterization of the *nirK* gene encoding the respiratory, Cu-containing nitrite reductase of *Bradyrhizobium japonicum. Biochim. Biophys. Acta, 1521*, 130-134.

Wang, S. P., and Stacey, G. (1991). Studies of the *Bradyrhizobium japonicum nodD1* promoter: a repeated structure for the *nod* box. *J. Bacteriol., 173*, 3356-3365.

Withey, J. H., and Friedman, D. I. (2002). The biological roles of trans-translation. *Curr. Opin. Microbiol., 5*, 154-159.

Wu, L. F., Chanal, A., and Rodrigue, A. (2000). Membrane targeting and translocation of bacterial hydrogenases. *Arch. Microbiol., 173*, 319-324.

Yang, W. (2000) Structure and function of mismatch repair proteins. Mutat Res 460, 245-256.

Yurgel, S., Mortimer, M. W., Rogers, K. N., and Kahn, M. L. (2000). New substrates for the dicarboxylate transport system of *Sinorhizobium meliloti. J. Bacteriol., 182*, 4216-4221.

CHAPTER 8

PSYMA *OF SINORHIZOBIUM MELILOTI*: NITROGEN FIXATION AND MORE

M. J. BARNETT[1] AND M. L. KAHN[2]
[1]*Department of Biological Sciences, Stanford University, Stanford, CA 94305;*
[2]*Institute of Biological Chemistry, Washington State University, Pullman, WA 99164, USA*

1. INTRODUCTION

pSymA, the smaller of the two megaplasmids of *Sinorhizobium meliloti*, contains a large number of the genes known to be needed for the development and productivity of the nitrogen-fixing symbiosis with legumes. However, pSymA contains no genes that are essential for cell viability and is, therefore, squarely in the category of accessory elements. Analysis of the genes contained in pSymA suggests many roles that the plasmid may play beyond symbiosis or to supplement the symbiotic interaction. But there are a large number of genes whose functions are either unknown or are difficult to specify precisely enough so that a specific role for the genes can be assigned. In this review, we attempt to set a context for further functional analysis of pSymA by discussing the content of this plasmid and the tools that can be brought to bear on its analysis.

Sinorhizobium meliloti is a nitrogen-fixing bacterial symbiont of several important forage crops, which include *Medicago* species, like alfalfa and the model legume *M. truncatula*, *Melilotus* species, like sweet clover, and *Trigonella* species, like fenugreek. The bacteria infect the roots of these plants and induce formation of an organ called a nodule, within which they are able to fix enough atmospheric N_2 to satisfy the host plant's fixed nitrogen requirements. This type of nitrogen-fixing symbiosis is ecologically important by providing a large fraction of the nitrogen available to natural ecosystems. It is also agronomically significant both by supporting optimal growth of many crop plants and by serving as an important nutrient input in many systems of sustainable agriculture. In addition, as an intimate cellular symbiont that acts by manipulating fundamental plant processes,

R. Palacios and W. E. Newton (eds.), Genomes and Genomics of Nitrogen-fixing Organisms, 113-132.

the study of *S. meliloti* provides a unique probe into the ecological and evolutionary context surrounding aspects of plant signal transduction and plant development.

S. meliloti belongs to the family *Rhizobiaceae* within the order *Rhizobiales* of the α-proteobacteria. *Rhizobiales* includes nitrogen-fixing symbionts, often referred to as rhizobia, that are members of the genera *Azorhizobium*, *Bradyrhizobium*, *Mesorhizobium*, *Rhizobium*, and *Sinorhizobium*. But it also includes photosynthetic genera, like *Rhodopseudomonas,* chemolithotrophs, like *Nitrobacter*, and animal pathogens, like *Brucella* and *Bartonella*. It is of considerable interest that these last two share with the rhizobia a prominent intracellular phase in their life histories. Members of this order generally possess large, often multipartite, genomes (Jumas-Bilak *et al.*, 1990).

There is some controversy concerning the phylogeny of the family *Rhizobiaceae.* Although the separation of genera, like *Sinorhizobium* and *Rhizobium*, is widely acknowledged, placement of species within previously described genera, like *Agrobacterium*, has been more controversial because taxonomies that rely heavily on grouping by 16S rRNA similarities sometimes have reassorted species within the family and have discounted other taxonomic and life history traits. Both the order and family have representatives that differ substantially in genome structure and lifestyles (Martínez-Romero and Caballero-Mellado, 1996; Young and Haukka, 1996; Young *et al.*, 2001). In addition to the *S. meliloti* strain 1021 genome (Barnett *et al.*, 2001; Capela *et al.*, 2001; Finan *et al.*, 2001; Galibert *et al.*, 2001), two other complete genome sequences of nitrogen-fixing symbiotic rhizobia have been published (as of June 2003), from *Mesorhizobium loti* and *Bradyrhizobium japonicum* (Kaneko *et al.*, 2000; Kaneko *et al.*, 2002). Complete sequences for *Brucella suis* (Paulsen *et al.*, 2002), *Brucella melitensis* (DelVecchio *et al.*, 2002), and *A. tumefaciens* C58 (Goodner *et al.*, 2001; Wood *et al.*, 2001) have also been determined and sequencing projects are underway for several other bacteria within this grouping. In this review, we discuss what is known to date about the *S. meliloti* pSymA plasmid.

2. HISTORY

In the late 1970s, researchers discovered that some bacterial nitrogen-fixing legume symbionts contained large plasmids (Casse *et al.*, 1979; Nuti *et al.*, 1977). Genetic characterization of *S. meliloti* was underway by this time (Kondorosi *et al.*, 1977; Meade and Singer, 1977) and, aided by advances in analytical methods (Eckhardt *et al.*, 1978), several *S. meliloti* strains were found to contain large ($> 300 \times 10^6$ daltons) "megaplasmids" (Banfalvi *et al.*, 1981; Rosenberg *et al.*, 1981). In *S. meliloti* 1021, the two circular megaplasmids present were later designated as pSymA and pSymB. Genes that hybridized to the *Klebsiella pneumoniae* nitrogenase genes, *nifD* and *nifH*, were shown to be on pSymA (Banfalvi *et al.*, 1981; Rosenberg *et al.*, 1981; Ruvkun and Ausubel, 1980), suggesting that this megaplasmid was essential for a functional symbiosis.

Later work demonstrated that spontaneous non-nodulating mutants (Nod$^-$) (Rosenberg *et al.*, 1981) and Nod$^-$ mutants generated by heat treatment (Banfalvi *et al.*, 1981) could be complemented by DNA linked to *nifHD*. Further analysis showed that symbiotically important genes on the megaplasmid included the *nod* genes that are required for nodulation and host specificity, other *nif* genes related to additional *Klebsiella* genes that encode nitrogenase-associated functions, and *fix* genes that, when mutated, lead to ineffective nodules unable to reduce (fix) N_2. Mutants "tagged" with Tn*5* were important in this work, by allowing DNA corresponding to mutations to be associated with sequences that could be easily mapped and retrieved (Buikema *et al.*, 1982; Long *et al.*, 1982; Meade *et al.*, 1982).

Several years later, the second megaplasmid of *S. meliloti* 1021, pSymB, was also shown to be important in symbiosis (Banfalvi *et al.*, 1985). Among its 1570 predicted proteins, pSymB carries polysaccharide specificity genes necessary for effective invasion of plant hosts (Finan *et al.*, 1985) and a gene used in dicarboxylate uptake that was needed for effectiveness (Bolton, *et al.*, 1986).

The findings that enzymes recognizing AT-rich target sites cut the GC-rich *S. meliloti* genome infrequently (Sobral *et al.*, 1991a) and the ability to separate the replicons with transverse alternating-field gel electrophoresis (Sobral *et al.*, 1991b) led to the generation of physical circular maps for all three *S. meliloti* replicons, including pSymA (Honeycutt *et al.*, 1993). Predictions of pSymA size ranged from 1.325 to 1.42 Mb (Honeycutt *et al.*, 1993). All strains of *S. meliloti* appeared to contain a pSymA megaplasmid and genes involved in nodulation and nitrogen fixation continued to be characterized at a rapid pace. However, these genes were all located within a 275-kb region on pSymA, so that, at the end of the 20th century, more than 1 Mb of pSymA was largely uncharted territory.

3. THE PSYMA SEQUENCE PROJECT

In 1997, one of us (Barnett), working in Sharon Long's laboratory, began to purify *Swa*I-linearized pSymA DNA by pulsed-field gel electrophoresis with the ultimate goal of determining the DNA sequence of pSymA. Sequence tagged site markers from a cosmid library made from this pSymA-enriched DNA were used to construct a detailed physical map (Barloy-Hubler *et al.*, 2000). As part of an international consortium to sequence the entire *S. meliloti* genome (Galibert *et al.*, 2001), these pSymA libraries and one total genome library were used for M13 shotgun sequencing of the entire pSymA replicon that used the physical map and BAC clones to guide correct assembly and to close sequence gaps (Barnett *et al.*, 2001).

4. GENERAL FEATURES OF PSYMA

The completed DNA sequence of pSymA confirms the prior length assessment. The exact size of 1,354,226 bp falls within the previous range of predictions (Burkhardt *et al.*, 1987; Honeycutt *et al.*, 1993) and is smaller than that of the other two replicons, pSymB (1.68Mb) and the chromosome (3.65 Mb). After manually editing the predictions of various bioinformatic tools, it was concluded that pSymA encoded about 1,293 proteins, which accounted for 83.6% of the total DNA. This usage was slightly lower than that of either the chromosome (85.9%) or pSymB (88.6%). Subsequent analyses, based on new sequence comparisons and proteomics data, suggest that there are several more potential protein-coding sequences in pSymA (Barnett and Kahn, unpublished data; Djordjevic *et al.*, 2003). The G+C content of pSymA is 60.4%, which is higher than previous estimates but still significantly lower than that of both pSymB (62.4%) and the chromosome (62.7%). This difference in G+C content led us to hypothesize that pSymA was acquired relatively recently by an ancestor of strain 1021 (Galibert *et al.*, 2001). The *nod* and *nif* genes have a notably lower G+C content than the rest of pSymA as do several regions with similarity to bacteriophage elements. In contrast, a pSymA region containing putative chemotaxis and pilus genes that are similar to chromosomal genes has a higher than average G+C content, suggesting that these genes may have been acquired by lateral transfer from the chromosome.

Our analysis of pSymA within the context of the total genome failed to find any pSymA genes that might be absolutely required for free-living growth. These results support previous data demonstrating that pSymA can be cured from a closely related strain, *S. meliloti* strain 2011, without drastic effects on growth in either rich or minimal-succinate-nitrate media (Oresnik *et al.*, 2000). However, pSymA is absolutely required for nodulation and nitrogen fixation.

5. COMPARATIVE GENOMICS

Of the complete and partial *Rhizobiaceae* genomes sequenced to date, *S. meliloti* is predicted to be most closely related to the broad-host-range symbiont, *Rhizobium* sp. NGR234. The NGR234 symbiosis plasmid (NGR234a) was completely sequenced (Freiberg *et al.*, 1997) and a partial shotgun sequence has been carried out for the remainder of the genome (Viprey *et al.*, 2000). Surprisingly, 54% of the predicted genes on NGR234a have no match anywhere in *S. meliloti*. However, of those with replicon-specific full-length matches (26%), more are found on pSymA (13%) than on either pSymB (9%) or the chromosome (4%) (Galibert *et al.*, 2001). A high proportion of the pSymA/NGR234a orthologs are genes related to nodulation and nitrogen fixation.

Comparison of the *S. meliloti* genome to the complete genome of the symbiont *Mesorhizobium loti* reveals a similar story: 35% of *M. loti* genes have no match in *S. meliloti* (Galibert *et al.*, 2001). Similar pSymA genes are dispersed throughout the *M. loti* genome, with a higher concentration in the *M. loti* "symbiotic island". As in the comparison with NGR234a, many of these are nodulation and nitrogen-fixation genes. The *Bradyrhizobium japonicum* 410-kb region that contains symbiotic genes (Göttfert *et al.*, 2001) has little similarity to pSymA other than the symbiotic genes themselves.

As previously mentioned, some *Rhizobiaceae* family members are not nitrogen-fixing symbionts, but plant pathogens that cause hypertrophies on plants. Recently, the complete genome of the crown gall pathogen, *A. tumefaciens* strain C58, was determined (Goodner *et al.*, 2001; Wood *et al.*, 2001). The genomic structure of *A. tumefaciens*, which consists of both circular and linear chromosomes plus the Ti and AT plasmids, is very different from that of *S. meliloti*. There is substantial congruence of the *A. tumefaciens* circular chromosome with the *S. meliloti* chromosome, which suggests that these units are either recently diverged or are subject to selection that conserves both the presence of particular genes and their relative order. In striking contrast, the pSymA genes also present in *A. tumefaciens* are dispersed throughout the *A. tumefaciens* genome with no apparent synteny in any of the replicons, although there is a somewhat higher proportion of orthologs on the linear and plasmid replicons than on the circular chromosome (Goodner *et al.*, 2001; Wood *et al.*, 2001).

6. PLASMID BIOLOGY

pSymA contains genes that encode a *repABC*-type of replication control system, like those commonly found in Gram-negative bacteria, including rhizobia (Thomas 2000; Turner *et al.*, 1996). *repABC* control systems are often used by large, low copy number plasmids. Although *repABC* replication control is often associated with other plasmid stabilization mechanisms, such as post-segregational killing, the latter are not obvious from inspection of the genes predicted on pSymA (Ramirez-Romero *et al.*, 2001). pSymA is considered quite stable by those working with *S. meliloti* 1021 but it was possible to cure pSymA by selecting for sucrose resistance as the result of the loss of an experimentally integrated *sacB* gene (Oresnik *et al.*, 2000). The loss of the plasmid was accomplished in two stages, which might reflect the relative frequency of eliminating the *sacB* gene by recombination or some other deletion process compared to loss of the whole plasmid. pSymB also contains a *repABC*-type replicon, but obviously from a different incompatibility group.

The megaplasmids in *S. meliloti* strain 41 have been reported to be able to transfer between bacteria (Banfalvi *et al.*, 1985). pSymA contains putative conjugative-transfer genes (*traACDG*) and a putative plasmid-transfer origin (*oriT*) sequence, but the previous analysis did not identify a full complement of *tra/trb* genes, such as those found on NGR234a (Freiberg *et al.*, 1997). However, the *A. tumefaciens avhB* genes have recently been shown to mediate conjugal transfer of pATC58 (Chen *et al.*, 2002). These *A. tumefaciens avhB* genes are homologs of the better studied *virB* genes, which transfer bacterial T-DNA into the plant cell (Kado, 2000). The ten *S. meliloti* pSymA genes described as homologous to *virB* and type-IV secretion systems are actually more similar to these *A. tumefaciens avhB* genes than they are to the *A. tumefaciens virB* genes. This suggests that the *S. meliloti* genes may be involved in conjugal transfer of pSymA. A mammalian pathogen, *Brucella abortus,* also has homologs to the *virB* genes and these are required for virulence and intracellular multiplication (Sieira *et al.*, 2000). However, mutants of *S. meliloti* 1021, which contain deletions of the pSymA *virB* operon, appear to nodulate and fix nitrogen normally (Wells, unpublished data), suggesting that these genes do not play a direct role in symbiosis.

7. ELEMENTS OF EXTERNAL ORIGIN

Consistent with data from many other bacterial plasmids, pSymA carries more than its share of phage and IS elements (3.6% of sequences *vs.* 2.2% for genome overall). pSymB has fewer of these elements (0.9%) and both plasmids fall far short of the remarkable 18% of NGR234a that is assigned to mobile elements. pSymA contains 12 of the 21 types of IS elements found in the *S. meliloti* genome (Table 1). Four of these 12 are pSymA-specific. The distribution of IS elements on pSymA is uneven with over half of the IS elements on pSymA being located in the region from *ca.* 208-kb to 590-kb, more than triple the density on the remaining three quarters of the replicon. The *nod-nif-fix* gene region in this sector is especially rich in IS elements and three of the four pSymA-specific elements are found only in this region. The two of these, ISRm7 and ISRm8, that are most closely linked to the *nod* genes are predicted to be nonfunctional. All of the ISRm11 elements are found in an approximately 345-kb region near the replication origin. The unique population of IS elements on pSymA and the bias in their distribution adds further credibility to the idea of a recent acquisition for pSymA and perhaps recent acquisition of symbiotic genes by a pSymA precursor. The role of IS elements in the *S. meliloti* genome evolution is not known but, in *Rhizobium* sp. strain NGR234, it was shown that intragenomic rearrangements may be mediated by reiterated sequences such as IS elements (Mavingui *et al.*, 2002).

pSymA carries two phage-related regions that have open reading frames (ORFs) similar to the lambda integrase/recombinase family (901to 904 kb and 1225 to 1228 kb). These phage regions are similar to Y4rABCD found on the *Rhizobium* species NGR234 symbiotic plasmid (Freiberg *et al.*, 1997).

Table 1. IS elements present on pSymA

IS	Family	Number of complete copies	Number of partial or nonfunctional copies
ISRm1	IS3	3	0
ISRm3	IS256	4	1
ISRm5	IS256	1	0
ISRm7*	IS3	0	1
ISRm8*	IS66	0	1
ISRm11‡	IS630	3	1
ISRm17	?	2	1
ISRm23*	IS3	1	1
ISRm24	IS30	3	0
ISRm25*	IS66	1	1
ISRm29	?	2	0
ISRm30	?	1	0

*pSymA specific elements; ‡ formerly ISRm2011-2

Members of the *Rhizobiaceae* contain repetitive palindromic sequences in intergenic regions; these are called *Rhizobium*-specific intergenic mosaic elements (RIME; Østeras *et al.*, 1995) as well as a related repeat with palindromic sequences (Østeras *et al.*, 1998). 476 of these repetitive sequences are found in the *S. meliloti* genome. Consistent with earlier observations (Østeras *et al.*, 1995), only 1.3% of these are found on pSymA, 6.7% on pSymB, and 92% on the chromosome (Galibert *et al.*, 2001). It has been proposed that the dispersion of RIMEs occurred on the chromosome before acquisition of the megaplasmids (Østeras *et al.*, 1995). A similar distribution of RIMEs has been observed in *Rhizobium* sp. NGR234 (Perret *et al.*, 2001).

8. TRANSFER RNA GENES

pSymA contains two tRNA genes. The first one, the proposed pSymA $tRNA_{met}$, is redundant with the chromosomal $tRNA_{met}$ and may be nonfunctional on the basis of secondary-structure predictions. The second (*selC*) has a UCA anticodon that specifies selenocysteine and its functionality is suggested by its location adjacent to genes encoding SelA (selenocysteine synthase), SelD (required to modify seryl tRNA to selenocysteine tRNA), and SelB (selenocysteine specific elongation factor), plus genes encoding formate dehydrogenase, the only selenoprotein identified thus far in the *S. meliloti* genome. tRNAs play a role in the horizontal transfer of bacterial pathogenicity islands; frequently, the 3' sequences of tRNA genes are used as recombination sites by P4 family recombinases (Hou, 1999). In *M. loti*, the symbiosis genes are present on a 610-kb island that is inserted into a phenylalanine tRNA gene with the 3' 17 nucleotides of the tRNA present as a direct repeat at the right end of the island (Kaneko *et al.*, 2000; Sullivan and Ronson, 1998). An adjacent P4 integrase-like gene may be responsible for integration of the *M. loti* island (Sullivan and Ronson, 1998). In *S. meliloti*, we found an 11-bp direct repeat of the selenocysteine tRNA 3' end about 518 kb from *selC*, but find no evidence for a closely-linked integrase gene.

9. NODULATION GENES

In *S. meliloti*, the early events in symbiosis are mediated by Nod factors, which are lipo-chito-oligosaccharides synthesized by the bacteria in response to chemical signals produced by the host plants (Lerouge *et al.*, 1990; Long, 1996). The genes necessary for biosynthesis of Nod factors are encoded by the *nod*, *nol*, and *noe* genes, which lie in six operons within 82 kb on pSymA. Transcription of the *nod* genes depends on three LysR-type activators, NodD1, NodD2, and NodD3 (Honma and Ausubel, 1987; Honma *et al.*, 1990; Mulligan and Long 1989). These activators bind to a conserved sequence in the *nod*-gene promoters, called the *nod* box (Fisher *et al.*, 1988; Fisher and Long, 1989; Fisher *et al.*, 1987). NodD1 is active in the presence of flavonoid inducers, such as luteolin and methoxychalcone (Maxwell *et al.*, 1989; Mulligan and Long, 1985; Peters *et al.*, 1986). NodD2 is activated by plant flavonoids and also by betaines, such as trigonelline (Phillips *et al.*, 1992). There is no known inducer for NodD3, but it is known that *nodD3* transcription is activated by another LysR-family member, SyrM (Barmett *et al.*, 1996; Swanson *et al.*, 1993). NodD3 activates *syrM* expression by binding to a degenerate *nod* box upstream of *syrM* (Barnett *et al.*, 1996; Swanson *et al.*, 1993), thus, establishing a positive feedback loop. In addition, SyrM activates transcription of an adjacent gene, *syrA*, which appears to affect acidic exopolysaccharide (EPSI) production (Barnett *et al.*, 1998). Because *syrM* and *nodD3* have been found to control synthesis of a unique class of nod factors that are *N*-acylated with (ω-1)-hydroxylated fatty acids (Demont *et al.*, 1994), they may also activate genes involved in the synthesis of these molecules.

The *nod* factor-pathway genes are well characterized and the complete genome did not contain any new *nod*-structural genes related to known *nod* genes from other species. pSymA does encode a second copy of the *ntrR* gene, which encodes a regulator thought to be responsible for *nod*-gene repression in the presence of nitrogen (Dusha *et al.*, 1989; Dusha and Kondorosi 1993; Oláh *et al.*, 2001). The sequence also revealed two additional copies of the *syrB* gene, a putative repressor of *syrM* expression (Barnett and Long, 1997); one of these, *syrB3*, is closely linked to the *nod* and *nif* genes. Functional tests are required to determine if *ntrR2*, *syrB2*, and *syrB3* are involved in *nod*-gene regulation and to determine if additional regulators in the *nod*-gene pathway are present on pSymA. Searching for additional *nod*-gene promoters (*nod* boxes) yielded no significant matches.

The chaperonins, GroES and GroEL, are required for *nod*-gene activation, presumably by facilitating the assembly of active NodD (Ogawa and Long, 1995). A gene, *groESL2,* encoding a functional chaperonin of this type, was previously shown to be present on pSymA (Ogawa and Long, 1995) and, based on mutational analysis, this is likely to be the only *groESL* in the genome able to substitute for the chromosomal *groESL1* (Ogawa, 1993). However, the pSymA sequence revealed a third complete *groESL* operon; additional experiments are necessary to determine the role of *groESL3*.

10. NITROGEN-FIXATION GENES

As indicated above, pSymA contains the only copy of the nitrogenase structural genes in *S. meliloti* 1021. In addition, pSymA contains homologs to most of the other nitrogenase-associated genes found in *Klebsiella*, including the primary transcriptional regulator of *nif*-gene expression, *nifA* (Fischer, 1994). A detailed discussion of *fix* and *nif* genes found on pSymA is part of the original publication of the sequence, so the following will emphasize points that we consider of special interest with regard to the metabolism related to nitrogen fixation.

First, neither pSymA nor the other replicons in *S. meliloti* 1021 include a homolog of *nifJ*, which codes for a pyruvate-flavodoxin oxidoreductase, or of *nifF*, which codes for the associated flavodoxin. These proteins are involved in transferring reductant to nitrogenase in *Klebsiella*. How electron transfer to nitrogenase is mediated in *S. meliloti* has not yet been established, but it has been suggested that the *fixABCX* operon, which is located next to the nitrogenase structural genes and is transcribed from a divergent promoter, may be involved

(Earl *et al.*, 1987). FixABC are homologs of electron transfer flavoproteins (ETF proteins) that carry electrons from various specific dehydrogenases to respiratory electron-transport chains and it is possible that, in *S. meliloti,* FixABC reduce FixX, a ferredoxin with an unusual structure. It has recently been shown that the *E. coli fixA* and *fixB* genes are needed for carnitine reduction during anaerobic growth (Walt and Kahn, 2002). A second set of ETF-related proteins of the FixAB type are encoded by *etfA2B2* on pSymA. An identical set of these genes, *etfA1B1*, is also present on the chromosome.

Second, also missing from the complement of genes in *S. meliloti* is a homolog of the *Klebsiella* NifL protein, which is involved in O_2-dependent regulation of nitrogenase expression. In *S. meliloti*, O_2-dependent regulation is carried out by the FixJ-FixL two component-regulatory system in conjunction with the FixK transcriptional regulator (Foussard *et al.*, 1997; Tuckerman *et al.*, 2001). Interestingly, the sequence of pSymA shows that these regulatory proteins are embedded in a large cluster of genes involved in denitrification and are not located near the nitrogenase genes. The primary function of these *fix* genes may be to regulate the microaerobic use of nitrate as a terminal electron acceptor rather than to control nitrogenase. Their use in controlling nitrogenase may be a secondary adaptation. It is clear that rhizobia use many different mechanisms to couple O_2 sensing with nitrogenase regulation (Fisher, 1994), suggesting that different rhizobia have recruited diverse control networks during the evolution of the different symbioses.

Third, three other classic *nif* genes, *nifQ, nifZ, and nifW*, are missing from the *S. meliloti* sequence, although they are found in the genomes of other rhizobia, including *M. loti, R. etli*, and *B. japonicum*. The functions of these proteins are not well established, although NifZ and NifW might form a complex. It is, therefore, difficult to know what other functions might substitute for them, but there are only two unassigned ORFs in the nitrogenase-related gene cluster in pSymA.

11. CARBON AND NITROGEN METABOLISM

A large fraction of the genes on pSymA appear to be involved in nitrogen metabolism. However, assigning specific functions to many of the pSymA genes that are related to known proteins can be difficult because, although a protein might clearly have a certain function (in transport, for example), the relationship to a protein with known specificity may not be close enough to convincingly indicate its substrate. The pSymA-cured derivative of *S. meliloti* 2011 had relatively few differences from wild-type in its "ordinary" physiology (Chen *et al.*, 2000). Moreover, only 29 of the 2000 protein spots that could be observed in the *S. meliloti* proteome disappear in the pSymA-cured strain (Chen *et al.*, 2000). Because about 20% of the predicted ORFs in *S. meliloti* are on pSymA, this result means that disproportionately few of these ORFs were expressed in free-living cells at a level detectable in these experiments. This result is also consistent with the idea that plasmid-encoded proteins are expressed only under special circumstances. Adding the *nod*-gene inducer luteolin significantly influenced expression of only 6 of the 29 proteins (Chen *et al.*, 2000).

The pSymA-cured strain, in contrast to wild type *S. meliloti* 1021, was unable to catabolize inosine, γ-aminobutyric acid (GABA), gluconate, trigonelline, glycine, or serine as sole carbon sources (Oresnik, *et al.*, 2000). The possible relationships of specific genes on pSymA to these phenotypes is discussed in Barnett *et al.* (2001), where it was noted that, although pSymA plausibly contains genes involved in catabolism of each of these compounds, many of these genes belong to families with representatives on the other replicons. Thus, the involvement of specific pSymA genes in catabolism of these compounds will require additional and more specific mutants to sort out the genetics and biochemistry of the use of these compounds.

Examination of the pSymA sequence suggests the involvement of plasmid genes in other aspects of metabolism. For example, *S. meliloti* 1021 grows well on formate as a carbon source. Genes needed for the synthesis of a selenocysteine-containing formate dehydrogenase, including those for both synthesizing selenocysteine and charging a pSymA-encoded selenocysteine tRNA, are located adjacent to the predicted pSymA origin of plasmid replication. Genes for an NAD-dependent formate dehydrogenase are present on the chromosome and it remains to be determined which of these formate dehydrogenases is used for growth on formate and under what circumstances, although selenium is presumably needed for the plasmid-encoded gene products to function.

pSymA also encodes an arginine deiminase pathway for the catabolism of arginine. This pathway, which is best described in anaerobes, releases carbamyl phosphate from arginine and then uses carbamate kinase to synthesize a molecule of ATP. Two copies of arginine deiminase are carried on pSymA: *arcA1*, which is linked to both ornithine carbamoyl transferase and carbamate kinase, and *arcA2*, which is linked to genes that are similar to basic amino acid-transport proteins. Other pathways for catabolism of basic amino acids are suggested by the presence of two related amino acid decarboxylases, SMA0680 and SMA0682, with specificity for basic (ornithine, lysine, arginine) amino acids.

Like many rhizobia, *S. meliloti* 1021 is able to use nitrate as a terminal electron acceptor. Within an approximately 50-kb region are genes for components of nitrate reductase (6 genes), a copper-containing nitrite reductase (2 genes), nitric oxide reductase (5 genes), and nitrous oxide reductase (1 gene) that allow the bacteria to convert nitrate to N_2. In addition, this cluster contains genes for copper transport, a *fixNOQP* gene cluster (containing a predicted Cu-cytochrome c, di-heme cytochrome c, and cytochrome c oxidase), a *fixGHIS* cluster of redox proteins (including a Cu-dependent ATPase), a *fixKTJL* group of regulatory genes involved in O_2 sensing, and such miscellaneous genes as *azu1*, which codes for the blue copper protein azurin, *cycB2*, which encodes a cytochrome c552, and *hemN*, which codes for coproporphyrinogen III oxidase, an enzyme involved in heme biosynthesis. Whether this gene cluster is actively involved in denitrification, even while nitrogen fixation is taking place, is an interesting question because simultaneous activity of both pathways would imply that the metabolic role of these activities was to deliver electrons to terminal electron acceptors rather than the synthesis of metabolically useful forms of nitrogen. Preliminary evidence (House and Kahn, unpublished observations) suggests that denitrification is operational in nodules. Furthermore, nodules formed by mutants, which are unable to convert NO into N_2O, are unusually sensitive to the addition of nitrate to the growth medium, whereas mutants, which are unable to make NO, are relatively resistant. As indicated above, the presence of the FixJ-FixL O_2 sensor in this cluster may indicate that expression of denitrification is under control of these proteins. An additional 25 proteins without a firmly assigned function are predicted in this region. It would be interesting to know if these also had a role in microaerobic use of either nitrogen oxides or other electron acceptors.

In *B. japonicum*, FixNOQP constitute a cytochrome cbb3 oxidase that functions as a proton pump and is essential for symbiosis (Arslan *et al.*, 2000). The *S. meliloti fixNOQP* gene cluster referred to above is one of three encoded on pSymA, but what distinguishes the products of these three sets of genes is unknown. Lack of all three sets of genes leads to an ineffective (Fix^-) phenotype, but it has not been determined whether the *fixN2O2Q2P2* or *fixN3O3Q3P3* genes are able to support fixation by themselves. *fixN2O2Q2P2* is associated with a second plasmid copy of homologs of the regulatory genes, *fixK* and *fixT*. The genes adjacent to these last two clusters do not encode an obvious alternative function that would need the high O_2-affinity cbb_3 cytochromes that are characteristically products of these genes.

12. CHEMOTAXIS AND PILUS FORMATION

Chemotaxis is an important adaptation to diverse environments. *S. meliloti* is chemotactic toward a wide variety of compounds, including *nod*-gene inducers in root exudates (Dharmatilake and Bauer, 1992), and possesses cellular machinery for efficient motility in the soil (reviewed in Armitage and Schmitt, 1997). pSymA contains a 14-kb cluster of putative chemotaxis and pilus-assembly genes, some of which are similar to genes on the *S. meliloti* chromosome. Genes in the cluster encode a putative CheB chemotaxis methylesterase, a CheR-like methyltransferase, a methyl-accepting chemotaxis protein, a CheA-like histidine kinase, a CheW-like protein, and pilus assembly proteins, PilA2, CpaA2, CpaB2, CpaE2, and CpaF2.

The pilus-assembly proteins are related to those of type-II secretion systems for biogenesis of a type-IV pilus. Similar proteins in the α-protobacterium, *Caulobacter crescentus*, form the polar pilus, which is a receptor for the bacteriophage C6K, and are regulated by the CtrA cell-cycle regulator (Skerker and Shapiro, 2000). A putative *pilQ* (SMa0163) homolog lies outside this cluster, but no other pilus genes are located in this region. Pili have not been characterized in *S. meliloti*. The function of these proteins is unknown but it is tempting to speculate that these genes play a role in attraction and adhesion to plant roots.

13. TRANSPORT

We predict that about one of every seven genes on pSymA is involved in transport. The 34 ABC transporter-gene clusters, which consist of genes encoding permease,

ATP-binding, and solute binding domains, are fairly evenly distributed on pSymA, except for a 100-kb region (732 to 833 kb) that contains eight clusters. The high degree of conservation between ABC transporters makes it difficult to predict the transported solute for the majority of these, although some of them were tentatively identified as transporting nitrate, sulfate/thiosulfate, amino acids, sugars, and polyamines. In the cation P-type ATPase subgroup of transporters, there are four transporters predicted to transport copper (*actP*, SMa1087, *fixI1, fixI2*), one predicted to transport potassium (*kdp*), one predicted to transport cadmium and/or magnesium (SMa1163), and one undetermined solute (SMa1155).

The second most abundant group of transporters comprises those in the major facilitator superfamily (MFS). This group includes amino acid antiporters (SMa0678, SMa0684, SMa1667, SMa1668), a Na^+/H^+ antiporter, a copper-export protein (SMa1198), three members of the RND transport family that function with associated membrane-fusion proteins (*nolG*, SMa1662, SMa1884), and many others for which no solute was provisionally assigned. Although *S. meliloti* has a large number (Bolton *et al.*, 1986) of MFS members in the RhtB amino acid efflux subclass, none are present on pSymA (Galibert *et al.*, 2001).

Iron is often scarce in the soil and is required for synthesis of symbiotically-important proteins, such as nitrogenase, cytochromes, and ferredoxins. In addition to the previously known regulon encoding a high-affinity siderophore iron-transport system *rhbABCDEFrhrArhtA* (Lynch *et al.*, 2001), there are two clusters of putative iron-transport genes at positions 283 kb and 985 kb.

In addition to the unique KdpABC-type of potassium transporter, pSymA carries the TrkH- and KUP-types that are also present on the chromosome. Potassium transport may be important for either pH adaptation or membrane potential during symbiosis because a mutant defective in potassium efflux was shown to be Fix^- (Putnoky *et al.*, 1998).

14. REGULATION AND SIGNAL TRANSDUCTION

The largest family of transcriptional regulators in *S. meliloti* is of the LysR/NodD-type. This group is over-represented on pSymA with 36 of the 85 total *S. meliloti* members. Of the 11 proteins that most closely resemble NodD, nine are on pSymA. The GntR family of small repressors is the second most abundant family of regulators with 15 members on pSymA. There are eight response regulators on pSymA and nine proteins with either global or local similarity to histidine kinases. None of the pSymA histidine kinases are similar to the family of seven sensor histidine kinases found on pSymB and the chromosome. We found neither SorC nor DeoR-type regulators on pSymA, consistent with other data suggesting that pSymA is not specialized for sugar metabolism (Galibert *et al.*, 2001). Nor did we identify any LuxR- or NtrC-like activators on pSymA. Three Crp/Fnr-like regulators are present on pSymA (SMa1067, 1141, 1245); these constitute a family distinct from the FixK family. We failed to discern any particular bias in the locations of the regulators as they are fairly evenly distributed about the replicon.

The *S. meliloti* genome contains more nucleotide cyclases than any other α-proteobacterial genome to date (Galibert *et al.*, 2001). Nine proteins on pSymA were annotated as having either global or partial similarity to adenylate cyclases. Two of these, *cyaF4* and *cyaF5*, belong to a family of five "type II" cyclases that contain a catalytic domain in the N-terminal part of the protein and C-terminal tetratricopeptide repeats and, thus far, have been found only in *S. meliloti* and *M. loti* (Galibert *et al.*, 2001; Sharypova *et al.*, 1999). Three (SMa0464, 1591, 1789) encode proteins of 1058-1159 amino acids with local similarity to both adenylate cyclases and transcriptional regulators. Of the remaining four, one (SMa1046) is most similar to *cyaE* and three (SMa0579, 1103, 2357) have local similarity to the C-terminal catalytic domain of classic adenylate cyclases.

It is difficult to determine specific functions of regulatory proteins from the sequence alone because subtle changes in substrate, DNA, and protein binding can strongly influence how a particular protein may act. However, the interesting work that has already been carried out with many of these proteins suggests that it will be very worthwhile to determine the targets of these new regulatory proteins.

15. STRESS RESPONSES

Plasmid-encoded genes often play a role in the exploitation of specific niches related to environmental stresses. Genes that are involved in cold shock (SMa0126, SMa0181, and SMa0738) and heat shock (SMa1118) responses are found on pSymA but they also have homologs on the chromosome.

It is important for a symbiotic soil bacterium to be able to withstand osmotic stress. Desiccation resistance might be mediated by a pSymA aquaporin, *aqpZ2*, which also has a chromosomal homolog. Several pSymA genes are potentially involved in metabolism of osmolytes, such as trehalose, glutamine, and glycine betaine. Trehalose synthase, *otsA*, is presumably required for synthesis of trehalose, an endogenous osmolyte in *S. meliloti* (Gouffi *et al.*, 1999), although several other trehalose-related proteins are encoded on pSymB. *S. meliloti* appears to lack OtsB, trehalose phosphatase, supporting the idea that trehalose synthesis occurs *via* an alternate pathway (Streeter and Bhagwat, 1999). Interestingly, *otsA* is linked to *gdhA*, which encodes the only glutamate dehydrogenase in the genome. GdhA is likely to play a role in metabolism of glutamate, another *S. meliloti* osmolyte. A pSymA copy of betaine aldehyde dehydrogenase, *betB2*, catalyzes the second step in betaine synthesis and presumably acts together with chromosomal genes involved in other steps of betaine synthesis. SMa1466 and SMa1467 are 44% identical to ABC-transport proteins for glycine, betaine, carnitine, and choline.

A hydroperoxidase (SMa2379) and two haloperoxidases (SMa1809 and SMa2031) may be part of protective mechanisms for dealing with either symbiotic or environmental oxidative stresses. SMa2389 is similar to other Ohr (organic hydroperoxide) stress-induced proteins from various bacteria, including *Xylella*, in which the homolog was recently shown to have thiol-dependent organic hydroperoxidase activity (Cussiol *et al.*, 2003).

Other predicted pSymA proteins may protect against specific damage caused by such agents. For example, the protein encoded by SMa1896 is predicted to be homologous to methionine sulfoxide reductase, an enzyme that reduces an oxidized form of methionine that is found both as the free amino acid and in oxidized proteins (Weissbach *et al.*, 2002). SMa1547 is the best match in the entire genome to *E. coli PimT*, L-isoaspartate protein carboxymethyl transferase. This protein initiates the restoration of damaged aspartate residues to aspartate and, therefore, plays a role in repairing damaged proteins (Li and Clarke, 1992).

Other putative pSymA proteins may confer resistance by exporting a broad spectrum of hydrophobic toxins. SMa1664 and SMa1662 encode homologs of AcrA and AcrB, the membrane and periplasmic subunits of a resistance complex. SMa1884 is also similar to AcrB and is adjacent to an AcrR-like regulator, SMa1882 (Nikaido and Zgurskaya, 2001).

Several gene products (ActP, NosD, NosF, FixI1, FixI2, SMA1198) are predicted to have copper-transport activity and, although some may be involved in acquiring copper for copper-containing proteins in the genome (and especially in pSymA), some may be involved in copper- (or other heavy metal-) export.

16. SULFUR METABOLISM

Sulfur metabolism is central to the *S. meliloti* symbiosis because a critical recognition step requires sulfated Nod factors. pSymA encodes the NodH enzyme responsible for the transfer of activated sulfate to the Nod factor (Ehrhardt *et al.*, 1995; Schwedock and Long, 1990). In addition, pSymA codes for a duplicate set of enzymes involved in the synthesis of activated sulfate, which is needed to meet the increased demand made by Nod factor synthesis. Analysis of the pSymA sequence identified additional enzymes putatively involved in sulfur metabolism, including a sulfite oxidase, sulfate/thiosulfate-transport proteins, two arylsulfatase-like proteins, sulfonate-binding proteins, and proteins similar to dibenzothiophene desulfurization enzymes (Barnett *et al.*, 2001). The desulfurization and sulfonate-binding proteins may be important for scavenging sulfur during either sulfate or cysteine starvation. Some desulfurization enzymes can remove covalently bound sulfur without breaking carbon-carbon bonds and are potentially important for desulfurization/ bioremediation of fossil fuels. The function of these putative enzymes in *S. meliloti* 1021 is unknown, but there is at least one *S. meliloti* strain, Orange 1, that can grow on dibenzothiophenes as a sole carbon and energy source (Frassinetti *et al.*, 1998). *A. tumefaciens* has orthologs of these putative desulfurization genes but, in *M. loti*, BLAST analysis returned only weak or no matches.

17. ORPHAN GENES

As of April 2002, 6.1% of the total predicted protein-coding genes in the *S. meliloti* genome have no database match. The number of these orphan genes on the plasmid replicons is higher, 10% for pSymA and 9.5% for pSymB. Some (and perhaps many) of these predicted open reading frames may not really be genes at all; experiments are necessary to determine which of these ORFs encode functional proteins. Of the 130 orphan genes on pSymA, 39 are shorter than 100 amino acids long. Many are either clustered together or in regions near transposons and transposon fragments. Preliminary data (Barnett and Toman, unpublished data) indicates that more than a third of the 130 pSymA orphans may be transcribed in free-living cells.

18. GENOME-WIDE ANALYSES

As this review is being written, we are in a situation now typical for microbiologists who are working with sequenced bacterial genomes, where the number of characterized genes is dwarfed by those that were discovered through the sequencing effort. With the focus on predictive bioinformatics that was necessary to organize the data in the sequence, it is useful to list/detail/describe the number of different predictions currently available on the internet. These include analyses at: CNRS site (http://sequence.toulouse.inra.fr/meliloti.html); Institute for Genome Research (www.tigr.org); EBI (www.ebi.ac.uk/proteome/index.html); at NIH (www.ncbi.nlm.nih.gov); Center for Biological Sequence Analysis (Denmark) (www.cbs.dtu.dk/services/GenomeAtlas/Bacteria/Sinorhizobium/meliloti/Rm1021); GeneQuiz EMBL-EBI (http://jura.ebi.ac.uk:8765/ext-genequiz/genomes/sme0108/); PromScan promoter scan site at the Sanger Institute (England) (http://www.promscan.uklinux.net/data.html); and the Munich Information Center for Protein Sequence site (http://pedant.gsf.de/).

These sites contain interesting analyses of different aspects of the sequence, from protein predictions and alignments, like those that were used in the original analysis, to structural analysis of the genome. To our knowledge, the only site that includes extensive human annotation of the pSymA sequence is the one at Toulouse but, especially because of subsequent publications of sequences from closely related bacteria such as *A. tumefaciens*, the automated analyses are of increasing interest.

19. A STRATEGY FOR ANALYZING PSYMA OF *S. MELILOTI*

Major projects are underway in several laboratories to adapt genomic technologies to the analysis of *S. meliloti* 1021. Data on pilot macroarrays, covering 214 *S. meliloti* genes, including 46 pSymA genes, have recently been published (Ampe *et al.*, 2003; Bergès *et al.*, 2003) and a major proteomics project has been referred to above (Chen *et al.*, 2000; Djordjevic *et al.*, 2003). Our efforts at Stanford University and Washington State University are aimed at developing something we call a "platform" for genetic analysis of this bacterium. This platform includes the construction of an Affymetrix oligonucleotide array for monitoring gene expression both from the postulated ORFs and from the intergenic regions. In addition to being able to observe the expression of mRNA that corresponds to ORFs identified in the sequence, an advantage of the Affymetrix approach is that the oligonucleotides have been designed both to detect intergenic RNA molecules that might have a regulatory role and to provide some indication of the possible validity of proteins too short to be uncovered by the statistical models used to predict "significant" ORFs.

In addition, we are in the process of cloning the postulated ORFs, using PCR technology and a strategy (House, Mortimer and Kahn, unpublished data) that incorporates lambda integrase site-specific recombination and then uses this, together with both homologous and site-specific recombination, to generate reporter-gene fusions to each promoter and deletion mutations for one or several genes. Preliminary work in this direction has yielded over 3000 clones, including almost all of pSymA, and we are working to extend this set to represent the entire genome. At some point in the near future, we envision collections of mutants, reporter genes, and protein overexpression plasmids that will allow investigators to carry out preliminary experiments to test hypotheses, at least in a crude way, without having to do the genetics *de novo*. Arrays of DNA from limited numbers of the plasmid clones should allow investigators to focus on the expression of subsets of genes identified by using the Affymetrix technology at a fraction of the cost of the oligonucleotide arrays. In this way, we hope to determine how the various unknown genes on pSymA influence both the survival of the free-living bacteria and symbiotic function.

It will also be instructive to use heterologous hybridization to see how many of the genes in pSymA are common to other *S. meliloti* strains and to other rhizobia. The results of such an analysis would help us to understand how many of the genes on pSymA are always needed for the bacteria to establish itself in its ecological niches, how many may be important in some specialized niches, and how many are just along for the ride. With so many genes on pSymA and in *S. meliloti*, it is tempting to try to use strategies for doing everything all at once. However, an alternative view is that preliminary surveys of all of the genes at one level of resolution may give us the ability to choose for further analysis those genes, which give us the best opportunity to investigate the critical properties of *S. meliloti*. In the next few years, we will learn how successful we have been in mining data that helps identify both the questions and the answers that define the essence of *S. meliloti*.

REFERENCES

Ampe, F., Kiss, E. Sabourdy, F., and J. Batut. (2003). Transcriptome analysis of *Sinorhizobium meliloti* during symbiosis. *Genome Biol., 4*, R15.

Armitage, J. P., and Schmitt, R. (1997). Bacterial chemotaxis: *Rhodobacter sphaeroides* and *Sinorhizobium meliloti*--variations on a theme? *Microbiol., 143*, 3671-3682.

Arslan, E., Kannt, A. Thony-Meyer, L., and Hennecke, H. (2000). The symbiotically essential cbb_3-type oxidase of *Bradyrhizobium japonicum* is a proton pump. *FEBS Lett., 470*, 7-10.

Banfalvi, Z., Kondorosi, E., and Kondorosi, A. (1985). *Rhizobium meliloti* carries two megaplasmids. Plasmid, 13, 129-138.

Banfalvi, Z., Sakanyan, V., Koncz, C., Kiss, A., Dusha, I., and Kondorosi, A. (1981). Location of nodulation and nitrogen fixation genes on a high molecular plasmid of *R. meliloti*. *Mol. Gen. Genet., 184*, 318-325.

Barloy-Hubler, F., Capela, D., Barnett, M. J., Kalman, S., Federspiel, N. A., Long, S. R., *et al.* (2000). High-resolution physical map of the *Sinorhizobium meliloti* 1021 pSyma megaplasmid. *J. Bacteriol., 182*, 1185-1189.

Barnett, M. J., Fisher, R. F., Jones, T., Komp, C., Abola, P. A., Barloy-Hubler, F., *et al.* (2001). Nucleotide sequence and predicted functions of the entire *Sinorhizobium meliloti* pSymA megaplasmid. *Proc. Natl. Acad. Sci. USA, 98*, 9883-9888.

Barnett, M. J., and Long, S. R. (1997). Identification and characterization of gene on *Rhizobium meliloti* pSyma, *syrB*, that negatively affects *syrM* expression. *Mol. Plant-Microbe Interact., 5*, 550-559.

Barnett, M. J., Rushing, B. G., Fisher, R. F., and Long, S. R. (1996). Transcription start sites for *syrM* and *nodD3* flank an insertion sequence relic in *Rhizobium meliloti*. *J. Bacteriol., 178*, 1782-1787.

Barnett, M. J., Swanson, J. A., and Long, S. R. (1998). Multiple genetic controls on *Rhizobium meliloti syrA*, a regulator of exopolysaccharide abundance. *Genetics, 148*, 19-32.

Bergès, H., Lauber, E., Liebe, C., Batut, J., Kahn, D., de Bruijn, F., *et al.* (2003). Development of *Sinorhizobium meliloti* pilot microarrays for transcriptome analysis. *Appl. Environ. Microbiol., 69*, 1214-1219.

Bolton, E., Higgisson, B., Harrington, A., and O'Gara, F. (1986). Dicarboxylic acid transport in *Rhizobium meliloti*: Isolation of mutants and cloning of dicarboxylic acid transport gene. *Arch. Microbiol., 144*, 142-146.

Buikema, W. R., Long, S. R., Brown, S. E., van de Bos, R. C., Earl, C., and Ausubel, F. M. (1982). Physical and genetic characterization of *Rhizobium meliloti* symbiotic mutants. *J. Mol. Appl. Genetics, 2*, 249-260.

Burkhardt, B., Schillik, D., and Pühler, A. (1987). Physical characterization of *Rhizobium meliloti* megaplasmids. *Plasmid, 17*, 13-25.

Capela, D., Barloy-Hubler, F., Gouzy, J., Bothe, G., Ampe, F., Batut, J., *et al.* (2001). Analysis of the chromosome sequence of the legume symbiont *Sinorhizobium meliloti*. *Proc. Natl. Acad. Sci USA, 98*, 9877-9882.

Casse, F., Boucher, C., Julliot, J. S., Michel, M., and Dénarié, J. (1979). Identification and characterization of large plasmids in *Rhizobium meliloti* using agarose gel electrophoresis. *J. Gen. Microbiol., 113*, 229-242.

Chen, H., Higgins, J., Oresnik, I., Hynes, M. F., Natera, S., Djordjevic, M. A., *et al.* (2000). Proteome analysis demonstrates complex replicon and luteolin interactions in pSymA-cured derivatives of *Sinorhizobium meliloti* strain 2011. *Electrophoresis, 21*, 3833-3842.

Chen, L., Chen, Y., Wood, D. W., and Nester, E. W. (2002). A new type IV secretion system promotes conjugal transfer in *Agrobacterium tumefaciens*. *J. Bacteriol., 184*, 4838-4845.

Cussiol, J. R., Alves, S. V., Oliveira, M. A., and Netto, L. E. (2003). Organic hydroperoxide resistance gene encodes a thiol-dependent peroxidase. *J. Biol. Chem.*, in press.

DelVecchio, V. G., Kapatral, V., Redkar, R. J., Patra, G., Mujer, C., Los, T., *et al.* (2002). The genome sequence of the facultative intracellular pathogen *Brucella melitensis*. *Proc. Natl. Acad. Sci. USA, 99*, 443-448.

Demont, N., Ardourel, M., Maillet, F., Promé, D., Ferro, M., Promé, J., *et al.* (1994). The *Rhizobium meliloti* regulatory *nodD3* and *syrM* genes control the synthesis of a particular class of nodulation factors *N*-acylated by (w-1)-hydroxylated fatty acids. *EMBO J., 13*, 2139-2149.

Dharmatilake, A. J., and Bauer, W. D. (1992). Chemotaxis of *Rhizobium meliloti* toward nodulation gene-inducing compounds from alfalfa roots. *Appl. Environ. Microbiol., 58*, 1153-1158.

Djordjevic, M. A., Chen, H. C., Natera, S., Noorden, G. V., Menzel, C., Taylor, S., *et al.* (2003). A global analysis of protein expression profiles in *Sinorhizobium meliloti*: Discovery of new genes for nodule occupancy and stress adaptation. *Mol. Plant-Microbe Interact., 16*, 508-524.

Dusha, I., Bakos, A., Kondorosi, A., de Bruijn, F. J., and Schell, J. (1989). The *Rhizobium meliloti* early nodulation genes (*nodABC*) are nitrogen-regulated: Isolation of a mutant strain with efficient nodulation capacity on alfalfa in the presence of ammonium. *Mol. Gen. Genet., 219*, 89-96.

Dusha, I., and Kondorosi, A. (1993). Genes at different regulatory levels are required for the ammonia control of nodulation in *Rhizobium meliloti*. *Mol. Gen. Genet., 240*, 435-444.

Earl, C. D., Ronson, C. W., and Ausubel, F. M. (1987). Genetic and structural analysis of the *Rhizobium meliloti fixA, fixB, fixC,* and *fixX* genes. *J. Bacteriol., 169*, 1127-1136.

Eckhardt, T. (1978). A rapid method for the identification of plasmid deoxyribonucleic acid in bacteria. *Plasmid, 11*, 584-588.

Ehrhardt, D. W., Atkinson, E. M., Faull, K. F., Freedberg, D. I., Sutherlin, D. P., Armstrong, R., *et al.* (1995). In vitro sulfotransferase activity of NodH, a nodulation protein of *Rhizobium meliloti* required for host-specific nodulation. *J. Bacteriol., 177*, 6237-45.

Finan, T. M., Hirsch, A. M., Leigh, J. A., Johansen, E., Kuldau, G. A., Deegan, S., *et al.* (1985). Symbiotic mutants of *Rhizobium meliloti* that uncouple plant from bacterial differentiation. *Cell, 40*, 869-877.

Finan, T. M., Weidner, S., Chain, P., Buhrmester, J., Wong, K., Vorhölter, F.-J., *et al.* (2001). The complete sequence of the 1,683 kilobase pSymB megaplasmid from the N_2-fixing endosymbiont *Sinorhizobium meliloti*. *Proc. Natl. Acad. Sci. USA, 98*, 9889-9894.

Fischer, H. M. (1994). Genetic regulation of nitrogen fixation in rhizobia. *Microbiol. Rev., 58*, 352-386.

Fisher, R. F., Egelhoff, T. T., Mulligan, J. T., and Long, S. R. (1988). Specific binding of proteins from *Rhizobium meliloti* cell-free extracts containing NodD to DNA sequences upstream of inducible nodulation genes. *Genes Dev., 2*, 282-293.

Fisher, R. F., and Long, S. R. (1989). DNA footprint analysis of the transcriptional activator proteins NodD1 and NodD3 on inducible *nod* gene promoters. *J. Bacteriol., 171*, 5492-5502.

Fisher, R. F., Swanson, J., Mulligan, J. T., and Long, S. R. (1987). Extended region of nodulation genes in *Rhizobium meliloti* 1021. II. Nucleotide sequence, transcription start sites, and protein products. *Genetics, 117*, 191-201.

Foussard, M., Garnerone, A.-M., Ni, F., Soupene, E., Boistard, P., and Batut, J. (1997). Negative autoregulation of the *Rhizobium meliloti fixK* gene is indirect and requires a newly identified regulator, FixT. *Mol. Microbiology, 25*, 27-37.

Frassinetti, S., Setti, L., Corti, A., Farrinelli, P., Montevecchi, P., and Vallini, G. (1998). Biodegradation of dibenzothiophene b a nodulating isolate of *Rhizobium meliloti*. *Can. J. Microbiol., 44*, 289-297.

Freiberg, C., Fellay, R., Bairoch, A., Broughton, W. J., Rosenthal, A., and Perret, X. (1997). Molecular basis of symbiosis between *Rhizobium* and legumes. *Nature, 387*, 394-401.

Galibert, F., Finan, T. M., Long, S. R., Pühler, A., Abola, A. P., Ampe, F., *et al.* (2001). The composite genome of the legume symbiont *Sinorhizobium meliloti*. *Science, 293*, 668-672.

Goodner, B., Hinkle, G., Gattung, S., Miller, N., Blanchard, M., Qurollo, B., *et al.* (2001). Genome sequence of the plant pathogen and biotechnology agent *Agrobacterium tumefaciens* C58. *Science, 294*, 2323-2328.

Göttfert, M., Röthlisberger, S., Kündig, C., Beck, C., Marty, R., and Hennecke, H. (2001). Potential symbiosis-specific genes uncovered by sequencing a 410-kilobase DNA region of the *Bradyrhizobium japonicum* chromosome. *J. Bacteriol., 183*, 1405-1412.

Gouffi, K., Pica, N., Pichereau, V., and Blanco, C. (1999). Disaccharides as a new class of nonaccumulated osmoprotectants for *Sinorhizobium meliloti*. *Appl. Env. Microbiol., 65*, 1491-1500.

Honeycutt, R. J., McClelland, M., and Sobral, B. W. S. (1993). Physical map of the genome of *Rhizobium meliloti* 1021. *J. Bacteriol., 175*, 6945-6952.

Honma, M., and Ausubel, F. M. (1987). *Rhizobium meliloti* has three functional copies of the *nodD* symbiotic regulatory gene. *Proc. Natl. Acad. Sci. USA, 84*, 8558-8562.

Honma, M. A., Asomaning, M., and Ausubel, F. M. (1990). *Rhizobium meliloti nodD* genes mediate host-specific activation of *nodABC*. *J. Bacteriol., 172*, 901-911.

Hou, Y. M. (1999). Transfer RNAs and pathogenicity islands. *Trends Biochem. Sci., 24*, 295-298.

Jumas-Bilak, E., Michaux-Charachon, S., Bourg, G., Ramuz, M., and Allardet-Servent, A. (1998). Unconventional genomic organization in the alpha subgroup of the Proteobacteria. *J. Bacteriol., 180*, 2749-2755.

Kado, C. I. (2000). The role of the T-pilus in horizontal gene transfer and tumorigenesis. *Curr. Opin. Microbiol., 3*, 643-648.

Kaneko, T., Nakamura, Y., Sato, S., Asamizu, E., Kato, T., Sasamoto, S., *et al.* (2000). Complete genome structure of the nitrogen-fixing symbiotic bacterium *Mesorhizobium loti*. *DNA Res., 7*, 331-338.

Kaneko, T., Nakamura, Y., Sato, S., Minamisawa, K., Uchiumi, T., Sasamoto, S., *et al.* (2002). Complete genome sequence of nitrogen-fixing symbiotic bacterium *Bradyrhizobium japonicum* USD110. *DNA Res., 9*, 189-197.

Kondorosi, A., Kiss, G. B., Forrai, T., Vincze, E., and Banfalvi, Z. (1977). Circular linkage map of the *Rhizobium meliloti* chromosome. *Nature, 268*, 525-527.

Lerouge, P., Roche, P., Faucher, C., Maillet, F., Truchet, G., Promé, J. C. *et al.* (1990). Symbiotic host-specificity of *Rhizobium meliloti* is determined by a sulphated and acylated glucosamine oligosaccharide signal. *Nature, 344*, 781-784.

Li, C., and Clarke, S. (1992). A protein methyltransferase specific for altered aspartyl residues is important in *Escherichia coli* stationary-phase survival and heat-shock resistance. *Proc. Natl. Acad. Sci. USA, 89*, 9885-9889.

Long, S. R. (1996). *Rhizobium* symbiosis: Nod factors in perspective. *The Plant Cell, 8*, 1885-1898.

Long, S. R., Buikema, W. E., and Ausubel, F. M. (1982). Cloning of *Rhizobium meliloti* nodulation genes by direct complementation of Nod$^-$ mutants. *Nature, 298*, 485-488.

Lynch, D., O'Brien, J., Welch, T., Clarke, P., Cuív, P., Crosa, J. H. *et al.* (2001). Genetic organization of the region encoding regulation, biosynthesis and transport of rhizobactin 1021, a siderophore produced by *Sinorhizobium meliloti*. *J. Bacteriol., 183*, 2576-2585.

Martínez-Romero, E., and Caballero-Mellado, J. (1996). *Rhizobium* phylogenies and bacterial genetic diversity. *Critical Rev. Plant Sci., 15*, 113-140.

Mavingui, P., Flores, M., Guo, X., Dávila, G., Perret, X, Broughton, W. J. *et al.* (2002). Dynamics of genome architecture in *Rhizobium* sp. strain NGR234. *J. Bacteriol., 184*, 171-176.

Maxwell, C. A., Hartwig, U. A., Joseph, C. M., and Phillips, D. A. (1989). A chalcone and two related flavonoids released from alfalfa roots induce *nod* genes of *Rhizobium meliloti*. *Plant Physiol., 91*, 842-847.

Meade, H. M., Long, S. R., Ruvkun, G. B., Brown, S. E., and Ausubel, F. M. (1982). Physical and genetic characterization of symbiotic and auxotrophic mutants of *Rhizobium meliloti* induced by transposon Tn*5* mutagenesis. *J. Bacteriol., 149*, 114-122.

Meade, H. M., and Signer, E. R. (1977). Genetic mapping of *Rhizobium meliloti*. *Proc. Natl. Acad. Sci. USA, 74*, 2076-2078.

Mulligan, J. T., and Long, S. R. (1989). A family of activator genes regulates expression of *Rhizobium meliloti* nodulation genes. *Genetics, 122*, 7-18.

Mulligan, J. T., and Long, S. R. (1985). Induction of *Rhizobium meliloti nodC* expression by plant exudate requires *nodD*. *Proc. Natl. Acad. Sci. USA, 82*, 6609-6613.

Nikaido, H., and Zgurskaya, H. I. (2001). AcrAB and related multidrug efflux pumps of *Escherichia coli*. *J. Mol. Microbiol. Biotechnol., 3*, 215-218.

Nuti, M. P., Ledeboer, A. M., Lepidi, A. A., and Schilperoort, R. A. (1977). Large plasmids in different *Rhizobium* species. *J. Gen. Microbiol., 100*, 241-248.

Ogawa, J. (1993). Ph.D. thesis, Stanford University, CA, USA.

Ogawa, J., and Long, S. R. (1995). The *Rhizobium meliloti groELc* locus is required for regulation of early *nod* genes by the transcription activator NodD. *Genes Dev., 9*, 714-729.

Oláh, B., Kiss, E., Györgypal, Z., Borzi, J., Cinege, G., Csanádi, G., *et al.* (2001). Mutation in the *ntrR* gene, a member of the *vap* gene family, increases the symbiotic efficiency of *Sinorhizobium meliloti*. *Mol. Plant-Microbe Interact., 14*, 887-894.

Oresnik, I. J., Liu, L.-L., Yost, C. K., and Hynes, M. F. (2000). Megaplasmid pRme2011a of *Sinorhizobium meliloti* is not required for viability. *J. Bacteriol., 182*, 3582-3586.

Østeras, M., Boncompagni, E., Vincent, N., Poggi, M.-C., and Le Rudulier, D. (1998). Presence of a gene encoding choline sulfatase in *Sinorhizobium meliloti bet* operon: Choline-o-sulfate is metabolized into glycine betaine. *Proc. Natl. Acad. Sci. USA, 95*, 11394-11399.

Østeras, M., Driscoll, B. T., and Finan, T. M. (1995). Molecular and expression analysis of the *Rhizobium meliloti* phosphoenolpyruvate carboxykinase (*pckA*) gene. *J. Bacteriol., 177*, 1452-1460.

Paulsen, I. T., Seshadri, R., Nelson, K. E., Eisen, J. A., Heidelberg, J. F., Read, T. D., *et al.* (2002). The *Brucella suis* genome reveals fundamental similarities between animal and plant symbionts. *Proc. Natl. Acad. Sci. USA, 99*, 13148-13153.

Perret, X., Parsons, J., Viprey, V., Reichwald, K., and Broughton, W. J. (2001). Repeated sequences of *Rhizobium* sp. NGR234 and *Sinorhizobium meliloti*: a comparative analysis through random sequencing. *Rev. Can. Microbiol., 47*, 548-558.

Peters, N. K., Frost, J. W., and Long, S. R. (1986). A plant flavone, luteolin, induces expression of *Rhizobium meliloti* nodulation genes. *Science, 233*, 917-1008.

Phillips, D. A., Joseph, C. M., and Maxwell, C. A. (1992). Trigonelline and stachydrine released from alfalfa seeds activate NodD2 protein in *Rhizobium meliloti*. *Plant Physiol., 99*, 1526-1531.

Putnoky, P., Kereszt, A., Nakamura, T., Endre, G., Grosskopf, E., Kiss, P., *et al.* (1998). The *pha* gene cluster of *Rhizobium meliloti* involved in pH adaptation and symbiosis encodes a novel type of K+ efflux system. *Mol. Microbiol., 28*, 1091-1101.

Ramirez-Romero, M. A., Tekkez-Sosa, J., Barrios, H., Perez-Oseguera, A., Rosas, V., and Cevallos, M. A. (2001). RepA negatively autoregulates the transcription of the *repABC* operon of the *Rhizobium etli* symbiotic plasmid basic replicon. *Mol. Microbiol., 42*, 195-204.

Rosenberg, C., Boistard, P., Dénarié, J., and Casse-Delbart, F. (1981). Genes controlling early and late functions in symbiosis are located on a megaplasmid in *Rhizobium meliloti*. *Mol. Gen. Genet., 184*, 326-333.

Ruvkun, G. B., and Ausubel, F. M. (1980). Interspecies conservation of nitrogenase genes. *Proc. Natl. Acad. Sci. USA, 77*, 191-195.

Schwedock, J., and Long, S. R. (1990). ATP sulphurylase activity of the *nodP* and *nodQ* gene products of *Rhizobium meliloti*. *Nature, 348*, 644-647.

Sharypova, L. A., Yurgel, S. N., Keller, M., Simarov, B. V., Pühler, A., and Becker, A. (1999). The eff-482 locus of *Sinorhizobium meliloti* CXM1-105 that influences symbiotic effectiveness consists of three genes encoding an endoglycanase, a transcriptional regulator and an adenylate cyclase. *Mol. Gen. Genet., 261*, 1032-44.

Sieira, R., Comerci, D. J., Sánchez, D. O., and Ugalde, R. A. (2000). A homologue of an operon required for DNA transfer in *Agrobacterium* is required in *Brucella abortus* for virulence and intracellular multiplication. *J. Bacteriol., 182*, 4849-4855.

Skerker, J. M., and Shapiro, L. (2000). Identification and cell cycle control of a novel pilus system in *Caulobacter crescentus*. *EMBO J., 19*, 3223-3234.

Sobral, B. W. S., Honeycutt, R. J., and Atherly, A. G. (1991a). The genomes of the family *Rhizobiaceae*: size, stability, and rarely cutting restriction endonucleases. *J. Bacteriol., 173*, 704-709.

Sobral, B. W. S., Honeycutt, R. J., Atherly, A. G., and McClelland, M. (1991b). Electrophoretic separation of the three *Rhizobium meliloti* replicons. *J. Bacteriol., 173*, 5173-5180.

Streeter, J. G., and Bhagwat, A. (1999). Biosynthesis of trehalose from maltooligosaccharides in Rhizobia. *Can J Microbiol., 45*, 716-721.

Sullivan, J. T., and Ronson, C. W. (1998). Evolution of rhizobia by acquisition of a 500-kb symbiosis island that integrates into a phe-tRNA gene. *Proc. Natl. Acad. Sci. USA, 95*, 5145-5149.

Swanson, J. A., Mulligan, J. T., and Long, S. R. (1993). Regulation of *syrM* and *nodD3* in *Rhizobium meliloti*. *Genetics, 134*, 435-444.

Thomas, C. M. (2000). Paradigms of plasmid organization. *Mol. Microbiol., 37*, 485-491.

Tuckerman, J. R., Gonzalez, G., and Gilles-Gonzalez, M. A. (2001). Complexation precedes phosphorylation for two-component regulatory system FixL/FixJ of *Sinorhizobium meliloti*. *J. Mol. Biol., 308*, 449-455.

Turner, S. L., Rigottier-Gois, L., Power, R. S., Amarger, N., and Young, J. P. W. (1996). Diversity of *repC* plasmid-replication sequences in *Rhizobium leguminosarum*. *Microbiology, 142*, 1705-13.

Viprey, V., Rosenthal, A., Broughton, W. J., and Perret, X. (2000). Genetic snapshots of the Rhizobium species NGR234 genome. *Genome Biology, 1*, 1-17.

Walt, A., and Kahn, M. L. (2002). The *fixA* and *fixB* genes are necessary for aerobic carnitine reduction in *Escherichia coli*. *J. Bacteriol., 184*, 4044-4047.

Weissbach, H., Etienne, F., Hoshi, T., Heinemann, S. H., Lowther, W. T., Matthews, B., *et al.* (2002). Peptide methionine sulfoxide reductase: Structure, mechanism of action, and biological function. *Arch. Biochem. Biophys., 397*, 172-178.

Wood, D. W., Setubal, J. C., Kaul, R., Monks, D. E., Kitajima, J. P., Okura, V. K., *et al.* (2001). The genome of the natural genetic engineer of *Agrobacterium tumefaciens* C58. *Science, 294*, 2317-2323.

Young, J. M., Kuykendall, L. D., Martínez-Romero, E., Kerr, A, and Sawada, H. (2001). A revision of *Rhizobium* Frank 1889, with an emended description of the genus, and the inclusion of all species of *Agrobacterium* Conn 1942 and *Allorhizobium undicola* de Lajudie *et al.* 1998 as new combinations: *Rhizobium radiobacter*, *R. rhizogenes*, *R. rubi*, *R. undicola* and *R. vitis*. *Int. J. Syst. Evol. Microbiol., 51*, 89-103.

Young, J. P. W., and Haukka, K. E. (1996). Diversity and phylogeny of rhizobia. *New. Phytol., 133*, 87-94.

CHAPTER 9

RHIZOBIUM ETLI GENOME BIOLOGY

G. DÁVILA, V. GONZÁLEZ, M. A. RAMÍREZ-ROMERO AND O. RODRÍGUEZ
Centro de Investigación sobre Fijación de Nitrógeno, UNAM, P.O. Box 565-A, Cuernavaca, Morelos 62170, México

1. INTRODUCTION

Rhizobium etli is a Gram-negative soil bacterium that forms nitrogen-fixing nodules on the roots of *Phaseolus vulgaris* L. Many genetic determinants from both symbionts are required for this association. The genome of bacteria belonging to this species is structured in several replicons; one circular chromosome and several large plasmids that could represent up to one third of the genome (Garcia-de los Santos *et al.*, 1996). Most of the genes for nodulation and nitrogen fixation are encoded in one of these plasmids, which is necessary, but not sufficient, for optimal symbiosis. The whole sequence of p42d, the symbiotic plasmid of *R. etli* CFN42, has recently been completed. Members of all the species belonging to the *Rhizobiaceae* family seem to have a single and ancient ancestral chromosome, even though the genetic diversity of the chromosome for any of such species is usually very high. To achieve symbiotic capability, these chromosomes must be associated with a specific symbiotic genome compartment that also determines the plant host range. Commonly, these Symbiotic Genome Compartments (SGC) occur either in replicons referred to as symbiotic plasmids (pSym) or as symbiotic islands (or regions) within the chromosome. Phylogenetic comparisons of these compartments from different species of the rhizobia have shown a mosaic structure, whereas compartments belonging to bacterial isolates from a single species are probably clonal and epidemic. A genomic project with CFN42, the type strain for *R. etli*, is currently in progress.

Some of the relevant structural qualities of the genomes of this bacterial family have been recently established by the complete genomic sequence of *Mesorhizobium loti* MAFF303099 (Kaneko *et al.*, 2000), *Sinorhizobium meliloti* (Barnett *et al.*, 2001; Capela *et al.*, 2001; Finan *et al.*, 2001; Galibert *et al.*, 2001),

R. Palacios and W. E. Newton (eds.), Genomes and Genomics of Nitrogen-fixing Organisms, 133-142.

and *Bradyrhizobium japonicum* USDA110 (Kaneko *et al.*, 2002). In addition, the pSym of *Rhizobium* sp. NGR234, called pNGR234a (Freiberg *et al.*, 1997), the chromosomal SGC of both *Bradyrhizobium japonicum* and *Mesorhizobium loti* R7A (Gottfert *et al.*, 2001; Sullivan *et al.*, 2002), and the pSym of *Rhizobium etli* p42d (González *et al.*, 2003) have been sequenced and analyzed. Analysis of complete genomes has revealed that they share many orthologous genes with a good level of chromosomal syntheny. However, the genetic relatedness of the symbiotic plasmids and the symbiotic chromosomal compartments has a different evolutionary history. This difference could be due either to the fact that they may have been acquired through horizontal transfer or to the frequent occurrence of genomic rearrangements, which is indicated by both the high concentration of insertion sequences and the large amount of reiterated DNA elements.

Native populations of *R. etli*, which were isolated from nodules on the common bean plant, *Phaseolus vulgaris*, in Mesoamerican soils, show a variation in both the size and the number of plasmids per strain. Most of the isolates contained four plasmids and only one out of 24 had a megaplasmid (Brom *et al.*, 2002). Interestingly, in these natural populations, the linkage distribution of genetic markers indicates a high level of conservation in the sequence integrity of their symbiotic plasmids. The pSym of *R. etli* strains fluctuates between two sizes as shown with plasmid-profile gels hybridized with a *nifH* probe. These plasmids differ in approximately 100 Kb and are equally distributed among *R. etli* strains isolated from bean-nodules from around the world. The larger pSym includes almost all the sequence present in the smaller one. The additional information present in the larger pSym appears to be conserved, clustered, and localized in a specific region. Moreover, the genetic diversity among the symbiotic plasmids of *R. etli* is exceptionally low, particularly when contrasted with that of their chromosomes. These characteristics support the existence of a unique common ancestor for the *R. etli* pSym.

2. *RHIZOBIUM ETLI* GENOME STRUCTURE

The *R. etli* CFN42 genome contains a circular chromosome with an estimated size of 4.5 Mb and an average GC content of 60%. In addition, 2 Mb are distributed in six large plasmids, p42a to p42f, whose sizes range from 184 to *ca.* 600 Kb. The symbiotic plasmid p42d has 371,255 bp (http://itzamna.cifn.unam.mx/retlidb/; González *et al.*, 2003). The analysis of plasmid-encoded functions indicates that:
(i) p42b, p42c, p42d and p42f are required for an efficient nodule occupancy;
(ii) p42b and p42d are essential for effective nodulation;
(iii) p42d and p42f participate in nitrogen fixation;
(iv) p42b, p42c and p42e are involved in the utilization of some carbon compounds;
(v) p42e and p42f participate in cellular growth and viability (Brom *et al.*, 2000).

Interestingly, in natural populations of *R. etli*, hybridization experiments, using an overlapping collection of cosmids (Girard *et al.*, 1991) that cover the whole p42d plasmid, show a high level of RFLP conservation (Figure 1). Moreover, Brom and coworkers found that the majority of *R. etli* strains share not only the presence but also the organization of some genetic regions associated with three replicons of R.

etli CFN42, namely p42b, p42d and p42f (Brom *et al.*, 2002). This information suggests a high level of structural and functional conservation between several plasmids of *R. etli* and, even though the whole genome is extremely variable, it constitutes a comprehensive functional unit structured in several replicons.

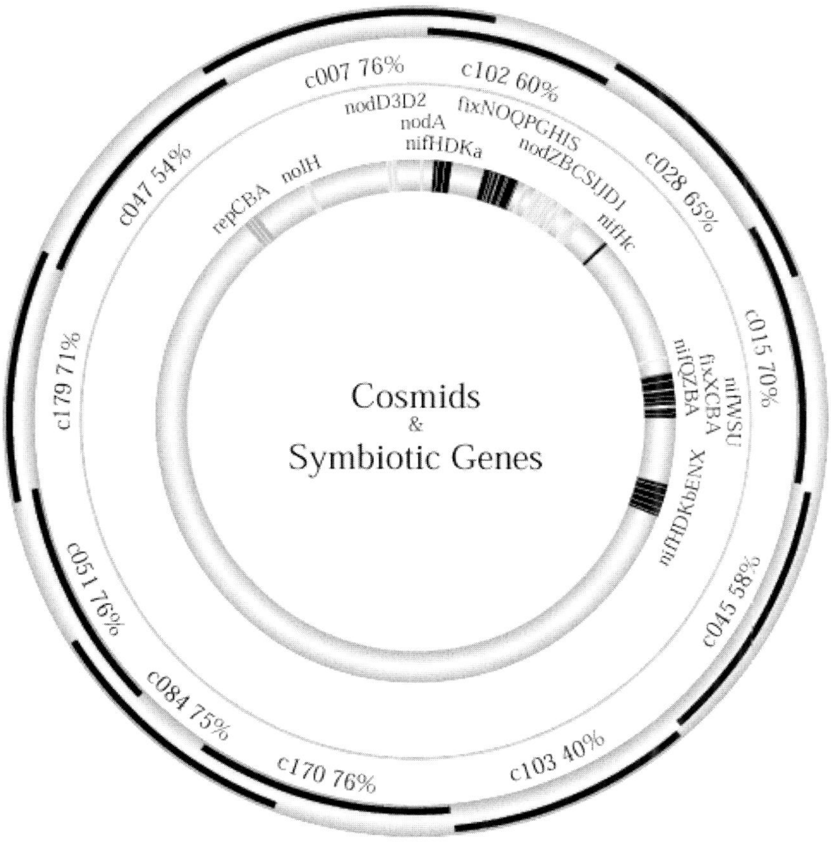

Figure 1. Conservation of regions covered by the complete p42d sequence in 50 natural isolates of Rhizobium etli.

Cosmid insert lengths and positions are shown in the outer ring, together with their names and conservation percentage of the RFLP patterns. The inner ring shows the nod *genes (light gre), the* nif *and* fix *genes (black), and the* rep *operon (dark grey).*

Another type of element relevant for the organization of these genomes corresponds to Elements Related to Insertion Sequences (ERIS). These elements are present in high numbers in the genomes of several rhizobia and they are particularly concentrated in the SGC. Even though in *R. etli* experimental evidence for functional ERIS is not available, a computational search for ERIS within the partial sequence of the CFN42 genome, suggests that the majority of the ERISs are distributed on two plasmids, p42a and p42d, but with only nine of them showing all the features required by a functional IS. The participation of ERIS in genome dynamics is currently under evaluation.

The CFN42 genome contains an unusually large number of reiterated sequences (RS), with about 700 elements (of 300-bp average size with more than 80% homology) belonging to about 200 different families. This assessment suggests that as much as 4.2 % of this genome is reiterated (Flores *et al.*, 1987). Reiterated sequences are scattered throughout the genome but, surprisingly, their distribution is not completely random. For example, all the elements from several RS families are exclusively shared by two replicons. The genetic nature of the RS comprises either structural genes, such as the two copies of *rpoN*, one of which is located in p42d and the other in the chromosome, or complete operons, like the two *nifHDK* regions present in p42d (Quinto *et al.*, 1985), or pseudo genes and ERIS shared between replicons. The participation of the RS in eliciting genomic rearrangements is reviewed below.

3. *RHIZOBIUM* GENOMIC PLASTICITY

Variability and plasticity of genomic structures are common phenomena among members of the *Rhizobiaceae* family. It is possible that these features are related to the capacity to both adapt and survive in a changing environment. Two processes participate in the modification of the bacterial genome structures; these are genomic rearrangements and horizontal gene transfer. In *R. etli* (see Chapter 12), it was initially shown that, when colonies derived from a single mother cell were analyzed, 3-6 % of them revealed some kind of genomic reorganization, implicating a high frequency of genomic rearrangements (Brom *et al.*, 1991; Flores *et al.*, 1988). The proportion of derivatives bearing rearrangements may increase up to 35 % after subculturing for one year under normal laboratory conditions (Romero *et al.*, 1998). The main process that produces these rearrangements involves homologous recombination between two RS and these rearrangements may generate cointegrations, deletions, amplifications, or inversions (Romero *et al.*, 1995).

The best-studied model of genomic plasticity in *R. etli* is the symbiotic plasmid p42d. The analysis of this plasmid revealed the presence of 29 RS (with at least 300 identical nucleotides) grouped in 12 families of two or three elements each. A segment of DNA bordered by a pair of direct repeats is known as an amplicon that may promote the amplification (or deletion) of the encompassed region. Figure 2 shows the 12 predicted amplicons of p42d; three of them have been experimentally confirmed (black arcs) and the remaining 9 are under study (grey arcs). The characterized amplicons, which are located in the symbiotic region of the plasmid, allow amplifications at frequencies that range from 10^{-3} to 10^{-4} (Romero *et al.*, 1991; Romero *et al.*, 1995).

Other types of rearrangements that occur correspond to cointegrations and translocations. The symbiotic plasmid p42d is able to form a transient cointegrate with p42a that is usually resolved so preserving the original structures and organization of the wild-type plasmids. However, in some cases, this cointegrate is spliced by recombination between the abundant RS shared by these plasmids yielding two chimerical replicons (Brom *et al.*, unpublished data). Another interesting observation is that, when a replicon that is incompatible with p42d is acquired by conjugation, instead of losing the resident plasmid, it recombines with p42b forming a hybrid replicon. Such a hybrid plasmid is stably maintained as long as the selective pressure exerted by the incompatible plasmid is present (Soberón and Cevallos unpublished data).

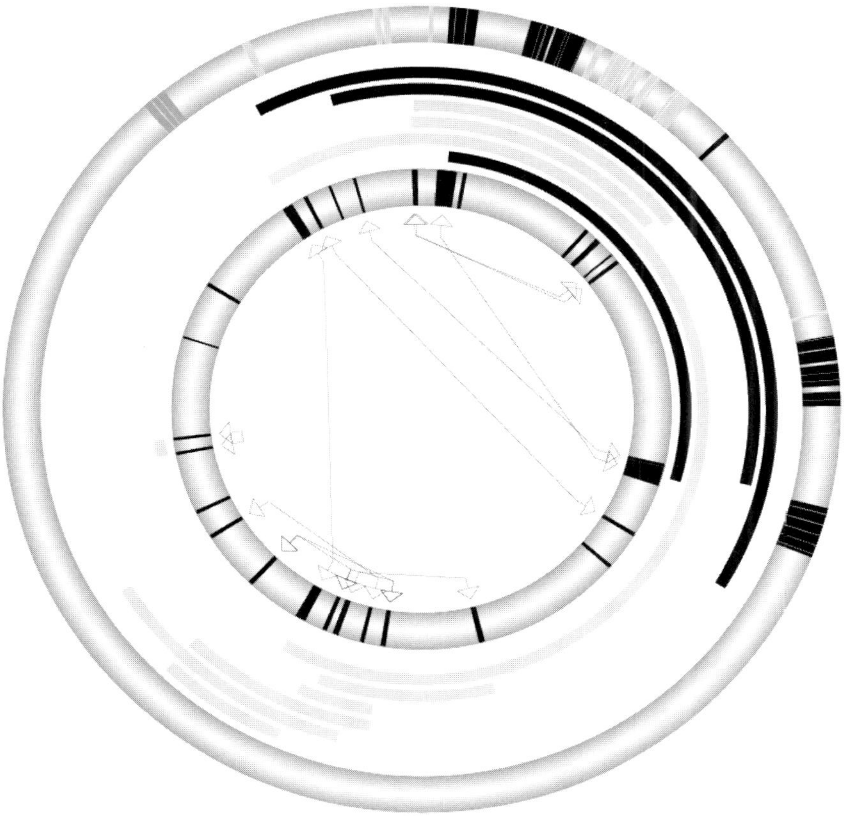

Figure 2. Amplicons of the symbiotic plasmid p42d of R. etli *CFN42.*
The external circle shows the replicator and some symbiotic genes as in Figure 1. Arcs in the second circle show the location and extension of putative (grey) and actual (black) amplicons. The inner circle illustrates the position and size of all RS present in the structure of p42d. The double-headed arrows in the internal circle connect all pairs of directly oriented repeats.

Conjugative transfer of the CFN42 plasmids can be detected in laboratory conditions. Studies by Brom *et al.* indicate that, of the six plasmids of the strain, only p42a is self-transmissible at high frequency; in addition, p42d transfers at low frequency depending on the presence of p42a. Furthermore, the mechanism that allows p42a-dependent transfer of pSym seems to involve the cointegration of these plasmids. Two mechanisms for cointegration have been proposed, one is RecA-dependent and the other involves a RecA-independent site-specific recombination (Brom *et al.*, unpublished data). In natural populations of *R. etli* isolated from Mexican and European soils, it is common to obtain strains containing one transmissible plasmid. Usually, many of the conjugative plasmids recovered are very similar to p42a. However, some self-transmissible plasmids with no similarity to p42a have been identified, and 10% of the analyzed strains simultaneously contain two conjugative plasmids (Brom *et al.*, 2002).

4. *RHIZOBIUM ETLI* TAXONOMY AND EVOLUTION

This group of nodule-forming bacteria belonging to the *Rhizobiaceae* family is heterogeneous with a high degree of genetic diversity (Martínez-Romero and Caballero-Mellado, 1996). Every attempt to classify the rhizobial species has confronted the fact that their most prominent feature is the symbiotic phenotype, even though it is mostly associated with either large plasmids or symbiotic islands. These genomic compartments may be transferred among different chromosomal backgrounds and this circumstance has complicated the understanding of *Rhizobium* phylogeny.

Strains able to engage in symbiosis with the common bean are not constrained to a single species (Martínez-Romero and Caballero-Mellado, 1996), indicating either a promiscuous host range of the plant or the frequent horizontal transfer of the pSym between different chromosomal backgrounds. *R. etli* has been isolated from México, South America, and Europe and is characterized by both its narrow host range and the reiteration of *nifH* gene (Segovia *et al.*, 1993). Population genetics, based on multi *locus* enzyme electrophoresis, has revealed that the genetic diversity of the chromosome of *R. etli* is exceptionally high (Piñero *et al.*, 1988; Souza *et al.*, 1992). In contrast, the *R. etli* pSym is remarkably conserved among isolates from different geographical regions (see Figure 1). These data support the idea that the evolution of this plasmid is both clonal and epidemic.

A detailed examination of the bacterial population of the bean rhizosphere has revealed that intermingled with the effective but scarce *R. etli* bacteria, there is a vast population of bacteria unable to nodulate beans, even though their chromosomes are similar to those of the effective ones (Segovia *et al.*, 1991). A possible explanation for this unequal distribution is a frequent loss of the pSym. This suggestion is supported by the fact that, when an ineffective bacterium (re)acquires the plasmid by *in vitro* conjugation, the symbiotic abilities are reestablished. Underlying this observation is the notion that a collection of different chromosomal backgrounds is able to endorse the symbiotic process and, therefore, these bacterial chromosomes have been chosen by association with a specific pSym. These bacteria were clustered in what is now named the species of *Rhizobium etli*.

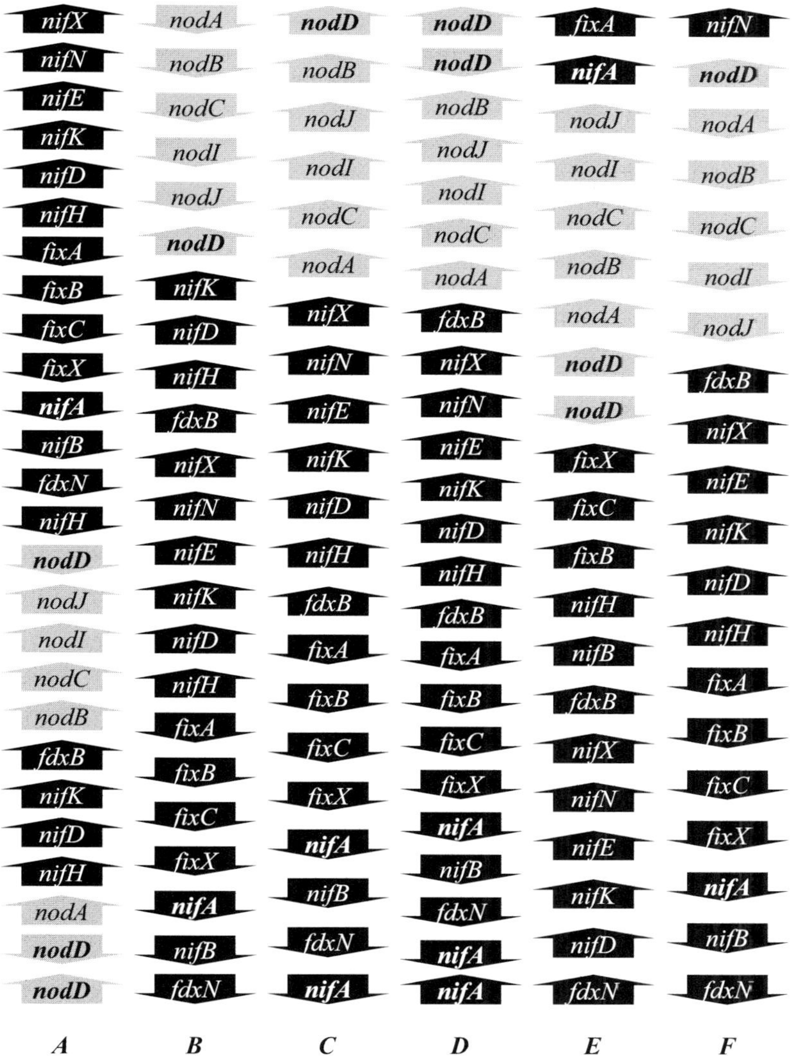

Figure 3. Order, orientation, and redundancy of the twenty orthologous genes shared by six SGC.

Only the symbiotic genes (neither in scale nor real spacing) are shown with nif, fix *and* fdx *as black arrows and* nod *as grey arrows. First and last gene delimit the smallest region that enclose the 20 orthologous genes in its respective SGC with the following sizes: A, p42d of* R. etli, *120 kb; B, pNGR234a of* R. *sp. NGR234, 250 kb; C, the symbiotic island of* M. loti *R7A, 300 kb; D, the symbiotic region of* M. loti *MAFF, 320 kb; E, the symbiotic region of* B. japonicum, *300 kb; and F, pSymA of* S. meliloti, *50 kb.*

Because not all soil bacteria can both replicate and express the symbiotic genes of the plasmid, then how diverse are these chromosomes? Could they be rationally classified into a single species? These are fundamental questions that still remain unsolved.

From the phylogenetic point of view, *R. etli* seems to share a common chromosomal ancestor with *R. tropicii*, *R. leguminosarum*, and *S. meliloti*. Notably, strains of *Brucella* and *Agrobacterium* also lie close to some rhizobial branches, highlighting the common origin of the α-proteobacteria group (Martínez-Romero *et al.*, 2000; Young *et al.*, 2001). On the contrary, the variability of the SGC in regard to the size, gene content, lack of synteny, and degree of distortion in the phylogeny of the 20 orthologous genes shared by them (see Figure 3) highlight the absence of a common ancestor for these compartments among the rhizobial species.

The genomic architecture of rhizobial species can be quite different in size and compartmentalization (Sobral *et al.*, 1991). For example, *R. etli* has one chromosome and several large plasmids, *S. meliloti* has one chromosome and two megaplasmids, both *M. loti* and *B. japonicum* contain large chromosomes with an integrated symbiotic region or island (Brom *et al.*, 2002; Gottfert *et al.*, 2001; Sobral *et al.*, 1991; Sullivan and Ronson, 1998). Moreover, *Agrobacterium tumefaciens* and several *Brucella* species have two chromosomes. Despite this variable organization, several common features have emerged from whole genome comparisons. We find, for example, that *A. tumefaciens* C58 chromosomes have extensive conservation in gene order with *S. meliloti*, but a limited synteny with *M. loti* (Goodner *et al.*, 2002; Wood *et al.*, 2002). On the other hand, the chromosome of *Brucella suis* shares more regions of synteny with the chromosome of *M. loti* than with that of either *A. tumefaciens* or *S. meliloti* (Paulsen *et al.*, 2002).

Taken as whole, these findings suggest a common chromosomal ancestor for all the rhizobial species. Nevertheless, it is clear that along the evolutionary history of these genomes, multiple events of genetic exchange have occurred as have DNA rearrangements, including whole replicon fusions. It has recently been shown that the three replicons composing the genomes of *Rhizobium* sp. NGR234 and *S. meliloti* can be stably fused without altering either bacterial growth or symbiotic proficiency (Mavingui *et al.*, 2002; Guo *et al.*, 2003; see Chapters 6 and 12).

ACKNOWLEDGEMENTS

We are indebted to Rafael Palacios for valuable discussions and encouragement through the *R. etli* genome project and to Susana Brom for a critical review of the manuscript. We wish to thank the technical support of Bustos, P., Santamaría, R. I., Canedo, M., Hernández, I. L., Gama, J. A., Gómez, R. E., Martínez, J. A., and Fernández, J. L. The computational assistance of José Espíritu and Víctor del Moral is also appreciated. This work was funded by grants from the National Council of Science and Technology (CONACyT, México, N-028) and the National Autonomous University of México (UNAM).

REFERENCES

Barnett, M. J., Fisher, R. F., Jones, T., Komp, C., Abola, A. P., Barloy-Hubler, F., *et al.* (2001). Nucleotide sequence and predicted functions of the entire *Sinorhizobium meliloti* pSymA megaplasmid. *Proc. Natl. Acad. Sci. USA, 98*, 9883-9888.

Brom, S., García-de los Santos, A., Girard, M. L., Dávila, G., Palacios, R., and Romero, D. (1991). High frequency rearrangements in *Rhizobium leguminosarum* bv. phaseoli plasmid. *J. Bacteriol., 173*, 1344-1346.

Brom, S., García-de los Santos, A., Stepkowski, T., Flores, M., Dávila, G., Romero, D., *et al.* (1992). Different plasmids of *Rhizobium leguminosarum* bv. phaseoli are required for optimal symbiotic performance. *J. Bacteriol., 174*, 5183-5189.

Brom, S., García-de los Santos, A., Cervantes, L., Palacios, R., and Romero, D. (2000). In *Rhizobium etli* symbiotic plasmid transfer, nodulation competitivity and cellular growth require interaction among different replicons. *Plasmad, 44*, 34-43.

Brom, S., Girard, L., García-de los Santos, A., Sanjuan-Pinilla, J. M., Olivares, J., and Sanjuan, J. (2002). Conservation of plasmid-encoded traits among bean-nodulating *Rhizobium* species. *Appl. Environ. Microbiol., 68*, 4978-4981.

Capela, D., Barloy-Hubler, F., Gouzy, J., Bothe, G., Ampe, F., Batut, J., *et al.* (2001). Analysis of the chromosome sequence of the legume symbiont *Sinorhizobium meliloti* strain 1021. *Proc. Natl. Acad. Sci. USA, 98*, 9877-9882.

Finan, T. M., Weidner, S., Wong, K., Buhrmester, J., Chain, P., Vorholter, F. J., *et al.* (2001). The complete sequence of the 1, 683-kb pSymB megaplasmid from the N2- fixing endosymbiont *Sinorhizobium meliloti*. *Proc. Natl. Acad. Sci. USA, 98*, 9889-9894.

Flores, M., González, V., Brom, S., Martínez, E., Piñero, D., Romero, D., *et al.* (1987). Reiterated DNA sequences in *Rhizobium* and *Agrobacterium* spp. *J. Bacteriol., 169*, 5782-5788.

Flores, M., González, V., Pardo, M. A., Leija, A., Martínez, E., Romero, D., *et al.* (1988). Genomic instability in *Rhizobium phaseoli*. *J. Bacteriol., 170*, 1191-1196.

Freiberg, C., Feilla, R., Bairoch, A., Broughton, W. J., Rosenthal, A., and Perret, X. (1997). Molecular basis of symbiosis between *Rhizobium* and legumes. *Nature, 387*, 394-401.

Galibert, F., Finan, T. M., Long, S. R., Pühler, A., Abola, P., Ampe, F., *et al.* (2001). The composite genome of the legume symbiont *Sinorhizobium meliloti*. *Science, 293*, 668-672.

García-de los Santos, A., Brom, S., and Romero, D. (1996). *Rhizobium* plasmids in bacteria-legume interactions. *World J. Microbiol. Biotechnol., 12*, 119-125.

Girard, M. L., Flores, M., Brom, S., Romero, D., Palacios, R., and Dávila, G. (1991). Structural complexity of the symbiotic plasmid of *Rhizobium leguminosarum* bv. phaseoli. *J. Bacteriol., 173*, 2411-2419.

Gonder, B., Hinkle, G., Gattung, S., Millar, N., Blanchard, M., Qurollo, B., *et al.* (2001). Genome sequence of the plant pathogen and biotechnology agent *Agrobacterium tumefaciens* C58. *Science, 294*, 2323-2328.

González, V., Bustos, P., Ramírez-Romero, M. A., Medrano-Soto, A., Salgado, H., Hernández-González, I., *et al.* (2003). The mosaic structure of the symbiotic plasmid of *Rhizobium etli* CFN42 and its relation to other symbiotic genome compartments. *Genome Biol., 4*, R36.

Gottfert, M., Rothlisberger, S., Kundig, C., Beck, C., Marty, R., and Hennecke, H. (2001). Potential symbiosis-specific genes uncovered by sequencing a 410-kilobase DNA region of the *Bradyrhizobium japonicum* chromosome. *J. Bacteriol., 183*, 1405-1412.

Guo, X., Flores, M., Mavingui, P., Fuentes, S., Hernández, G., Dávila, G., *et al.* (2003). Natural genomic design in *Sinorhizobium meliloti*: Novel genomic architecture. *Genome Res., 13*, 1810-1817.

Kaneko, T., Nakamura, Y., Sato, S., Asamizu, E., Kato, T., Sasamoto, S., *et al.* (2000). Complete genome structure of the nitrogen-fixing symbiotic bacterium *Mesorhizobium loti*. *DNA Res., 7*, 331-338.

Kaneko. T., Nakamura, Y., Sato, S., Minamisawa, K., Uchiumi, T., Sasamoto, S., *et al.* (2002) Complete genomic sequence of nitrogen-fixing symbiotic bacterium *Bradyrhizobium japonicum* USDA110. *DNA Res., 9*, 189-197.

Martínez-Romero, E., and Caballero-Mellado, J. (1996). *Rhizobium* phylogenies and bacterial genetic diversity. *Crit. Rev. Plant Sci., 15*, 113-140.

Martínez-Romero, E., Caballero-Mellado, J., Gándara, B., Rogel, M. A., López-Merino, A., Wang, T., *et al.* (2000). Ecological, phylogenetic and taxonomic remarks on diazotrophs and related genera. In F. O. Pedroza, M. Hungria, M. G. Yates, and W. E. Newton (Eds.), *Nitrogen fixation: From molecules to crop productivity* (pp. 155-160). Dordrecht, The Netherlands: Kluwer Academic Publishers.

Mavingui, P., Flores, M., Guo, X., Dávila, G., Perret, X., Broughton, W. J., *et al.* (2002). Dynamics of genome architecture in *Rhizobium* sp. strain NGR234. *J. Bacteriol., 184*, 171-176.

Paulsen, I. T., Seshadri, R., Nelson, K. E., Eisen, J. A., Heidelberg, J. F., Read, T. D., *et al.* (2002). The *Brucella suis* genome reveals fundamental similarities between animal and plant pathogens and symbionts. *Proc. Natl. Acad. Sci. USA, 99*, 13148-13153.

Piñero, D., Martínez, E., and Selander; R. K. (1988). Genetic diversity and relationships among isolates of *Rhizobium leguminosarum* bv. phaseoli. *Appl. Environ. Microbiol., 54*, 2825-2832.

Quinto, C., De la Vega, H., Flores, M., Leemans, J., Cevallos, M. A., Pardo, M. A., *et al.* (1985). Nitrogenase reductase: A functional multi-gene family in *Rhizobium phaseoli. Proc. Natl. Acad. Sci. USA, 82*, 1170-1174.

Romero, D., Brom, S., Martínez-Salazar, J., Girard, M. L., Palacios, R., and Dávila, G. (1991). Amplification and deletion of a *nod-nif* region in the symbiotic plasmid of *R. phaseoli. J. Bacteriol., 173*, 2435-2441.

Romero, D., Dávila, G., and Palacios, R. (1998). The dynamic genome of *Rhizobium*. In F. J. de Bruijn, J. R. Lupski, and G. M. Winstock (Eds.), *Bacterial genomes. Physical structure and analysis* (pp. 153-161). New York: Chapman and Hall.

Romero, D., Martínez-Salazar, J., Girard, L., Brom, S., Dávila, G., Palacios, R., *et al.* (1995). Discrete amplifiable regions (Amplicons) in the symbiotic plasmid of *Rhizobium etli* CFN42. *J. Bacteriol., 177*, 973-980.

Romero, D., and Palacios, R. (1997). Gene amplification and genomic plasticity in prokaryotes. *Annu. Rev. Genet., 31*, 91-111.

Segovia, L., Piñero, D., Palacios, R., and Martínez, E. (1991). Genetic structure of a population of non symbiotic *Rhizobium leguminosarum* bv. phaseoli isolated from soil. *Appl. Environ. Microbiol., 57*, 426-433.

Segovia, L., Young, J. P., and Martínez-Romero, E. (1993). Reclassification of American *Rhizobium leguminosarum* biovar phaseoli type I strains as *Rhizobium etli* sp. nov. *Int. J. Syst. Bacteriol., 43*, 374-377.

Sobral, B. W., Honeycutt, R. J., Atherly, A. G., and McClelland, M. (1991). Electrophoretic separation of the three *Rhizobium meliloti* replicons. *J. Bacteriol., 173*, 5173-5180.

Souza, V., Nguyen, T. T., Hudson, R. R., Pinero, D., and Lenski, R. E. (1992). Hierarchical analysis of linkage disequilibrium in *Rhizobium* populations: Evidence for sex? *Proc. Natl. Acad. Sci. USA, 89*, 8389-8393.

Sullivan, J. T., and Ronson, C. W. (1998). Evolution of rhizobia by acquisition of a 500-kb symbiosis island that integrates into a phe-tRNA gene. *Proc. Natl. Acad. Sci. USA, 95*, 5145-5149.

Young, J. M., Kuykendall, L. D., Martínez-Romero, E., Kerr, A., and Sawada, H. (2001). A revision of *Rhizobium* Frank 1889, with an emended description of the genus, and the inclusion of all species of *Agrobacterium* Conn 1942 and *Allorhizobium undicola* de Lajudie *et al.* 1998 as new combinations: *Rhizobium radiobacter*, *R. rhizogenes*, *R. rubi*, *R. undicola* and *R. vitis*. *Int. J. Syst. Bacteriol., 51*, 89-103.

Wood, D. W., Setubal, J. C., Kaul, R., Monks, D. E., Kitajima, J. P., Okura, V. K., *et al.* (2001). The genome of the natural genetic engineer *Agrobacterium tumefaciens* C58. *Science, 294*, 2317-2323.

CHAPTER 10

THE DAWN OF FUNCTIONAL GENOMICS IN NITROGEN FIXATION RESEARCH

S. ENCARNACIÓN

Programa de Ingeniería Metabólica, Centro de Investigación sobre Fijación de Nitrógeno, Universidad Nacional Autónoma de México, Cuernavaca, Morelos CP62210, México

1. INTRODUCTION

Functional genomics is changing our understanding of biology and our approach to biological research. It brings about concerted high-throughput genetics with analyses of gene transcripts, proteins, and metabolites to answer the ultimate question posed by all genome-sequencing projects: what is the biological function of each and every gene? Functional genomics is stimulating a change in the research paradigm - away from the analysis of single genes, proteins, or metabolites towards the analysis of each of these parameters on a global scale. By identifying and measuring several, if not the entire, molecular group of actors that take part in a given biological process, functional genomics offers the panorama of obtaining a truly holistic representation of life.

Functional genomics is based on high-throughput methods which are not necessarily hypothesis-dependent. It offers insights into mRNA expression, protein expression, protein localization, and protein interactions and may cast light on the flow of information within signaling pathways. At its beginning, biology involved observing nature and experimenting on its isolated parts. Genomic research now generates new types of complex observational data derived from nature. During the early twenty-first century, this new biology will have positive consequences in all fields, including agriculture, and will hasten the development of design-based biological engineering of cells and organisms to perform new functions. This chapter describes the tools that are currently being used for functional-genomics work and considers the impact of this new discipline on nitrogen-fixation research.

Although the earth's atmosphere is 78% N_2, free gaseous N_2 cannot be utilized by either animals or higher plants. Instead, they depend on reduced-nitrogen

R. Palacios and W. E. Newton (eds.), Genomes and Genomics of Nitrogen-fixing Organisms, 143-167.

sources that are present in the soil. To be incorporated by a living organism, nitrogen must be fixed (usually combined with either oxygen or hydrogen) into compounds, such as either nitrate or ammonia, that plants can utilize.

The nitrogenase enzyme complex catalyzes biological nitrogen fixation, a process present only in a relatively few types of prokaryotes. Bacteria containing nitrogenase occupy different ecological niches and include: (i) strict anaerobes, like *Clostridium* spp., *Desulfovibro*, *Desulfotomaculum*, and *Methanococcus*; (ii) phototrophic anaerobes, like *Chromatium*, *Chlorobium*, *Thiopedia*, and *Ectothiospira*; (iii) facultative heterotrophs, which grow aerobically when not fixing N_2, like *Klebsiella*, *Bacillus*, *Enterobacter*, *Citrobacter*, and *Proprionibacterium*; (iv) facultative phototrophs, like *Rhodospirillum* and *Rhodopseudomonas*; (v) microaerophiles, which are normally aerobes when not fixing N_2, like *Mycobacterium*, *Thiobacillus*, *Spirillum*, *Aquaspirillum*, *Methanosinus*, and *Rhizobium*; (vi) microaerophilic phototrophs, such as *Lyngbya*, *Oscillatoria*, *Plectonema*, and *Spirulina*; (vii) aerobic heterotrophs, such as *Azotobacter*, *Azotococcus*, *Azomonas*, *Beijerinckia*, and *Derxia*; and (viii) aerobic phototrophs, like *Anabena*, *Calothrix*, *Gloeocapsa*, *Nostoc*, and other genera of cyanobacteria.

All of these supply fixed nitrogen to the global nitrogen cycle. Due to their fundamental position in the nitrogen cycle, diazotrophs are present in practically all ecosystems, with representatives in environments as varied as aerobic soils (*e.g.*, *Clostridium*, *Azotobacter* and *Azospirillum* species), the ocean surface layer (*e.g.*, *Trichodesmium*), and specialized nodules on legume roots (*e.g.*, *Allorhizobium, Azorhizobium*, *Bradyrhizobium*, *Mesorhizobium*, *Rhizobium, and Sinorhizobium*). In any ecosystem, diazotrophs must respond to varied environmental conditions to regulate the nitrogen-fixation process.

The complete DNA sequence of a number of nitrogen-fixing microorganisms has been obtained and several genomic projects are now in progress (as described in this volume). The genome information from all these nitrogen-fixing organisms will allow researchers to rapidly apply information obtained from genome sequencing to the developing area of functional genomics, which will provide new insights into the complex molecular relationships that underpin both symbiotic and non-symbiotic nitrogen fixation.

The general pictures of biological function revealed by coding sequences are suggestive but not conclusive. The next great goals in the biology of nitrogen-fixing bacteria is to understand the function of the encoded proteins, to reconstruct the metabolic pathways in which these participate, and to determine when and under what conditions these proteins are actively expressed.

2. FUNCTIONAL GENOMICS - THE ROLE OF GENE-EXPRESSION STUDIES

We are witnessing a remarkable change in the scale of molecular microbiological research and are entering an era of integrative 'genomic science'. In the past decade, we have moved from the time when entire research papers were based on the sequencing of either a single gene or operon to single papers describing the sequence of a whole genome. The number of completed microbial genomes continues to increase and the availability of this level of genetic information has spawned the terms 'functional genomics', 'transcriptomics' (Schena *et al.*, 1996; Velculescu *et al.*, 1997), and 'proteomics' (Dutt, Lee 2000; Quadroni, James 1999; Wasinger *et al.*, 1995), which describe the large-scale application of mass mutagenesis, gene-expression profiling, and global protein analysis. However, there has also been a concern that this genome-wide approach might signal a move towards 'non-hypothesis-driven' or 'data-driven' science (Brent 1999), a term that has been used rather pejoratively. It is clear that scientific inference uses a combination of both deductive and inductive reasoning (Kell and King, 2000). Functional genomics allows us to make new and unexpected links between the function of unrelated and hitherto uncharacterized genes and to suggest hypotheses that must subsequently be tested by more traditional methods of molecular genetics and biochemistry (Hughes, 2000). An example lies with the rising numbers of proteins that have unexpected dual cellular functions, such as aconitases, which, in addition to their catalytic role in the TCA cycle, have been shown to act as a post-transcriptional regulators by binding mRNA (Tang and Guest, 1999). Genomic-scale research may be termed 'non-hypothesis-driven' science, but we suggest it should be viewed positively because it is likely to reveal the function of many genes that have been missed by more conventional approaches. The need for this approach is apparent when we consider that the genome sequences of different organisms still contain a large amount of FUN (*i.e.*, function unknown) genes, which remain to be functionally characterized (Hinton, 1997). The role of FUN genes will not be discovered without the application of functional genomic technologies combined with creative experiments.

A new technology, called 'DNAarray', has attracted tremendous interest among biologists. This technology promises to monitor the expression of a whole genome in a single experiment so that researchers can have a better picture of the interactions among thousands of genes simultaneously. On the other hand, the understanding of probably 10-fold more proteins than genes in a organism is still a long way away, and the hard work to unravel the complexity of biological systems is yet to come. A new fundamental concept called proteome (PROTEin complement of a genOME) has recently emerged that should drastically help to unravel biochemical and physiological mechanisms at the functional molecular level (Dutt and Lee, 2000). A new discipline, "proteomics", has been initiated that complements structural genomic research. Proteomics can be defined as "the qualitative and quantitative comparisons of proteomes under different conditions to further unravel biological processes".

3. THE TRANSCRIPTOME

The development of arrays of either DNA or oligonucleotides offers a tool for the analysis of the expression of the mRNA under specific cellular conditions. DNAarrays allow us to produce a "signature" or "gene-expression profile" for a particular organism under defined environmental conditions. Since the first reports of DNAarray technology in 1995 (DeRisi *et al.*, 1996; Lockhart *et al.*, 1996; Schena *et al.*, 1995), the potential of DNAarrays has definitely captured the minds of biologists. The term DNAarray essentially includes a mixture of technologies that share common characteristics. In general, it refers to high-density arrangements of DNA fragments that represent gene sequences obtained from various sources, including either those previously identified in public databases or those from expressed-sequence tags, called ESTs, which are randomly selected clones sequenced from cDNA libraries, bonded to a structural support. These techniques offer extreme flexibility because targeted arrays can be designed to suit specific needs and applications. The methodology is based on the principle that complementary sequences of DNA can be used to probe immobilized DNA molecules. The classical Southern and Northern blotting approaches for the recognition of specific DNA and mRNA species (Alwine *et al.*, 1977; Southern 1975) provided the basis for microarray hybridization with fluorescently labeled cDNA. As a replacement for detecting and studying one or a few genes at a time, DNAarrays allow thousands or tens of thousands of specific DNA or RNA sequences to be detected simultaneously in a single experiment.

For DNA microarrays, either presynthesized oligonucleotides or PCR-fragments are immobilized by using a robotic arrayer and capillary printing tips, which can print more than 10,000 cDNAs or 250,000 different oligonucleotides (Wilkins and Gooley, 1997) per square centimeter onto a glass slide.

To measure relative gene expression using cDNA microarrays, RNA is prepared from the two samples to be compared, and labelled cDNA is made by reverse transcription, incorporating any Cy3 (green) or Cy5 (red) fluorescent dye. Then, the two labeled cDNAs are mixed and hybridised to the microarray, and the slide is scanned. In cases where the green Cy3 and red Cy5 signals are overlaid, yellow spots indicate equal intensity for the dyes (Schena, 1995). With the use of image-analysis software, signal intensities are determined for each dye at each component of the array, and the logarithm of the ratio of Cy5 intensity to Cy3 intensity is calculated. Positive log (Cy5/Cy3) ratios indicate relative excess of the transcript in the Cy5-labeled sample, and negative log (Cy5/Cy3) ratios indicate relative excess of the transcript in the Cy3-labeled sample. After several such experiments have been performed, either clustering or some other computational analysis is used to identify the overall gene-expression profile - or lists of up- or down-regulated genes can be used analyze the data set. These techniques are detailed in Chapter 11.

4. TRANSCRIPTOMICS IN NITROGEN-FIXATION RESEARCH.

In the area of symbiotic nitrogen-fixation research, the application of DNA array technology to bacteria has just started. *S. meliloti, K. peneumoniae, Rhizobium* sp NGR234, and *R. etli* are the only nitrogen-fixing bacteria which have been analyzed using DNAarrays. The slow application of DNAarrays to bacterial research is probably due to the fact that bacterial mRNA is less stable than eukaryotic mRNA and it is not polyadenylated, which complicates its purification. The isolation of sufficient, good-quality bacterial mRNA is especially complicated from complex environments, such as nodules, and the relatively few completed genome sequences in nitrogen-fixing organisms also contribute to the low number of projects which involve this methodology. This situation will change in the future because the genome sequencing of more nitrogen-fixing bacteria is in progress. We hope that work in our own and in other laboratories worldwide will soon produce a push of informative data concerning bacterial gene expression.

The first massive approach to transcriptional analyses of entire symbiotic replicons was based on a high-resolution transcriptional analysis of the symbiotic plasmid of Rhizobium sp. NGR234 (Perret *et al.*, 1999) at the Universite de Geneve, which developed methods to study the regulation of bacterial genes during symbiosis. To analyze the transcription of all putative ORFs, the nucleotide sequence of pNGR234a, sequenced previously by Freiberg *et al.*, 1997, was divided into 441 segments designed to represent all coding and intergenic regions. Each of these segments was amplified by the polymerase chain reaction, transferred to filters, and probed with radioactively labeled RNA either extracted from bacterial cultures grown under various experimental conditions or from bacteroids of determinate and indeterminate nodules.

Genes involved in the synthesis of Nod factors were induced rapidly after the addition of flavonoids, whereas others thought to act within the plant (*e.g.*, those encoding the type-III secretion system) responded more slowly. Many insertion (IS) and transposon (Tn)-like sequences were expressed strongly under all conditions tested. In addition, a number of loci other than those known to encode *nif*, *fix, nod*, *noe*, *nol*, and *nod* genes were also transcribed in nodules. A greater diversity of transcripts was found in bacteroids of determinate as opposed to those from indeterminate nodules.

In *S.meliloti*, Batut and colleges (Cabanes *et al.*, 2000) attempted to determine the expression of a significant portion of the genes expressed by this organism. In these experiments, RNA fingerprinting by arbitrarily primed PCR was used to isolated genes regulated during the symbiotic interaction with alfalfa (*Medicago sativa*). Sixteen partial cDNAs, whose corresponding genes were differentially expressed between symbiotic and free-living conditions, were isolated. Thirteen sequences corresponded to genes up-regulated during symbiosis, whereas the remaining three were instead repressed during establishment of the symbiotic interaction. Seven cDNAs corresponded to either known or predicted *fix, nod*, and *nif* genes. Four presented high sequence similarity with genes not yet identified in *S. meliloti*, including genes that encoded a component of the pyruvate dehydrogenase complex, a cell-surface protein component, a copper transporter, and an argininosuccinate lyase. Finally, five cDNAs did not exhibit any similarity with sequences present in databases.

A detailed expression analysis of the nine *nod-nif-fix* genes provided evidence for an unexpected variety of regulatory patterns, most of which had not been described previously. Using nylon macroarrays, Batut and colleagues have explored the potential of a transcriptome approach to dissect the establishment of the *S. meliloti-M.truncatula* symbiosis. They have also probed free-living *S. meliloti* under a variety of environmental conditions, such as microaerobiosis, in the presence of the nod-gene inducer, luteolin (Ampe *et al.*, 2003).

The direct application of microarrays in nitrogen-fixing systems has already begun to receive extensive examination, as evidenced by work like that performed by Dong and colleagues (2001) at the University of Wisconsin with *Klebsiella pneumoniae*, a common diazotrophic endophyte of maize. *K. pneumoniae* and *E. coli* are closely related enteric bacteria. DNA from this strain was hybridized to a microarray containing 96% (n = 4,098) of the annotated open reading frames from *E. coli* K-12 (Richmond *et al.*, 1999). Using a criterion of 55% identity or greater, 3,020 (70%) of the *E. coli* K-12 open reading frames were also found to be present in *K. pneumoniae* strain 342. Approximately 24% (n = 1,030) of the *E. coli* K-12 open reading frames were absent from strain 342. Genes with high identity between the two organisms are those involved in energy metabolism, amino-acid metabolism, fatty-acid metabolism, cofactor synthesis, cell division, DNA replication, transcription, translation, transport, and regulatory proteins. Genes that were less highly conserved included those involved in carbon-compound metabolism, membrane proteins, structural proteins, putative transport proteins, cell processes, such as adaptation and protection, and central intermediary metabolism. Open reading frames of *E. coli* K-12 with either little or no identity in strain 342 included putative regulatory proteins, putative chaperones, surface structure proteins, mobility proteins, putative enzymes, hypothetical proteins, and proteins of unknown function, as well as genes presumed to have been acquired by lateral transfer from sources, such as phage, plasmids, or transposons.

These microarray results were also compared to the genome sequence of *K. pneumoniae* MGH78578, a clinical isolate. Of the 4,290 ORFs in *E. coli* K-12, 3,053 (71%) were found in *K. pneumoniae* MGH78578. Only 183 (4.3%) of the *E. coli* K-12 genes that were missing in *K. pneumoniae* MGH78578 were present in *K. pneumoniae* strain 342 and 113 (2.6%) of the *E. coli* K-12 genes that were missing in *K. pneumoniae* 342 were present in *K. pneumoniae* MGH78578. Nitrogen-fixation genes were present in *K. pneumoniae* strain 342 but absent in both *K. pneumoniae* MGH78578 and *E. coli* K12 (Dong *et al.*, 2001). As shown by this work, the main advantage the microarray approach is that it permits the identification of thousands of genes in an organism without any need for sequencing. The inconvenience is that it indicates only the genes in common between strains of interest, whereas genes that are exclusive to a *Klebsiella* strain remain unknown.

The genome of *S. meliloti* 1021 has been sequenced by an international consortium (Galibert *et al.*, 2001). A comprehensive genome-wide functional analysis of the *S. meliloti* genome is now in progress (see Chapter 11). Other genomic projects that are using transcriptomics analysis include those of *Rhodopseudomonas palustris* (http://www.jgi.doe.gov/), *Mesorhizobium loti*, *B. japonicum* (see Chapter 7) and *R. etli* (see below).

5. TRANSCRIPTOMICS IN PLANTS DURING SYMBIOTIC NITROGEN FIXATION.

Plant-gene DNAarrays have recently been used to study plant-microbe interactions from the viewpoint of the host by determining the effects of the interaction on the mRNA-expression profile of plant cells. Moreover, some groups are using the same approach by comparison of the plant cellular transcriptional signatures in response to bacterial strains carrying well-defined mutations.

Recent advances in plant genomics are sure to accelerate progress in understanding plant-controlled aspects of the legume-rhizobia nitrogen-fixing symbiosis. Udvardi recently used an array of 2,304 cDNA clones derived from nitrogen-fixing nodules of *Lotus japonicus* to detect differences in relative gene-transcript abundance between nodules and uninfected roots (Colebatch *et al.*, 2002; Udvardi, 2002). Transcripts of 83 different genes were found to be more abundant in nodules than in roots. More than 50 of these have never before been identified as nodule-induced in any species. Expression of 36 genes was detected in nodules but not in roots. Several known nodulin genes were included among the nodule-induced genes. Also included were genes involved in sucrose breakdown and glycolysis, CO_2 recycling, and amino-acid synthesis, which are processes that are accelerated in nodules compared with roots. Genes involved in membrane transport, both cell-wall and protein synthesis, and signal transduction and regulation of transcription were also induced in nodules. Genes that may subvert normal plant defense responses, including two genes that encode enzymes involved in detoxification of active oxygen species and one that may block phytoalexin synthesis, were also identified. The data represent a rich source of information for hypothesis building and future exploration of symbiotic nitrogen fixation. The goal is to use high-density arrays of non-redundant ESTs to monitor global changes in *L. japonicus* gene transcription both during nodule development and following the onset of nitrogen fixation in wild type as well as to monitor the effects of specific mutations.

Using *M. truncatula* as a model plant, Endre and colleagues constructed several cDNA libraries representing different time points during nodule initiation. Partial sequencing of ~11,000 clones from these libraries has contributed to the >150,000 ESTs from *M. truncatula* (http://www.medicago.org/MtDB). Their goals are to establish a unigene set of cDNA clones representing the breadth of the transcriptome and to investigate thoroughly the genome-wide patterns of gene expression by using hybridisation of probes to cDNA microarrays. As a first step, they have assembled a pilot array of ~1,000 non-redundant cDNA clones (called the 'kiloclone set') that contains nodulation markers as well as root-specific clones. The kiloclone set also contains clones encoding proteins with putative functions in signal transduction, transcriptional regulation, control of cell division and cell death, pathogen responses, secondary metabolism, and a number of genes of unknown function. The identity of the clones composing the kiloclone set has been verified by resequencing. Hybridization experiments have monitored differences in gene expression during the early steps of the symbiotic process.

In addition to its use in regard to gene expression, the DNAarray methodology can be used to compare the gene content of an organism with a model organism whose nucleotide sequence is known. Furthermore, strain comparison by the hybridization of genomic DNA to microarrays (genomotyping) is a more realistic approach than the whole-genome sequencing of dozens of strains. This high-definition evolutionary picture could explain the strategies used for development and survival of different bacteria strains. Gene-specific microarrays are being used by Jaime Mora at University of Mexico (personal communication) to compare the entire genome of *R. etli* CE3 (of low nitrogen-fixation capability) with the closely related *R. etli CR*652 (of high nitrogen-fixation capability).

6. THE PROTEOME

Many researchers consider the next key landmark to be an overview of the characteristics and activity of every protein that an organism can synthesize in its lifetime; its 'proteome'. Protagonists argue that proteomics is one of the most important of the so-called 'post-genomic' approaches to understanding gene function because it is the proteins expressed by genes that are ultimately responsible for all processes that take place within the cell. Proteomics is a term generally used to describe analysis of the levels of individual proteins. The technology is more complex than that used for measurement of mRNA, but there are major reasons for studying proteomics; these include the fact that levels of expression of proteins are not particularly well-correlated with mRNA levels and because many important regulatory signals involve posttranslational changes in proteins (*e.g.*, phosphorylation and oxidation-reduction changes). The major technology used today involves separation (and quantification) of proteins by 2D gel electrophoresis and rapid identification by mass spectrometry, particularly matrix-assisted laser desorption ionization methods.

The concept of the proteome, like the transcriptome, is fundamentally different from that of the genome because the genome is virtually static and can be well defined for an organism. In contrast, the proteome frequently changes in response to external and internal events. For example, a rhizobial organism will express different proteins, and so have a different proteome, when cultivated in minimal media as compared to being in symbiosis with a plant. In some ways, this situation paints a dark picture for biochemistry. If we cannot yet fully understand how the proteins in the simplest organisms operate, how can we ever hope to understand the functions of the 6,000 to 7,000 polypeptides thought to occur in bacteria, which after post-translational modifications may total more than 70,000? The response to this question is simply that more effort must be placed into the investigation of proteins and, on a larger scale, proteomes.

The overall picture, highlighted by proteomics, will permit cell biologists to start building a complex map of cell function by discovering how changes in one signaling pathway, the cascade of molecular events sparked by a signal, affect other pathways and also how proteins within one signaling pathway interact with each other. This whole representation also allows researchers to look at the multiplicity of factors involved in biological processes, very few of which are caused by a single gene. The field of proteomics has emerged as a robust and global approach to analyze protein expression. There are hundreds of "proteome projects" (see examples in http://au.expasy.org/ch2d/) that impact all areas of biology from agriculture to medical sciences.

Some years ago, a protein chemist would have been happy to identify two or three proteins a year. Now, increasing volumes of genome data, combined with mass-spectrometry technologies, permit a researcher to identify hundreds of proteins within a week. The term `proteomics' was first formalized in 1996, whereas the primary experimental tool to monitor genome-wide protein expression, 2-D electrophoresis (2DE), has been available since 1975 (O'Farrell 1975) and is used to separate proteins by isoelectric focusing in a first dimension and then by their apparent molecular weight in an SDS-PAGE second dimension (Leymarie, 1996; Tsugita *et al.*, 1996). Current methods can resolve as many as 1,500 proteins on a 22 x 24 cm gel so, in principle, it is possible to resolve the entire proteome of an organism like *R. etli* or *S. meliloti* on a single two-dimensional gel. 2-D PAGE is also one of the most efficient and powerful methods for purifying proteins in small quantities (Kamo *et al.*, 1995; Klose and Kobalz, 1995). Immobilized pH gradients (IPG) can now be used for the pH range 3 to 12 and have become the method of choice for isoelectric focusing. IPG gels do not suffer from cathodic drift and focus proteins to equilibrium, thus, providing very high reproducibility (Bjellqvist *et al.*, 1982; Sanchez *et al.*, 1997).

The visualization of the separated proteins by different staining techniques is limited due to the low dynamic range of most staining techniques (Merril *et al.*, 1979). The recent development of fluorescent dyes for proteins (Patton, 2000) may overcome this limitation. Microgram quantities of protein can be studied in an initial experiment to identify proteins whose expression change in a significant manner. Subsequently, the same sample can be applied in milligram quantities to purify individual peptides for amino-acid analysis, mass spectrometry, either amino-terminal or internal amino-acid sequencing, and other techniques (Rouquie *et al.*, 1997; Touzet,*et al.*, 1996; Towbin *et al.*, 1979; Wilkins and Gooley, 1997).

Aebersold *et al.* (1987) showed that amino-terminal and internal protein sequence information could be obtained from 2DE-separated proteins. Direct amino-acid sequencing provided an important connection between proteomic and genomic information - yielding a genetic basis for as yet uncharacterized proteins; however, equipment and reagent costs, as well as the limited sensitivity associated with direct sequencing, have been restrictive. More recently, peptides (from digested proteins obtained from 2DE gels or blots) are characterized by mass spectrometry (MS) (Haynes *et al.*, 1998; Quadroni and James, 1999.).

Among the more impressive recent developments is the use of tandem mass spectrometry (MS) to sequence gel-separated proteins. The two common ionization techniques for proteins are either electrospray (or nanospray for smaller quantities) ionization (ESI) or matrix-assisted laser desorption ionisation (MALDI) (Karas and Hillenkamp 1988; Patterson and Aebersold, 1995). MS has become a powerful, rapid and sensitive tool for the analysis of proteins. An even more sensitive approach is peptide mass fingerprinting; here, low nanogram quantities of proteins are enzymatically cleaved, the peptide masses are determined by mass spectrometry, and then are used for database searches (Aebersold *et al.*, 1987; Bairoch and Apweiler, 1999; Henzel *et al.*, 1993). All mass spectrometers have three components: an ionisation source, such as either MALDI or ESI; a mass analyzer, such as quadrupoles, ion trap, or time-of-flight (TOF) tube; and an ion detector, such as electron multipliers or photomultipliers. The flexibility of combining ionisation sources with mass analysers has brought forward numerous types of MS instruments with varying mass accuracy, sensitivity, and, importantly, applications.

Ions produced in the ion source are separated in the mass analyzer by their mass-to-charge (*m*/*z*) ratio. MS data are recorded as "spectra", which display ion intensity versus the *m*/*z* value. The two techniques that have become preferred methods for ionization of peptides and proteins are ESI and MALDI, due to their effective application on a wide range of proteins and peptides (Fenn *et al.*, 1989; Karas *et al.*, 1989; Porubleva *et al.*, 2001). In general, MALDI/time-of-flight MS is more effective for the analysis of higher molecular-weight proteins, whereas ESI/ion-trap MS offers better sensitivity of detection, down to the femtomole level.

A key feature of MS analysis of gel-separated proteins and peptides is the ability to generate different types of structural information about a particular peptide of interest. For example, the mass spectrometer can directly provide information on the mass of a particular peptide but can also generate *de novo* amino-acid sequence information from the tandem mass spectra obtained either by post-source decay or collision-induced dissociation (as described in Keough *et al.*, 1999). A new approach to post-source decay matrix-assisted laser desorption ionization mass spectrometry provides easy to interpret ion-mass spectra (Southern 1975). After proteins in two-dimensional gels are identified, studies can be initiated to evaluate post-translational modifications. Recently, tandem MS has been used to identify proteins in macromolecular complexes (Link *et al.*, 1999).

There is, however, a limit to how many proteins a single gel can separate and the sensitivity is still not adequate to detect proteins appearing at very low levels. Some classes of protein, particularly hydrophobic membrane-bound proteins, will not run on 2D gels. On the other hand, the 2D gel is the only method that can resolve large numbers of proteins in a quantitative way.

7. PROTEOMICS AND NITROGEN-FIXATION RESEARCH

Differential proteome analysis compares the expression profile of 2DE-separated proteins from an arbitrary reference state of a cell (or organism) to the profile of a non-standard condition, such as free-living with either symbiosis or following the addition of an inducer to the system. The differences between two proteomes give an indication of the response of the system to the perturbations. There are an increasing variety of applications of differential proteome analysis to interesting problems.

A pioneer proteome study on nitrogen-fixing bacteria was undertaken five years ago by Guerreiro and colleagues who used derivatives of *R. leguminosarum* biovar *trifolii* strain ANU843, which had been cured of indigenous plasmids, to investigate plasmid-encoded functions (Guerreiro *et al.*, 1998). The level of synthesis of thirty-nine proteins was affected after curing of plasmid A. The differences observed upon plasmid curing included: protein loss, either up- or down-regulation of specific proteins, and the novel synthesis of new proteins. The results suggested that normally a complex interplay between the cured plasmid and the remaining replicons occurred. Twenty-two proteins appeared to be absent in the cured strains and plasmid-borne genes presumably encoded these. Of these, a small heat-shock protein, a cold-shock protein, a hypothetical protein, and the alpha and beta subunits of the electron transfer flavoprotein were identified by *N*-terminal microsequencing and were predicted to be encoded on plasmid e. Four of the sequenced proteins that were putatively encoded on plasmid e and two others encoded on plasmid c were novel. In addition, curing of plasmid e and c consistently decreased the levels of 3-isopropylmalate dehydratase and malate dehydrogenase, respectively, suggesting that levels of these proteins may be influenced by plasmid-encoded functions. A protein with homology to 4-oxalocrotonate tautomerase, which is involved in the biodegradation of phenolic compounds, was found to be newly synthesized in the strain cured of plasmid e. Proteome analysis provides a sensitive tool to examine the functional organization of the *Rhizobium* genome and the global gene interactions that occur among the different replicons.

S. meliloti was also analysed for the contribution of plasmid-encoded functions to the intracellular regulation of this bacterium (Chen *et al.*, 2000). Protein profiles of strain 2011 were compared with those from mutant strains, which were either cured of pRme2011a (pSyma) or contained an extensive deletion of this plasmid (strain SmA146). pSyma is 1.4 Mbp with an estimated coding potential of 1,400 proteins and contains both the nodulation and nitrogen-fixation genes. Under the growth conditions used, they detected 60 differences between the parent strain and its pSyma-cured derivative. Although the majority of these differences were due to regulatory changes, such as up- and down-regulation, some proteins were totally missing in some strains. The 60 proteins were classified into 21 subgroups, based on their protein levels when the cells were grown in the presence or absence of luteolin. Comparisons were made between the different strains to assess the possible interactions of the different proteins of the subgroups and plasmid pSyma. The results suggested that pSyma has a role in the regulation of the expression of genes from the other replicons (3.5 Mbp chromosome and the 1.7 Mbp pSymB plasmid) present in the *S. meliloti*.

Because cell cycle and environmental changes are multi-factorial, involving many metabolic changes, it is difficult to unravel the role played by specific global gene regulators. The definition of important regulons by the use of appropriate regulatory mutants provides the framework for a better understanding of complex cellular responses. This approach has led to the global characterization of AniA (*R. etli*), NolR (*S. meliloti*), and NifA and FixK2 (*B. japonicum*). Our group is currently analysing the role of NiFA, OxyR, FnR, FixK, NtrC, and others in *R. etli* when in either the free-living or symbiotic state.

Proteome analysis has yielded clues to carbon flux in *R. etli* based on the analysis of a regulon, AniA (Encarnación *et al.*, 2002). We reported previously that the oxidative capacity and ability to grow on carbon sources, such as pyruvate and glucose, were severely diminished in a *Rhizobium etli phaC* mutant strain, which is unable to synthesize poly-ß-hydroxybutyric acid (PHB) (Cevallos *et al.*, 1996). By random Tn*5* mutagenesis of the *phaC* strain, we isolated mutants containing single Tn*5* insertions that had recovered the ability to grow on either pyruvate or glucose (VEM58). Nucleotide sequencing of the region surrounding the Tn*5* insertions showed that they had interrupted an open reading frame that was designated as *aniA*. An *aniA*::Tn*5* mutant (VEM5854) was constructed and found to synthesize only 40% the wild-type level of PHB. Both VEM58 and VEM5854 produced significantly more extracellular polysaccharide than the wild type. Both organic-acid excretion and levels of intracellular reduced nucleotides were lowered to wild-type levels in VEM58 and VEM5854, in contrast to those of strain CAR1, which were significantly elevated. Proteome analysis of VEM58 showed a drastic alteration of protein expression, including the absence of a protein identified as PhaB. We propose that the *aniA* gene product plays an important role in directing carbon flow in *R. etli*. Clearly, the fact that AniA modulates the pattern of protein synthesis, carbon flux, and energy is extremely important in metabolism, as demonstrated by the fact that cells lacking AniA failed to produce 795 proteins, including the *phaB*-gene product.

Another regulon analysed with this methodology was described by Rolfe and colleagues, who identified NolR-regulated proteins in *S. meliloti* (Cevallos *et al.*, 1996). Analysis of Coomassie-stained preparative two-dimensional (2D) gels showed that at least 52 of the altered proteins could be reproducibly detected and isolated from a *nolR* mutant. These 52 altered proteins were classified into five groups based on both abundance and the effect of the presence (or absence) of luteolin addition in the growth medium. *N*-terminal microsequencing of 38 proteins revealed that the most striking feature of the *nolR* mutation was the number and broad spectrum of cellular functions that were affected by its loss. These included proteins involved in the tricarboxylic acid (TCA) cycle, heat shock and cold shock functions, translation elongation, oxidative stress, and cell growth and maintenance. They proposed that the NolR repressor is a global regulatory protein that responds to environmental factors.

Dainese-Hatt and colleagues (1999) used this approach to study the regulation of microaerobically and anaerobically induced genes, which included genes involved in nitrogen fixation, in the symbiotic bacterium, *B. japonicum*. Their results showed that, in addition to the two known regulons controlled by the transcription factors NifA and FixK2, a third set of proteins may exist in *B. japonicum*, which are induced by anaerobic conditions and are regulated independently. Further, 19 heat shock proteins (Hsp) were induced when *B. japonicum*, was shifted from 28°C to 43°C (Munchbach *et al.*, 1999). Up-regulated proteins included the small Hsp (sHsp; HspB, C, D, E, and H) and three others with strong sequence similarity to the sHsp family. Two other low molecular-mass proteins, which corresponded to GroES1 and GroES2, and five novel proteins were found. Four proteins of approximately 60 kDa were identified as GroEL2, GroEL4, GroEL5, and DnaK. An analysis of the heat-shock induction of DnaK, of four of the most strongly induced GroESL proteins, and six of the sHsp revealed that the proteins could be placed into four distinct regulatory groups based on the kinetics of protein appearance.

Similar to regulon analysis, this methodology was used to described stimulons proposed by Rolfe and colleagues in *R. leguminosarum* strains, where protein-expression profiles in response to specific genetic perturbations in exopolysaccharide (EPS) biosynthesis genes were examined, using two-dimensional gel electrophoresis (Guerreiro *et al.*, 2000). Lesions in either *pssA*, *pssD*, or *pssE* of *R. leguminosarum* bv. viciae VF39 or in *pssA* of *R. leguminosarum* bv. trifolii ANU794 not only abolished the capacity of these strains to synthesize EPS but also had a pleiotropic effect on protein-synthesis levels. Twenty-two protein differences were observed for the two-*pssA* mutant strains. The differences identified in the *pssD* and *pssE* mutants of strain VF39 were a distinct subset of the protein synthesis changes that occurred in the *pssA* mutant. The *pssD* and *pssE* mutant strains shared identical alterations in the proteins synthesized, suggesting that they share a common function in the biosynthesis of EPS. In contrast, a *pssC* mutant that produces 38% as much EPS as the parental strain showed no differences in its protein-synthesis patterns, suggesting that the absence of EPS was contributing to the changes in protein synthesis and that there may be a complex interconnection of the EPS-biosynthetic pathway with other metabolic pathways. Genetic complementtation of *pssA* restored wild-type protein-synthesis levels, indicating that many of the observed differences in protein synthesis were also a specific response to a non-functional PssA. It is evident that enzymatic pathways and regulatory networks are more interconnected and more sensitive to changes in the cell than is often appreciated. In these cases, linking the observed phenotype directly to the mutated gene can be misleading because the phenotype could be attributable to downstream effects of the mutation.

The differential study of *Sinorhizobium meliloti* highlights the application of proteomics in identifying markers during both early and late exponential-phase growth (Guerreiro *et al.*, 1999). Changes in gene expression of cells grown in a defined medium occur at different growth phases. Fifty-two reproducible changes in protein-expression levels were detected when early exponential-phase cells were compared to late exponential-phase cells. *N*-terminal microsequencing of eighteen constitutive proteins along with four more proteins, induced in late exponential phase, allowed the identification of proteins not previously described in rhizobia. As a follow-up to experiments with plant root exudates, *R. leguminosarum* bv. trifolii strain ANU843 was treated with the flavonoid, 7,4'-dihydroxyflavone (Guerreiro *et al.*, 1997), and proteome analysis of the flavonoid-treated cells revealed that the global expression pattern of proteins was largely unaltered by the treatment. Four inducible proteins and 20 constitutively expressed proteins were subjected to sequence analysis and the identity of 12 proteins was established. NodE was present during all phases of growth but was diminished in stationary phase cells, whereas NodB was not detected in the later stages of growth.

Djordjevic and colleagues, using the symbiosis between *S. meliloti* and *Melilotus alba* (Natera *et al.*, 2000), characterized novel symbiosis proteins and determined how the two symbiotic partners alter their respective metabolisms as part of the interaction. Proteome maps from control *M. alba* roots, wild-type nodules, and *S. meliloti* cultures and bacteroids were generated and compared. They found that over 250 proteins were up-regulated in the nodule as compared with the root and over 350 proteins were down-regulated in the bacteroid form of the rhizobia as compared with cultured cells. They identified nearly 100 nodule, bacterial, and bacteroid proteins by using *N*-terminal amino-acid sequencing and mass-fingerprint analysis, in conjunction with data-base searching, including leghemoglobin and NifH as well as proteins involved in nitrogen and carbon metabolism. Bacteroid cells showed down-regulation of several proteins involved in nitrogen acquisition, including glutamine synthetase, urease, a urea-amide binding protein, and a P_{II} isoform, indicating that the bacteroids were nitrogen proficient. The down-regulation of several enzymes involved in polyhydroxybutyrate synthesis and cell division were also observed.

This same group, using *R. leguminosarum* bv. trifolii, also recognized proteins that are implicated in the early stages of nodulation between strains ANU843 and ANU794 and the subterranean clover cultivar, Woogenellup (Morris and Djordjevic, 2001). On the roots of cv. Woogenellup, strain ANU843 induces nitrogen-fixing nodules, whereas strain ANU794 forms abnormal nodules, which fail to develop beyond an early stage. Proteome maps from control and inoculated roots were generated and compared at 24-h and 48-h post inoculation. The 16 proteins that were differentially regulated included an alpha-fucosidase, several ethylene-induced proteins, a Cu/Zn superoxide dismutase, a hypothetical 16.5 kDa protein, tubulin alpha-chain, chaperonin 21-precursor, and triosephosphate isomerase. The 22 constitutively expressed proteins included several pathogenesis and stress-related proteins. These results might suggest that ethylene levels are up-regulated during the early stages of infection but that this does not result in the induction of common pathogenesis-related proteins. Further, the specific induction of alpha-fucosidase by ANU794 might be important in the nodulation-failure phenotype.

8. PROTEOMICS IN PLANTS DURING SYMBIOTIC NITROGEN FIXATION

The legume-*Rhizobium* symbiosis leads to the formation of a new compartment in the plant cell, the symbiosome (Saalbach *et al.*, 2002). This compartment harbours the bacteroids surrounded by a peribacteroid membrane (PBM) that originates from the plant plasma membrane. The PBM and the space between the PBM and the bacteroid membrane, called peribacteroid space (PS), mediate the exchange of metabolites between the symbionts. One of the focal points for proteomics in legumes is the peribacteroid membrane that separates the microsymbionts from the plant cytoplasm. Proteins bound to this symbiotic membrane interface are expected to control the traffic of nutrients and signals between the symbionts. Proteins in the PBM and PS fractions obtained from symbiosomes were separated by two-dimensional gel electrophoresis and 89 spots were analysed by tandem mass spectrometry. Interestingly, endomembrane proteins, including V-ATPase, BIP, and an integral membrane protein known from COPI-coated vesicles, were found in the PBM fraction, supporting the role of the endomembrane system in PBM biogenesis.

Other researchers identified 17 putative PBM proteins, six of these were homologous to proteins of known function (Panter *et al.*, 2000). These included chaperones and serine and thiol proteases, all of which are involved in some aspect of protein processing in plants. The PBM homologs of these proteins may play roles in protein translocation, folding, maturation, or degradation in symbiosomes. Two proteins were homologous to known nodule-specific proteins from soybean; nodulin 53b and nodulin 26B. Although the function of these nodulins is unknown, nodulin 53b has independently been shown to be associated with the PBM. All of the eight proteins with identifiable homologs are likely to be peripheral rather than integral membrane proteins. The identification of homologs of HSP70 and HSP60 associated with the PBM is the first evidence that the molecular machinery for co- or post-translational import of cytoplasmic proteins is present in symbiosomes. This observation has important implications for the biogenesis of this unique nitrogen-fixing organelle.

In *M. truncatula*, Mathesius *et al.*, (2001) have established a proteome reference map for root proteins using two-dimensional gel electrophoresis combined with peptide mass fingerprinting to aid in the dissection of nodulation and root developmental pathways. *M. truncatula* has been chosen as a model legume for the study of nodulation-related genes and proteins. From 2,500 proteins displayed on 2D gels, they analysed 485 proteins by peptide mass fingerprinting, and 179 of those were identified by matching against the current *M. truncatula* expressed sequence tag (EST) database, which contains DNA sequences of approximately 105,000 ESTs. The majority of proteins identified were metabolic enzymes and stress-response proteins, but two nodulins were also identified in uninoculated root tissue, supporting other evidence for a role of nodulins in normal plant development.

9. PROTEOMICS IN CONCERT WITH TRANSCRIPTOMICS

Biological systems are comprised of protein components found at a wide variety of abundances, from millions of molecules of a single species per cell to less than one copy per cell. Because of this wide range of concentrations, a full accounting of all proteins in a cell is presently unavailable. Two-dimensional gel electrophoresis permits the separation and detection of many, but not all, protein species and as yet undetected proteins may, in fact, constitute the majority of the proteome. Conventional separation and analytical methods (two-dimensional gel electrophoresis and mass spectrometry) allow identification and quantification of many of the most abundant gene products, however, the greater part of the gene products (found at low abundance) can be neither identified nor measured in complex mixtures at present. The gene products that are found at low levels can be characterized by transcriptomics, which can yield important biological information about what genes are turned on and when. It has the disadvantage that, although the snapshot it provides will reflects the genome's strategy for protein synthesis, it does not represent the realization of those plans, unlike protein-based analysis, which has the ability to show post-transcriptional control as well as post-translational modifications of proteins (Kaufmann *et al.*, 1999; Nyman, 2001).

Experimental evidence clearly shows a disparity between the relative expression levels of mRNA and their corresponding proteins (Anderson and Seilhamer, 1997; Gygi *et al.*, 1999). Additionally, it has been proven mathematically that expression information from both mRNA and proteins is required to understand a gene network (Hatzimanikatis and Lee, 1999). A nonlinear stability analysis shows that a combination of gene-expression information at both the message level and at the protein level is required to describe even simple models of gene networks. A further informatics challenge is to establish effective connections between protein level and nucleic-acid level information about genes and gene networks.

10. GLOBAL APPROACHES TO STUDY THE *R. ETLI -P.VULGARIS* INTERACTION

At the Nitrogen Fixation Research Center of the National University of Mexico, we are developing an integrated program, including genomics, proteomics, and transcriptomics to aid in our understanding of symbiotic nitrogen fixation. These approaches have been used to study *R. etli* in its free-living state and during its symbiotic interaction with *P. vulgaris*. Combining the strengths and advantages of both the genome-based global approaches and conventional tools will allow us to attain a global picture of this biological interaction.

Our initial goal is gene products that are differentially present between the symbiotic and non-symbiotic states. Proteome maps from *R. etli* cultures and bacteroids have been generated and compared. Using proteome analysis, we have identified proteins that are essential for *Rhizobium* during nitrogen fixation in association with legume plants. To discover bacterial patterns of proteins involved in the infection and differentiation stages of symbiosis, we obtained 2D-gels of proteins expressed at the appropriate time points in the nodule. The protein maps of the nodule-expressed proteins will help us to understand the complex metabolism during nitrogen fixation in *R. etli*.

Over 300 proteins were either induced or up-regulated in the bacteroid compared with the free living bacteria and over 400 proteins were down-regulated in the bacteroid form of the rhizobia as compared with cultured cells. *N*-terminal amino-acid sequencing, in conjunction with database searching, were used to assign a putative identity to nearly 80 bacteroid proteins. These included the previously identified nodule proteins, GroEL and NifH, as well as proteins involved in carbon and nitrogen metabolism in *R. etli*. The down-regulation of a cell-division protein was also observed.

When free-living, *R. etli* undergoes a transition from an aerobic to a fermentative metabolism during successive subcultures in minimal medium. This metabolic transition does not occur in cells sub-cultured in either rich medium or in minimal medium containing either biotin or thiamine (Encarnación *et al.*, 1995). We characterized the aerobic and fermentative metabolism of *R. etli* using proteome analysis. According to their synthesis patterns in response to aerobic (rich medium, minimal medium with biotin or minimal medium with thiamine) or fermentative (minimal medium without supplements) growth conditions, proteins were assigned to five different classes: (i) proteins produced only in aereobic conditions, *e.g.*, catalase-peroxidase KatG and the E2 component of pyruvate dehydrogenase; (ii) proteins produced under both conditions but strongly induced in aerobic metabolism, *e.g.*, malate dehydrogenase and the succinyl-CoA synthetase α subunit; (iii) proteins that were induced equally under all conditions tested, *e.g.*, AniA, DnaK, and GroEL; (iv) proteins down-regulated during aerobic metabolism; and (v) proteins specific to only one of the conditions analyzed. Northern-blotting studies of *katG* expression confirmed the proteome data for this protein. The negative regulation of carbon metabolism proteins observed in fermentative metabolism is consistent with the drastic physiological changes which occur during this process (Encarnación *et al.*, 2003).

The metabolic alteration observed during the change from aerobic to fermentative growth is similar to that observed in *R. etli* when these bacteria undergo differentiation prior to nitrogen fixation in symbiosis with legume-plants. We proposed that the fermentative metabolism, which occurs when free living is similar to the physiological condition observed in the bacteroids during symbiosis. During the fermentative-like response, the bacteria grow less, increase their size and become pleiomorphic, poly-β-hydroxybutyrate is accumulated, and a symbiotic oxidase is induced just as occurs during symbiotic nitrogen fixation (Encarnacion *et al.*, unpublished data).

To investigate also the expression of symbiotic functions in free-living conditions, we have grown *R. etli* in minimal medium without a nitrogen source and analyzed the genome expression with proteome methodology. Differentially expressed proteins, which were also detected in 2D gels in fermentative metabolism, were identified. Surprisingly, some of the proteins expressed by *R. etli* under fermentative conditions of growth were found to be present in the differentiated bacteroid stage of *R. etli* (nitrogen-fixation condition), suggesting that the fermentative metabolism may be related to the first stages of the establishment of the symbiotic relationship between *R. etli* and the plant (*Phaseolus vulgaris*). We are currently identifying the proteins by mass spectrometry, using both electrospray ionization and MALDI.

The sequencing of the *R. etli* genome is in progress and we already have the complete sequence of one replicon, pSym (González *et al.*, 2003; see Chapter 9). This replicon contains 359 genes, which are mainly involved in symbiotic processes. With this sequence, we constructed two microarrays, the first of which contains 372 PCR products and represents, kilobase by kilobase, the coding and non-coding regions. The second microarray was design to contain open reading frames. With those microarrays, we have recently performed time-course studies of the *R. etli-P. vulgaris* symbiosis during nitrogen fixation. Symbiotic plasmid gene expression at various time points during nitrogen fixation with *P. vulgaris* was compared by quantification of hybridization signals, background subtraction, and calculation of normalized intensity values of the individual spots, as described by Eymann *et al.* (2002). Expression level ratios of two or greater that were measured in two independent experiments were considered significant. Final evaluation of the macroarray data included the consideration of putative operons, which were derived from the symbiotic plasmid sequence, using (http://www.cifn.unam.mx/retlidb/), as well as previously known transcriptional units. An ontology (gene functional classification) analysis revealed gene-expression patterns of different subsets of genes within the same functional class. Coordinated up-regulation of specific mRNAs clustered into functional groups. Nearly all genes showed some regulation over the course of development. These data will help to identify genes with a critical role during the nitrogen-fixing interaction between *R. etli* and *P. vulgaris*.

In addition to expression profiling at the transcriptome level, protein patterns of certain *R. etli* regulatory mutants were compared by high-resolution 2D gel electrophoresis to learn more about the global effects of these mutation during symbiosis and in the free-living state. With this combination of functional genomics tools (proteomics and transcriptomics), we intend to elucidate the genome expression from *R .etli* in both the free-living and symbiotic states.

11. PROTEIN-PROTEIN INTERACTIONS: APPLICATIONS OF MODULAR MAPS

An emerging paradigm in protein biology involves the systematic identification of proteins that interact with each other at a "biological level" (Lueking *et al.*, 1999; Lueking *et al.*, 2001; MacBeath and Schreiber, 2000; Rain *et al.*, 2001). Proteins can be defined in the context of their interactions to provide a unique characterization of associations within whole proteomes. Protein domains and their interactions and associations increase the complexity of higher organisms. Many biological regulatory events result through protein-protein interactions (Schwikowski *et al.*, 2000, Walhout and Vidal, 2001; Walhout *et al.*, 2000; Walter *et al.*, 2000). Establishing protein-interaction maps is still difficult but not impossible. The complexity of this task can be appreciated by the *ca.* 300,000 interactions possible within a single yeast cell (Bader and Hogue, 2002; Fromont-Racine *et al.*, 1997; Gygi *et al.*, 1999; von Mering *et al.*, 2002).

Modular proteomics relies on many technologies because of (i) the relatively low physiological concentration of proteins inside the cells and (ii) the difficulty of attaining a multi-protein state in its "biological form". Classical methods for identifying such interactions include immunoprecipitation, cross-linking, and pull-down strategies. However, for some proteins, these methods are not effective, not available, or not sufficiently sensitive. Furthermore, it is not feasible to use low-throughput techniques to identify potentially interacting partners in entire proteomes. Alternatively, high-density protein filters and microarrays offer the required high-throughput capability for whole-genome analyses.

To study protein-protein interactions, we are employing the combination of techniques previously used to identify proteins and peptides (Fields and Song, 1989; Puig *et al.*, 2001; Rigaut *et al.*, 1999), which includes the use of antibodies specific for proteins, 2D chromatography and MS for the direct analysis of protein complexes from mixtures, to characterize the role of chaperons, such as DnaK-DnaJ and GroEL-GroES. We are also starting the analysis of protein complexes by characterization of a tricarboxylic acid cycle metabolon. Here, we are using methods involving "co-precipitation/mass spectrometry", the bait proteins for which are purified on an affinity column and then separated by SDS-PAGE.

12. TRANCRIPTOMICS, PROTEOMICS AND BIOINFORMATICS

Genomics coupled with proteomics and trasncriptomics are now important high throughput techniques for qualifying and analyzing both gene and protein expression, discovering new gene or protein products, and understanding gene and protein functions. The microarray technology has made it possible to collect vast amounts of data on gene expression. The bottleneck within transcriptomics thereby lies in analyzing and managing data. One solution to this problem is to use software that digitizes microarray images, making it possible to let the computer search for patterns and trends in the data collected. Software can also assist in the visualization of gene expression in cells and making comparisons between levels of expression in different metabolic conditions.

Although many bioinformatics companies sell software, which assists in microarray analysis, there are several freely available software packages that can be used to perform the analytical techniques. A standard software for hierarchical clustering is Cluster and TreeView (http://rana.lbl.gov/EisenSoftware.htm), which also creates self-organizing maps and performs principal-components analysis. GeneCluster 2.0 was initially used for constructing self-organizing maps (http://www-genome.wi.mit.edu/cancer/software/genecluster2/gc2.html), now the latest version now also finds nearest neighbours and performs other supervised methods. MultiExpression Viewer (http://www.tigr.org/software/) is software that creates self-organizing maps and performs hierarchical clustering as well as finding principal components. This package also includes a component for support vector machines. MAExplorer (http://maexplorer.sourceforge.net) is a tool that performs many aspects of microarray processing, including the raw image analysis. It contains a few analytical techniques, including hierarchical clustering. RELNET (http://www.chip.org/relnet) and other software packages, like GETools software (http://www.cifn.unam.mx/Computational_Genomics/GETools/), which is a set of integrated programs linked to RegulonDB, uses as input a set of genes and their expression values from a transcriptome experiment and then extracts all information associated with them from RegulonDB, which is a database on transcriptional regulation and operon organization in *Escherichia coli* K-12 (http://www.cifn.unam.mx/Computational_Genomics/regulondb/). In this way, GETools provides ways to compare a microarray dataset directly with approximately 670 mapped promoters, 300 operons, 165 transcriptional regulators and 400 regulatory sites.

The following web sites also contain large amounts of microarray data that are of good quality and are freely available for academic use: Stanford Microarray Database (http://genome-www5.stanford.edu/MicroArray/SMD/) with 3,290 microarrays measured across 11 species, from 80 publications; and the National Center for Biotechnology Information Gene Expression Omnibus (http://www.ncbi.nlm.nih.gov/geo/) with 2,354 microarrays from 105 types of microarray, measured across 78 experiments.

Bioinformatics is also indispensable for the examination of the data obtained in proteome analysis. An excellent resource of Internet-accessible proteome databases is the Expert Protein Analysis System (ExPASy), available online at http://www.expasy.ch/. Furthermore, there is the desire to develop software packages that can take multiple protein-expression profiles and automatically identify quantitative changes of interest (Melanie 3 2D Gel Analysis Software on World Wide Web URL: http://www.expasy.ch/melanie/). The key features of Melanie 3 include the ability to perform multivariate statistics on datasets, make comparisons with online databases, and interface with a MS spectrum. A set of proteomic tools (Wilkins *et al.*, 1998a; Wilkins *et al.*, 1998b) is available at the

ExPASy website. The Swiss Institute for Bioinformatics maintains this database and has released a summary of the protein sequence data bank (Bairoch and Apweiler, 1999).

Numerous proteome projects are now in progress, resulting in the generation of two-dimensional electrophoresis databases that are accessible on the Internet and can be browsed with interactive software and integrated with in-house results. Clusters of Orthologous Groups of proteins (COG) is a new database search and represents an attempt at a phylogenetic classification of proteins from complete genomes (http://www.ncbi.nlm.nih.gov/COG). It is to serve as a platform for functional annotation of newly sequenced genomes and for studies on genome evolution. Proteome Analysis @ EBI database (http://www.ebi.ac.uk/proteome) from SWISS-PROT/TrEMBL contains complete non-redundant proteome sets that are constructed by selecting entries from SPTR, which is a comprehensive protein sequence database consisting of SWISS-PROT, TrEMBL, and TrEMBLnew (Arenkov *et al.*, 2000).

The construction of databases from nitrogen-fixing organisms that will be able to correlate genes identified from cDNA microarrays with their corresponding protein functions is an important goal to be achieved. The first protein map constructed for a nitrogen-fixing organism was developed by the Genomic Interaction Group at the Australian National University and was called ANU-2DPAGE (http://semele.anu.edu.au/2d/2d.html). The proteome maps of *S. meliloti* and *Medicago truncatula* will be updated continuously and will be a powerful tool for investigating the molecular mechanisms of root symbioses in legumes.

Post-translational modification of proteins is important for the regulation in many cellular processes (Kaufmann *et al.*, 1999; Oda *et al.*, 1999), including recognition, signaling, targeting and metabolism. The post-translational modification databases and resources can be found on the Internet at websites: *O*-GlycBase, which is the *O*-glycosylated protein database (http://www.cbs.dtu.dk/databases/OGLYCBASE); PhosphoBase, which is the phosphorylation site database (http://www.cbs.dtu.dk/database/PhosphoBase); GlycoSuiteDB, which forms a database of glycan structures (http://www.glycosuite.com); RESID, which is the database of amino-acid modifications (http://www-nbrf.georgetown.edu/pirwww/dbinfo/resid.html); and DSDBASE, which forms the disulfide database derived from 3D data (http://www.ncbs.res.in/~faculty/mini/dsdbase/dsdbase.html).

13. CONCLUSIONS

The genomic sequences now accessible and those now on-going, along with DNAarrays and proteome analysis, will provide a powerful set of tools to study the functional genomics of many nitrogen-fixing species. A unification of genomics, proteomics, and trancriptomics technologies is needed if we are to start to understand the complexity of biological function in nitrogen fixation. The high-throughput technologies of functional genomics are opening our eyes to the true complexity of cellular differentiation that underpins symbiotic nitrogen fixation. The detailed understanding of cellular function offered by proteomics and transcriptomics analysis, under both free-living and nitrogen-fixing conditions, has potentially staggering implications for nitrogen-fixation research. Integration of the results of transcriptome and proteome analyses will undoubtedly lead to a better understanding and increased efficiency of nitrogen-fixing symbioses. Eventually, this understanding may possibly lead to an expansion of the symbiotic fixation of nitrogen to agriculturally essential non-legumes.

ACKNOWLEDGEMENTS

This manuscript is dedicated to Jaime Mora. We gratefully thank Michael Dunn for critically reviewing the manuscript.

REFERENCES

Aebersold, R., Leavitt, J., Saavedra, R., Hood, L., and Kent, S. (1987). Internal amino acid sequence analysis of proteins separated by one- or two- dimensional electrophoresis after "*in situ*" protease digestion on nitrocellulose. *Proc. Natl. Acad. Sci. USA, 84*, 6970-6974.

Alwine, J., C., Kemp, D. J., and Stark, G. R. (1977). Method for detection of specific RNAs in agarose gels by transfer to diazobenzyloxymethyl-paper and hybridization with DNA probes. *Proc. Natl. Acad. Sci. U S A, 74*, 5350-5354.

Ampe F, Kiss E, Sabourdy F, Batut J. (2003). Transcriptome analysis of *Sinorhizobium meliloti* during symbiosis. *Genome Biol., 4*, R15.

Anderson, L., and Seilhamer, J. (1997). A comparison of selected mRNA and protein abundances in human liver. *Electrophoresis, 18*, 533-537.

Arenkov, P., Kukhtin, A., Gemmell, A., Voloshchuk, S., Chupeeva, V., and Mirzabekov, A. (2000). Protein microchips: Use for immunoassay and enzymatic reactions. *Anal. Biochem., 278*, 123-131.

Bader, G. D., and Hogue, C., W. (2002). Analyzing yeast protein-protein interaction data obtained from different sources. *Nature Biotechnol., 20*, 991-997.

Bairoch, A., and Apweiler, R. (1999). The SWISS-PROT protein sequence data bank and its supplement TrEMBL in 1999. *Nucleic Acids Res., 27*, 49-54.

Bjellqvist, B., K. Ek., Richetti, P., Gianazza, E., Gorg, A., Westermeir, R., and Postel W. (1982). Isoelectric focusing in immobilized pH gradients: Principle, methodology and some applications. *J. Biochem. Biophys. Methods, 6*, 317-339.

Brent, R. (1999). Functional genomics: Learning to think about gene expression data. *Curr. Biol., 9*, 338-341.

Büssow, K., Nordhoff, E., Lübbert, C., Lehrach, H., and Walter, G. (2000). A human cDNA library for high-throughput protein expression screening. *Genomics, 65*, 1-8.

Cabanes, D., Boistard, P., and Batut J. (2000). Identification of *Sinorhizobium meliloti* genes regulated during symbiosis. *J. Bacteriol., 182*, 3632-3637.

Cevallos, M. A., Encarnación, S., Leija, A., Mora, Y., and Mora, J. (1996). Genetic and physiological characterization of a *Rhizobium etli* mutant strain unable to synthesize poly-beta-hydroxybutyrate. *J. Bacteriol., 178*, 1646-1654.

Chen, H., Higgins, J., Kondorosi, E., Kondorosi, A., Djordjevic, M. A., Weinman, J. J., *et al.* (2000). Identification of *nolR*-regulated proteins in *Sinorhizobium meliloti* using proteome analysis. *Electrophoresis, 21*, 3823-3832.

Chen, H., Higgins, J., Oresnik, I. J., Hynes, M. F., Natera, S., Djordjevic, M. A., *et al.* (2000). Proteome analysis demonstrates complex replicon and luteolin interactions in pSyma-cured derivatives of *Sinorhizobium meliloti* strain 2011. *Electrophoresis, 21*, 3833-3842.

Colebatch, G., Kloska, S., Trevaskis, B., Freund, S., Altmann, T., and Udvardi, M. K. (2002). Novel aspects of symbiotic nitrogen fixation uncovered by transcript profiling with cDNA arrays. *Mol. Plant-Microbe Interact., 15*, 411-420.

Dainese-Hatt, P., Fischer, H. M., Hennecke, H., and James, P. (1999). Classifying symbiotic proteins from *Bradyrhizobium japonicum* into functional groups by proteome analysis of altered gene expression levels. *Electrophoresis, 20*, 3514-3520.

DeRisi, J., Penland, L., Brown, P. O., Bittner, M. L., Meltzer, P. S., Ray, M., *et al.* (1996). Use of a cDNA microarray to analyse gene expression patterns in human cancer. *Nature Genet., 14*, 457-460.

Deiwick, J., and Hensel, M. (1999). Regulation of virulence genes by environmental signals in *Salmonella typhimurium. Electrophoresis, 20*, 813-817.

Dong, Y., Glasner, J. D., Blattner, F. R., and Triplett, E. W. (2001). Genomic interspecies microarray hybridization: rapid discovery of three thousand genes in the maize endophyte, *Klebsiella pneumoniae* 342, by microarray hybridization with *Escherichia coli* K-12 open reading frames. *Appl. Environ. Microbiol., 67*, 1911-1921.

Dutt, M. J., and Lee, K, H. (2000). Proteomic analysis. *Curr. Opin. Biotechnol., 11*, 176-179.

Encarnacion, S., Dunn, M., Willms, K., and Mora J. (1995). Fermentative and aerobic metabolism in *Rhizobium etli. J. Bacteriol., 177*, 3058-3066.

Encarnacion, S., Vargas, M del C., Dunn, M. F., Davalos, A., Mendoza, G., Mora, Y., *et al.* (2002). AniA regulates reserve polymer accumulation and global protein expression in *Rhizobium etli. J. Bacteriol., 184*, 2287-2295.

Eymann, C., Homuth, G., Scharf, C., and Hecker, M. (2002). *Bacillus subtilis* functional genomics: Global characterization of the stringent response by proteome and transcriptome analysis. *J Bacteriol., 184*, 2500-2520.

Fellay, R., Perret, X., Viprey, V., Broughton, W. J., and Brenner, S. (1995) Organization of host-inducible transcripts on the symbiotic plasmid of *Rhizobium sp.* NGR234. *Mol. Microbiol, 16*, 657-667.

Fenn, J. B., Mann, M., Meng, C.K., Wong, S. F. and Whitehouse, C. M. (1989). Electrospray ionization for mass spectrometry of large biomolecules. *Science, 246*, 64-71.

Fields, S., and Song, O. (1989). A novel genetic system to detect protein-protein interactions. *Nature, 340*, 245-246.

Fromont-Racine, M., Rain, J. C., and Legrain, P. (1997). Toward a functional analysis of the yeast genome through exhaustive two hybrid screens. *Nature Genet., 16*, 277-282.

Guerreiro, N., Djordjevic, M., and Rolfe B. (1999). Proteome analysis of the model microsymbiont *Sinorhizobium meliloti*: Isolation and characterization of novel proteins. *Electrophoresis, 20*, 818-825.

Guerreiro, N., Redmond, J. W., Rolfe, B. G., and Djordjevic, M. A. (1997). New *Rhizobium leguminosarum* flavonoid-induced proteins revealed by proteome analysis of differentially displayed proteins. *Mol. Plant-Microbe Interact., 10*, 506-516.

Guerreiro, N., Ksenzenko, V. N., Djordjevic, M. A., Ivashina, T. V., and Rolfe, B. G. (2000). Elevated levels of synthesis of over 20 proteins results after mutation of the *Rhizobium leguminosarum* exopolysaccharide synthesis gene *pssA*. *J. Bacteriol., 182*, 4521-4532.

Guerreiro, N., Stepkowski, T., Rolfe, B. G., and Djordjevic, M. A. (1998). Determination of plasmid-encoded functions in *Rhizobium leguminosarum* biovar trifolii using proteome analysis of plasmid-cured derivatives. *Electrophoresis, 19*, 1972-1979.

González, V., Bustos, P., Ramírez-Romero, M. A., Medrano-Soto, A., Salgado, H., Hernández-González, I., *et al.*, (2003). The mosaic structure of the symbiotic plasmid of *Rhizobium etli* CFN42 and its relation to other symbiotic genome compartments. *Genome Biol., 4*, R36.

Gygi, S., Rist, B., Gerber, S., Turecek, F., Gelb, M., and Aebersold, R. (1999). Quantitative analysis of complex protein mixtures using isotope-coded affinity tags. *Nature Biotechnol., 17*, 994-999.

Gygi, S. P., Rochon, Y., Franza, B. R., and Aebershold, R. (1999). Correlation between protein and mRNA abundance in yeast. *Mol. Cell. Biol., 19*, 1720-1730.

Haynes, P., Gygi, S., Figeys, D., and Aebersold, R. (1998). Proteome analysis: Biological assay or data archive? *Electrophoresis, 19*, 1862-1871.

Hatzimanikatis, V., and Lee, K. (1999). Dynamical analysis of gene networks requires both mRNA and protein expression information. *Metab. Eng., 1*, 275-281

Henzel, W. J., Billeci, T. M., Stults, J. T., Wong, S. C., Grimley, C., and Watanabe, C. (1993). Identifying proteins from two-dimensional gels by molecular mass searching of peptide fragments in protein sequence databases. *Proc. Natl. Acad. Sci. USA, 90*, 5011-5015.

Hinton, J. C. (1997). The *Escherichia coli* genome sequence: the end of an era or the start of the FUN? *Mol. Microbiol., 3*, 417-422.

Hughes, D. (2000). Evaluating genome dynamics: The constraints on rearrangements within bacterial genomes. *Genome Biol., 1*, REVIEWS0006

Kamo, M., Kawakami, T., Miyatake, N., and Tsugita, A. (1995). Separation and characterization of *Arabidopsis thaliana* proteins by two-dimensional gel electrophoresis. *Electrophoresis, 16*, 423-300.

Kaneko T., Nakamura, Y., Sato, S., Asamizu, E., Kato, T., Sasamoto, S., *et al.* (2000). Complete genome structure of the nitrogen-fixing symbiotic bacterium *Mesorhizobium loti. DNA Res., 7*, 331-338.

Karas, M., Bahr, U., Ingendoh, A., and Hillenkamp, F. (1989). Laser desorption-ionization mass spectrometric of proteins with masses 100,000 to 250,000 dalton. *Angew Chem. Int. Ed. Engl., 28*, 760-761.

Karas, M., and Hillenkamp, F. (1988). Laser desorption ionization of proteins with molecular masses exceeding 10,000 daltons. *Anal. Chem., 60,* 2299-2301.

Kaufmann, H., Mazur, X., Fussenegger, and M., Bailey, J. (1999). Influence of low temperature on productivity, proteome and protein phosphorylation of CHO cells. *Biotechnol. Bioeng., 63*, 573-582.

Kell, D. B., and King, R. D. (2000). On the optimization of classes for the assignment of unidentified reading frames in functional genomics programmes: The need for machine learning. *Trends Biotechnol., 18*, 93-98.

Keough, T., Youngquist, R., and Lacey, M. A (1999). Method for high-sensitivity peptide sequencing using postsource decay matrix-assisted laser desorption ionization mass spectrometry. *Proc. Natl. Acad. Sci. USA, 96*, 7131-7136.

Klose, J., and Kobalz, U. (1995). Two-dimensional electrophoresis of proteins: An updated protocol and implications for a functional analysis of the genome. *Electrophoresis, 16*, 1034-1059.

Leymarie, J., Damerval, C., Marcotte, L., Combes, V., and Vartanian, N. (1996). Two-dimensional protein patterns of *Arabidopsis* wild-type and auxin-insensitive mutants, axr1, axr2, reveal interactions between drought and hormonal responses. *Plant Cell Physiol., 37*, 966-975.

Link, A., Eng, J., Schieltz, D., Carmack, E., Mize, G., Morris, D., *et al.* (1999). Direct analysis of protein complexes using mass spectrometry. *Nature Biotechnol., 17*, 676-682.

Lockhart, D. J., Dong, H., Byrne, M. C., Follettie, M. T., Gallo, M. V., Chee, M. S., *et al.* (1996). Expression monitoring by hybridization to high-density oligonucleotide arrays. *Nature Biotechnol., 14*, 1675-1680.

Lueking, A., Horn, M., Eickhoff, H., Büssow, K., Lehrach, H., and Walter, G. (1999). Protein microarrays for gene expression and antibody screening. *Anal. Biochem., 270*, 103-111.

Lueking, A., Konthur, Z., Eickhoff, H., Büssow, K., Lehrach, H., and Cahill, D.J. (2001). Protein microarrays a tool for the post- genomic era. *Curr. Genomics, 2*, 151-159.

MacBeath, G., and Schreiber, S., L. (2000). Printing proteins as microarrays for high-throughput function determination. *Science, 289*, 1760-1763.

Malmqvist, M. (1993). Surface plasmon resonance for detection and measurement of antibody-antigen affinity and kinetics. *Curr. Opin. Immunol., 5*, 282-286.

Mathesius, U., Keijzers, G., Natera, S. H., Weinman, J. J., Djordjevic, M. A., and Rolfe, B. G. (2001). Establishment of a root proteome reference map for the model legume *Medicago truncatula* using the expressed sequence tag database for peptide mass fingerprinting. *Proteomics, 1*, 1424-1440.

Melanie 3 2D Gel Analysis Software on World Wide Web URL: http://www.expasy.ch/melanie/

Merril, C., Switzer, R., and VanKeuren, M. (1979). Trace polypeptides in cellular extracts and human body fluids detected by two-dimensional electrophoresis and a highly sensitive silver stain. *Proc. Natl. Acad. Sci. USA, 76*, 4335-4339.

Morris, A., C., and Djordjevic, M. A. (2001). Proteome analysis of cultivar-specific interactions between *Rhizobium leguminosarum* biovar trifolii and subterranean clover cultivar Woogenellup. *Electrophoresis, 22*, 586-598.

Munchbach, M., Dainese, P., Staudenmann, W., Narberhaus, F., and James P. (1999). Proteome analysis of heat shock protein expression in *Bradyrhizobium japonicum. Eur. J. Biochem., 264*, 39-48.

Natera, S., H., Guerreiro, N., and Djordjevic, M., A. (2000). Proteome analysis of differentially displayed proteins as a tool for the investigation of symbiosis. *Mol. Plant-Microbe Interact., 13*, 995-1009.

Nyman, T., A. (2001). The role of mass spectrometry in proteome studies. *Biomol. Eng., 18*, 221-227.

Oda, Y., Huang, K., Cross, F., Cowburn, D., and Chait, B. (1999). Accurate quantitation of protein expression and site-specific phosphorylation. *Proc. Natl. Acad. Sci. USA, 96*, 6591-6596.

O'Farrell, P.H. (1975). High resolution two-dimensional electrophoresis of proteins. *J. Biol. Chem., 250*, 4007-4021.

Ouellet, F., Overvoorde, P. J., and Theologis, A. (2001). IAA17/AXR3: Biochemical insight into an auxin mutant phenotype. *Plant Cell, 13*, 829-842.

Panter, S., Thomson, R., de Bruxelles, G., Laver, D., Trevaskis, B., and Udvardi, M. (2000). Identification with proteomics of novel proteins associated with the peribacteroid membrane of soybean root nodules. *Mol. Plant-Microbe Interact., 13*, 325-333

Patterson, S., and Aebersold, R. (1995). Mass spectrometric approaches for the identification of gel-separated proteins. *Electrophoresis, 16*, 1791-1814.

Patton, W.F. (2000). A thousand points of light: The application of fluorescence detection technologies to two-dimensional gel electrophoresis and proteomics. *Electrophoresis, 21*, 1123-1144.

Perret, X., Freiberg, C., Rosenthal, A., Broughton, W. J., and Fellay, R. (1999). High-resolution transcriptional analysis of the symbiotic plasmid of *Rhizobium sp.* NGR234. *Mol. Microbiol., 32*, 415-425.

Porubleva, L., Vander Velden, K., Kothari, S., Livier, D. J., and Chitnis, P. R. (2001). The proteome of maize: Use of gene sequence and expressed sequence tag data for identification of proteins with mass fingerprints. *Electrophoresis, 22*, 1724-1738.

Puig, O., Caspary, F., Rigaut, G., Rutz, B., Bouveret, E., Bragado-Nilsson, E., *et al.* (2001). Seraphin, B. The tandem affinity purification (TAP) method: A general procedure of protein complex purification. *Methods, 24*, 218-229.

Quadroni, M., and James, P. (1999). Proteomics and automation. *Electrophoresis, 20*, 664-677.

Quail, P. H. (2000). Phytochrome-interacting factors. *Semin. Cell Dev. Biol., 11*, 457-466.

Rain, J. C., Selig, L., De Reuse, H., Battaglia, V., Reverdy, C., Simon, S., *et al.* (2001). The protein-protein interaction map of *Helicobacter pylori*. *Nature, 409*, 211-215.

Raftery, M., J., and Geczy, C. L. (2002). Electrospray low energy CID and MALDI PSD fragmentations of protonated sulfinamide cross-linked peptides. *J. Am. Soc. Mass Spectrom., 13*, 709-718.

Rigaut, G., Shevchenko, A., Rutz, B., Wilm, M., Mann, M., and Seraphin, B. (1999). A generic protein purification method for protein complex characterization and proteome exploration. *Nature Biotechnol., 17*, 1030-1032.

Richmond, C. S., Glasner J. D., Mau, R., Jin, H., and Blattner, F. R. (1999). Genome-wide expression profiling in *Escherichia coli* K-12. *Nucleic Acids Res., 27*, 3821-3835.

Rouquie, D., Peltier, J. B., MarquisMansion, M., Tournaire, C., Doumas, P., and Rossignol, M. (1997). Construction of a directory of tobacco plasma membrane proteins by combined two-dimensional gel electrophoresis and protein sequencing. *Electrophoresis, 20*, 705-711.

Saalbach, G., Erik, P., and Wienkoop, S. (2002). Characterisation by proteomics of peribacteroid space and peribacteroid membrane preparations from pea (*Pisum sativum*) symbiosomes. *Proteomics, 2*, 325-337.

Sanchez, J., Rouge, V., Pisteur, M., Ravier, F., Tonella, L., Moosmayer, M., *et al.* (1997). Improved and simplified in-gel sample application using reswelling of dry immobilized pH gradients. *Electrophoresis, 18*, 324-327.

Schena, M., Shalon., D., Davis., R. W., and Brown, P.O. (1995). Quantitative monitoring of gene expression patterns with a complementary DNA microarray. *Science, 270*, 467-470.

Schena, M., Shalon, D., Heller, R., Chai, A., Brown, P. O., and Davis, R. W. (1996). Parallel human genome analysis: microarray-based expression monitoring of 1000 genes. *Proc. Natl. Acad. Sci. USA, 93*, 10614-10619.

Schwikowski, B., Uetz, P. and Fields, S. (2000). A network of protein-protein interactions in yeast. *Nature Biotechnol., 18*, 1257-1261.

Southern, E. M. (1975). Detection of specific sequences among DNA fragments separated by gel electrophoresis. *J. Mol. Biol., 98*, 503-517.

Swiderek, K., Davis, M., and Lee, T. (1998). The identification of peptide modifications derived from gel-separated proteins using electrospray triple quadrupole and ion trap analyses. *Electrophoresis, 19*, 989-997.

Tang, Y., and Guest, J, R. (1999). Direct evidence for mRNA binding and post-transcriptional regulation by *Escherichia coli* aconitases. *Microbiology, 145*, 3069-3079.

Touzet, P., de Vienne, D., Huet, J.C., Ouali, C., Bouet, F., and Zivy, M. (1996). Amino acid analysis of proteins separated by two-dimensional electrophoresis in maize: isoform detection and function identification. *Electrophoresis, 17*, 1393-1401.

Towbin, H., Staehelin, T., and Gordon, J. (1979). Electrophoretic transfer of proteins from polyacrylamide gels to nitrocellulose sheets: procedure and some applications. *Proc. Natl. Acad. Sci. USA, 76*, 4350-4354.

Tsugita, A., and Kamo, M. (1999). 2-D electrophoresis of plant proteins. *Meth. Mol. Biol., 112*, 95-97.

Tsugita, A., Kamo, M., Kawakami, T., and Ohki, Y. (1996). Two-dimensional electrophoresis of plant proteins and standardization of gel patterns. *Electrophoresis, 17*, 855-865.

Tsugita, A., Kawakami, T., Uchiyama, Y., Kamo, M., Miyatake, N., and Nozu, Y. (1994). Separation and characterization of rice proteins. *Electrophoresis, 15*, 708-720.

Udvardi, M. K. (2002). Legume genomes and discoveries in symbiosis research. *Genome Biol., 21*, REPORTS 4028.

Velculescu, V. E., Zhang, L., Zhou, W., Vogelstein, J., Basrai, M. A., Bassett, D. E. Jr., *et al.* (1997). Characterization of the yeast transcriptome. *Cell, 88*, 243-251.

von Mering, C., Krause, R., Snel, B., Cornell, M., Oliver, S. G., Fields, S., *et al.* (2002). Comparative assessment of large-scale data sets of protein-protein interactions. *Nature, 417*, 399-403.

Walhout, A. J., and Vidal, M. 2001. High throughput yeast two- hybrid assays for large-scale protein interaction mapping. *Methods 24*, 297-306.

Walhout, A. J., Sordella, R., Lu, X., Hartley, J. L., Temple, G. F., Brasch, M. A., *et al.* (2000). Protein interaction mapping in *C. elegans* using proteins involved in vulva development. *Science, 287*, 116-122.

Walter, G., Büssow, K., Cahill, D., Lueking, A., and Lehrach, H. (2000). Protein arrays for gene expression and molecular interaction screening. *Curr. Opin. Microbiol., 3*, 298-302.

Wasinger, V. C., Cordwell, S. J., Cerpa-Poljak, A., Yan, J. X., Gooley, A. A., Wilkins, M. R., *et al.* (1995). Progress with gene-product mapping of the Mollicutes: *Mycoplasma genitalium*. *Electrophoresis, 16*, 1090-1094.

Wilkins, M., Gasteiger, E., Tonella, L., Ou, K., Tyler, M., Sanchez, J., *et al.* (1998a). Protein identification with *N* and *C*-terminal sequence tags in proteome projects. *J. Mol. Biol., 278*, 599-608.

Wilkins, M., Gasteiger, E., Wheeler, C., Lindskog, I., Sanchez, J., Bairoch, A., *et al.* (1998b). Multiple parameter cross-species protein identification using MultiIdent - a world-wide web accessible tool. *Electrophoresis, 19*, 3199-3206.

Wilkins, M., and Gooley, A. (1997). Protein identification in proteome projects. In M. R. Wilkins, K. Williams., R. Appel., and D. Hochstrasser (Eds.), *Proteome research: New frontiers in functional genomics* (pp. 35-64). Tokyo: Springer.

Ye, R. W., Wang, T., Bedzyk, L., and Croker, K. M. (2001). Applications of DNA microarrays in microbial systems. *J. Microbiol. Methods, 47*, 257-272.

CHAPTER 11

TRANSCRIPTOMICS IN *SINORHIZOBIUM MELILOTI*

A. BECKER[1] AND F. J. DE BRUIJN[2]

[1]*Lehrstuhl für Genetik, Fakultät für Biologie, Universität Bielefeld, P.O. Box 100131, D-33501 Bielefeld, Germany, and* [2]*UMR INRA-CNRS 2594/441 Laboratoire des Interactions Plantes-Microorganismes, Chemin de Borde Rouge, BP27 31326 Castanet-Tolosan cedex, France*

1. INTRODUCTION TO TRANSCRIPTOMICS

The availability of complete bacterial genome sequences has greatly changed the face of microbiology. Value has to be added to the nucleotide sequence to convert genome data into gene-function data. Knowing the complete sequence of a genome is only the first step towards understanding how all components of a bacterial cell work together. Information about a predicted gene can be deduced from sequence similarities to either genes or sequence motifs of known function, from the location of a gene in either an operon or a gene cluster, from expression patterns, and from mutant phenotypes. Gene-expression analyses comprise monitoring changes in levels of both RNA and protein. In the post-genomic era, experimental approaches have moved from the targeted investigation of individual genes to the investigation of thousands of genes in a single untargeted experiment. Recent developments of novel tools allow the analysis of a whole genome in a single highly parallel experiment. High-throughput analysis of differential gene expression is a powerful tool for gaining information about biological processes on a genomic scale. The transcriptome is the RNA complement of all transcribed genes of an organism. Arrays have developed into the most popular tool for transcriptome analysis. These tools allow the measurement of concentrations of RNA complementary to defined genomic sequences in a highly parallel way.

In prokaryotes, gene expression is predominantly controlled at the level of transcription. Transcription is the first step towards protein synthesis. In particular, in prokaryotes, where transcription and translation are spatially and chronologically coupled, correlation of mRNA concentrations and protein abundance is expected. Most of the studies that addressed this issue have found only modest correlations (Gmuender *et al.*, 2001). However, combined analyses of both transcript and

R. Palacios and W. E. Newton (eds.), Genomes and Genomics of Nitrogen-fixing Organisms, 169-181.

protein content in response to a specific stimulus resulted in similar patterns (Yoshida *et al.*, 2001; Petersohn *et al.*, 2001). RNA abundance is affected both by transcription activity and by the stability of the RNA. It is not only important to identify the factors affecting the expression of a gene and to understand the regulatory mechanisms, but also to identify co-regulated genes to improve our knowledge about the context in which a gene is expressed. Because coordinated increases or decreases in the abundance of RNA derived from many different genes can be detected, DNA micro- and macro-arrays represent ideal tools for high-throughput analysis of gene regulation on the transcriptional level (see Lockhard and Winzeler, 2000).

Both micro- and macro-array experiments rely on the principle of hybridization, which is the pairing of complementary strands of nucleic acids (either DNA or RNA) in a sequence-specific manner. Single-stranded DNA fragments that represent either specific genes or other regions of a genome are coupled in an ordered pattern (array) to a solid substrate. Labeled nucleic acids are added that bind to the complementary sequence of the DNA molecules on the surface. Both the site and strength of the signals derived from labeled nucleic acids bound to the array allow the quantity of different sequence species in the unknown mixture of nucleic acids to be measured. Thus, the relative amounts of thousands of transcripts under various conditions can be compared in a single experiment (Figure 1).

In microarray experiments, RNA from two different cell populations is prepared and converted into targets that are labeled by two different fluorophores either by direct labeling or by reverse transcription into cDNA. Labeled RNA or cDNA is mixed and hybridized to a microarray. The experimental steps of a microarray experiment for gene-expression analysis comprise the design and generation of probe DNA, array fabrication, preparation and labeling of target DNA or RNA, array hybridization, and signal detection, followed by image and data analysis. These steps are discussed in detail by Rhodius *et al.* (2002). The nomenclature of probes and targets that is commonly used reflects the similarity between arrays and reverse dot-blot technologies. The DNA with a defined identity coupled to a solid substrate is referred to as "probe" and the labeled nucleic acid as "target". Moreover, "array" refers to the whole printed area on a substrate, a "spot" is the smallest unit of an array area that contains molecules of one species of a DNA probe, a "grid" is a sub-area of the array, *i.e.*, a group of spots usually produced by one print tip, and a "metagrid" is a group of grids. If an array consists only of one metagrid, the metagrid is identical to the array. Currently, the most frequently used arrays can be categorized into macroarrays and microarrays.

Macroarrays are usually fabricated by spotting probes on either a nylon or nitrocellulose membrane-based matrix. The term macroarray reflects the lower spot density on these arrays in comparison to microarrays. Targets for these filter arrays are usually radioactively labeled and the bound target can be detected using phosphoimagers. Therefore, different samples are not hybridized simultaneously to one array, but in a consecutive way, as described below.

Glass slides are usually used for the fabrication of microarrays. These glass substrates are coated to enable the coupling of DNA to the surface. In contrast to macroarrays, microarrays contain spots at a high spot density.

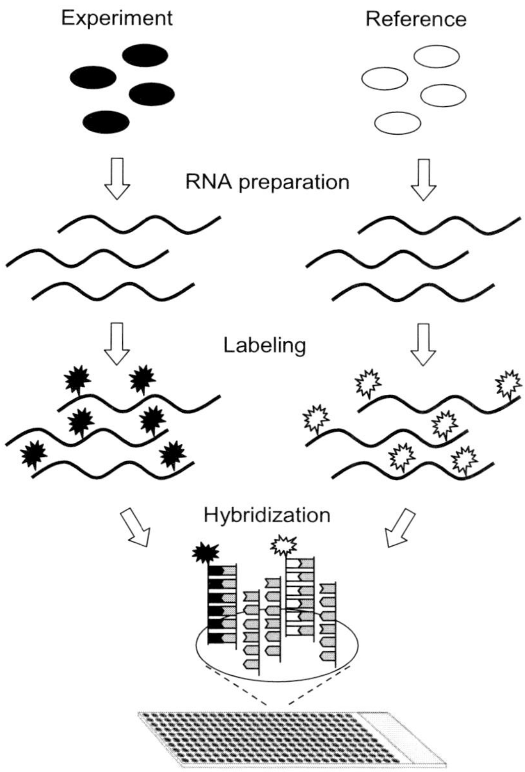

Figure 1. Microarray hybridization experiment for gene-expression analysis.

Either PCR products or pre-synthesized oligonucleotides are usually spotted. Targets are prepared from two RNA populations that are each labeled with a different fluorophore by either direct labeling or reverse transcription into fluorescently labeled cDNA. Gene-expression patterns can be compared by determining the ratio of fluorescence intensities of the two fluorophores after hybridization.

2. INTRODUCTION TO THE BIOLOGICAL SYSTEM

Rhizobia are soil bacteria capable of establishing a N_2-fixing symbiosis with plants of the family Leguminosae. During this symbiosis, new specialized organs are formed, the nitrogen-fixing root nodules, as a result of an elaborated developmental program directed by an exchange of signals between the two partners. Plant flavonoids secreted by the roots trigger Nod-factor production by the bacteria (Denarie *et al.*, 1996). In turn, Nod factors induce on specific host-plants a transduction pathway leading to nodule development (Catoira *et al.*, 2000).

When bacteria that are growing either in the rhizosphere or on the root surface become trapped between two epidermal root-hair cell walls, alteration and invagination of a root-hair cell wall initiates the development of an infection thread, a tubular structure in which the bacteria penetrate the plant and propagate further towards the inner cortex where the infection threads ramify. Among the few known bacterial genes required for initiation and elongation of infection threads, those directing the production of different types of cell-surface polysaccharides play a major role, possibly by protecting the invading bacteria against plant-defense mechanisms (Pellock *et al.*, 2000). Oxidative stress protection through the production of detoxifying enzymes is also essential for the survival of symbiotic bacteria during infection (Santos *et al.*, 2000). Bacterial cells at the tip of the infection threads eventually enter the nodule cells through a process that appears to involve binding to the root-cell cytoplasmic membrane and uptake into the cells by endocytosis (Vasse *et al.*, 1990; Brewin, 1998).

Once inside the plant cells, the bacteria differentiate into non-dividing cells, called bacteroids. The genetic control of both endocytosis and bacteroid differentiation is essentially unknown with the exception of BacA, an inner-membrane protein affecting the cell envelope composition (Ferguson *et al.*, 2002). *S. meliloti bacA* mutants are released from the infection threads but senesce rapidly before bacteroid differentiation, possibly because of their increased sensitivity to stress (Ferguson *et al.*, 2002; Glazebrook *et al.*, 1993). Finally, bacteroids synthesize proteins essential for symbiotic nitrogen fixation, such as those responsible for dicarboxylic-acid transport (*dct* genes) (Watson *et al.*, 1988), nitrogenase synthesis (*nif* genes), and microoxic respiration (*fix* genes) (Kaminski *et al.*, 1998). In nodules, *nif* and *fix* gene expression is driven by O_2-limitation (Soupene *et al.*, 1995).

In addition to this specific symbiotic lifestyle, the physiology of rhizobia has been investigated under a wide variety of environmental conditions, such as either nutrient deprivation or resistance to environmental stress, which is relevant to their survival in soils (*e.g.*, Milcamps *et al.*, 1998; Trzebiatowski *et al.*, 2001; Ricillo *et al.*, 2000; 2001). Over the past several years, *Sinorhizobium meliloti*, the symbiont of alfalfa, has become one of the primary model organisms to study microbial persistence, competition, and stress response in soil, as well as plant-microbe interactions. Moreover, one of its hosts, *Medicago truncatula*, has become a model organism to study the plant partner of symbiotic nitrogen fixation (Cook *et al.*, 1977).

Despite the above-mentioned information, we lack a global view on how the *Rhizobium*-legume symbiosis develops. In particular, we need to learn more about the bacterial genes that control plant-cell adhesion, colonization of infection threads, plant-cell invasion, and bacteroid differentiation. Genomics is, of course, of particular interest in this respect (see Ampe *et al*, 2003).

3. *SINORHIZOBIUM MELILOTI* MICROARRAYS

The basis upon which we are able to study rhizobia is changing due to the availability of complete genomic sequences that allow the study of gene expression on a global scale. DNA macro- and micro-arrays represent an ideal tool for global analyses of gene expression on the level of transcription in rhizobia. Perret *et al.* (1999) first reported both the complete sequence of the symbiotic plasmid of *Rhizobium* sp. NGR234 and analyses of transcripts derived from this plasmid in the symbiosis. This study demonstrated that such an approach improves our understanding of regulatory mechanisms in nodule symbiosis. Recently, the complete genome sequence of *S. meliloti* 1021 became available (Barnett *et al.*, 2001; Capela *et al.*, 2001; Finan *et al.*, 2001; Galibert *et al.*, 2001).

As a tool for systematic genome-wide gene-expression analysis in *S. meliloti* 1021, whole-genome microarrays and macroarrays were generated. Two sets of experiments with pilot macroarrays containing 34 and 214 *S. meliloti* gene probes, respectively, were carried out both to validate the experimental conditions to be used and to show that valid biological results can be obtained with genes whose function or mode of regulation is known (Berges *et al.*, 2003; Ampe *et al.*, 2003; Ampe and Batut, 2003). These pilot experiments are described below.

In the case of the microarrays, a set of primer pairs, which were specific for all 6207 predicted protein-coding genes of the *S. meliloti* 1021 genome (Barnett *et al.*, 2001; Capela *et al.*, 2001; Finan *et al.*, 2001; Galibert *et al.*, 2001, http://sequence.toulouse.inra.fr/rhime/Consortium/home.html), was generated using Primer3 (Rozen and Skaletsky, 1996; 1997; 1998). These primer pairs were designed to amplify internal gene-specific DNA fragments of 80-350 bp. Amplification, using the specific primer pairs, and re-amplification, using standard primers directed against 5' extensions of the specific primers, resulted in 6046 gene-specific PCR fragments for microarray production. This set was supplemented with 161 70-mer oligonucleotides, which substituted for the ORF-specific probes that were not obtained by PCR, to generate microarrays containing probes for all 6207 predicted protein-coding genes of the *S. meliloti* 1021 genome. In addition to the PCR fragment-based microarrays, microarrays containing 6208 70-mer oligonucleotides (synthesized by Operon-Qiagen), which were specific for the predicted protein-coding genes, were generated (for details, see Becker, 2003).

Probe specificity is a very important issue in the design of microarrays. In particular, repeat sequences, multi-copy genes, and paralogous genes have to be considered. Usually, probes, which contain sequence stretches of more than 15 bp and are more than 80% similar, cannot be differentiated by hybridization of an array with a complex labeled target-molecule mixture (Kane *et al.*, 2000). Whereas the *S. meliloti* 1021 PCR-fragment-based arrays contain 201 probes that probably cross-hybridize on this basis, the oligonucleotide-based arrays contain only 177 probes that may cross-hybridize. 100 of these probes were derived from transposase-gene sequences. The higher specificity of the oligonucleotide-based arrays is often beneficial due to shorter probe sequences, but it makes them less applicable for expression analyses of strains not very closely related to the reference strain 1021.

As controls, DNA probes from an unrelated organism, alien DNA, and both stringency-control and spiking-control probes are included in the arrays. Alien DNA fragments have artificial sequences that are not found in any known genome sequence. Stringency controls are 80% identical to selected gene-specific probes. Probes from an unrelated organism, alien probes, and stringency controls are used to control the specificity of hybridization. For a spiking control, an *in vitro* transcribed RNA is added to the target preparation and labeled together with the target molecules to verify the labeling procedure. This spiking RNA is complementary to a probe on the array which is unrelated to the genome sequence to be analyzed.

Table 1 summarizes characteristics and sources of the different arrays that were generated for transcription profiling in *S. meliloti* 1021 (see also Becker, 2003).

Table 1. S. meliloti *arrays*

	PCR fragment-based microarray	Oligonucleotide-based microarray	PCR fragment-based macroarray
Number of gene-specific probes	6046 PCR fragments 161 oligonucleotides	6208	6061
Length of gene-specific probes	80-350 bp	70 N	80-350 bp
Number of gene-specific probes that are likely to cross-hybridize	201	177	201
Alien control probes	3	19	5
Alien probes used as spiking controls	3	3	5
Stringency control probes	0	2	0
Source	Institute for Genome Research, Bielefeld University (A. Becker)	Institute for Genome Research, Bielefeld University (A. Becker)	INRA-CNRS Toulouse (J. Batut)

In order to validate gene expression studies in *S. meliloti* using the PCR fragment-based whole-genome microarrays, hybridizations based on RNA samples obtained from cells cultured under identical conditions were performed. In each of these experiments, transcript abundance of two independent cultures of *S. meliloti* 1021 in TY complex medium (Beringer, 1974) was compared. In all cases, signal intensities differed less than 1.6-fold (Figure 2A).

Phosphorus is an essential nutrient, which is taken up in the form of inorganic phosphate and organic phosphate compounds. In most soils, soluble phosphate is present at low concentrations. *S. meliloti* possess two different phosphate-transport systems, a high-affinity system that is encoded by *phoCDET* and a low-affinity system encoded by the *orfA-pit* genes (Voegele *et al.*, 1997; Bardin *et al.*, 1998). To assess the PCR-fragment-based whole-genome microarrays, the response of *S. meliloti* to different phosphate concentrations was analyzed. *S. meliloti* 1021 was cultured in minimal medium with either a low (0.1 mM) or high (2 mM) phosphate concentration. Microarray hybridizations revealed *ca.* 100 genes that displayed at least two-fold lower transcript abundance and *ca.* 150 genes that showed at least two-fold higher transcript levels in low phosphate medium (Figure 2B). Among the genes that displayed higher transcript levels in low phosphate medium were the *phoCDET* genes. Induction of these genes in response to phosphate starvation was previously reported by Bardin *et al.* (1996), so confirming the results from microarray hybridizations.

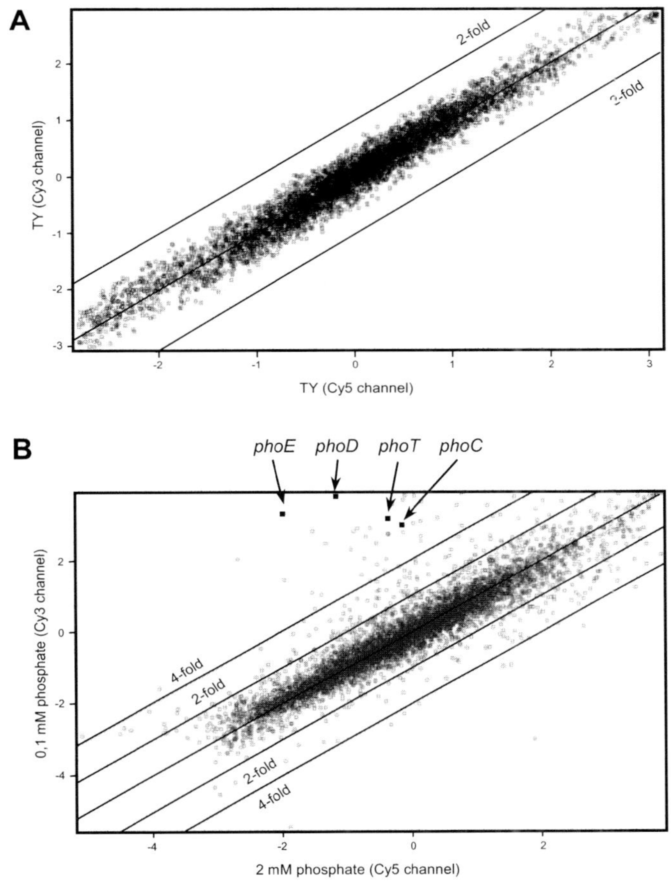

Figure 2. A. Comparison of transcript abundance in RNA preparations from two independent S. meliloti 1021 cultures in complex medium. B. Comparison of transcript abundance in RNA preparations from S. meliloti *1021 cells cultured in minimal medium with low and high phosphate concentrations.*
Mean-centered log intensities are shown for the Cy3 and Cy5 channel of each experiment.

4. *S. MELILOTI* MACROARRAYS

As pointed out above, microarrays based on glass-slide supports are becoming the most popular form of arrays. Miniaturisation allows the spotting of several thousand genes on a glass-slide-cover surface, hence allowing transcriptome analysis of large (regions of) genomes (see section above). Yet, the production and use of microarrays are technically challenging and require specific and expensive equipment, including both an accurate robotic spotter and a laser scanner. Therefore, in parallel to our experiments with micro-arrrays, we investigated the use of macroarrays, which use a nylon membrane as a support, as a worthwhile alternative because their manufacture and analysis are easier than those of glass microarrays. Moreover, most laboratories have the required equipment and are trained in nylon-membrane handling, plus a relatively simple robot (if any) is sufficient, and phosphor-imager screens are frequently available. In addition, nylon membranes are cheap as compared to glass slides and the spotted membranes are both stable over time and reusable (Ampe and Batut, 2003). In fact, macroarrays can be designed for other applications, such as either comparative genomics, through hybridisation of the membranes with labelled genomic DNA from different microbial isolates, or for population studies, using environmental DNA as a probe community (meta-genome libraries from soil; Goodman and Liles, personal communication).

As described above, we used *S. rhizobium* as our model soil and symbiotic bacterium because its entire genomic-DNA sequence is known and it forms nodules on, and fixes N_2 with, the model legume *M. truncatula*, which is also being studied intensively in our laboratories. In principle, the development and use of macro-gene arrays is very simple. In practice, however, many strategic and technological choices have to be made. In preparation for whole genome analysis, therefore, we carried out a number of pilot studies, using membranes carrying first 34, and subsequently 214, *S. meliloti* genes that are either known or predicted to either carry out specific functions in the free-living state, or be regulated in a known fashion or be potentially involved in critical steps of the establishment of the *Rhizobium*-legume symbiosis. For our first set of experiments, *S. meliloti* pilot DNA macroarrays were designed carrying probes corresponding to 34 genes of known regulation (see Berges *et al.*, 2003). We were interested in evaluating the reproducibility of DNA-nylon arrays, as well as identifying the main sources of experimental errors. Our experimental design intended to include four experimental parameters (RNA labeling, type of PCR spotted, repetition of experiments, and replication of spots) using RNA generated under two biological conditions (either aerobic or micro-aerobic growth) and to examine their influence on the expression of the 34 genes. Normalized results of all experiments were analysed by performing both Anova and Principal Component Analysis (PCA) in order both to identify the main sources of experimental variability and to select optimal experimental conditions (Berges *et al.*, 2003).

Variance and Principal Component Analysis showed that the most important non-biological parameter was the labeling method. Anova yielded a high Fischer ($F = 29.66$) value associated with a low *p*-value ($<10^{-6}$), which indicated that the type of protocol used to label RNA strongly influenced the results. When DIG-labeled probes were used, the background was strongly increased when membranes were rehybridized, thus, preventing their multiple use. As a result, the signal/background ratio was lower with DIG-labeled than with ^{33}P-labeled probes. In conclusion, despite the higher amount of RNA required, we prefer to use ^{33}P-labeling for the reverse transcription.

The size of the PCR products was also found to be important. In several studies, full-length genes were amplified (Richmond *et al.*, 1999), whereas other studies used gene fragments of constant size (Loos *et al.*, 2001). We amplified either the genes from the predicted start codon to the stop codon, which yielded PCR products from 335 bp to 2888 bp, or internal fragments of 238-387 bp. When the Anova analysis was performed with all the data generated with tagged-primers, the PCR size appeared to be a crucial factor influencing the results (high Fischer value ($F = 20.22$) associated to a low p-value ($<10^{-6}$; Berges *et al.* 2003). The ratio between hybridisation signals obtained with full-length and constant-size PCR products globally increased with the corresponding ratios of the PCR products sizes (size of full-length PCR product / size of constant PCR products).

Specific sequences referred to as 'tags' are often included in the primers used to amplify genes or gene fragments in transcriptome studies (Richmond *et al.*, 1999). These tags are most useful for both enabling re-amplification using a single pair of primers corresponding to the tags and generating restriction sites facilitating further cloning of PCR product. Tags also make it possible to quantify the amount of PCR products spotted on DNA arrays, an important quality control. Results showed that the influence of the tag sequence was minimal and lower than that yielded by the repetition of a single experiment. Therefore, the presence of the 19-mer tag in all the primers did not influence the transcriptome results and the use of tagged primers can be highly recommended (Berges *et al.*, 2003).

The last factor considered was a biological one; gene expression under aerobic and microaerobic conditions. A local Anova analysis performed as described by Didier *et al.*, (2001) enabled us to identify those genes being more responsive to changes in the O_2 status. Genes, which showed both high variation in expression depending on the O_2 input and low p-values testifying the significance of the observed effect, were considered to be specifically regulated by O_2. As expected (Soupene *et al.*, 1995), the *fix* genes involved in microoxic respiration were significantly induced by O_2 limitation with induction factors ranging from 9 to 11 for the *fixKNP* genes. The induction by O_2 limitation of other genes was also detected by the macroarrays experiments. Those induced 2- to 3-fold by O_2 limitation, which is in agreement with the literature (Davey and de Bruijn, 2001; David *et al.*, 1988; Fousard *et al.*, 1997), included: *fixT,* which encodes an antikinase under the control of FixK; *nifA,* the nitrogen-fixation transcriptional activator; *tspO*, a tryptophan-rich sensory protein; and *ndiB*, a nutrient-deprivation induced protein. The variability between replicated spots on a membrane was found to be low (Berges *et al.*, 2003), thus supporting the idea that variations observed here are essentially due to O_2 limitation.

For the second, more extensive, pilot experiment, nylon macroarrays featuring 214 genes (3.5% of the total genome content), which were selected by mining *the S. meliloti* genome sequence, were produced. Genes potentially related to adhesion and attachment to the host plant, oxidative-stress protection, iron metabolism, invasion, calcium binding, toxin and protease production, cell-surface organization, and regulation were selected based on either their overall similarity with known genes or the presence of specific motifs. Thirty-four control genes from *S. meliloti* were included, the regulation of which was known under at least some of the conditions studied. Finally, five genes from *Corynebacterium* with no homologues in *S. meliloti* were included to assess the quality of the hybridizations. The comprehensive list of the genes included in the custom nylon macroarrays as well as the complete expression results is available on the Internet (Ampe *et al.*, 2003). (http://sequence.toulouse.inra.fr/rhime/Infection/Infection_macroarray_2002.htm).

The expression of 214 *S. meliloti* genes was monitored under ten environmental conditions, including free-living aerobic and microaerobic conditions, addition of the plant symbiotic elicitor luteolin, and a variety of symbiotic conditions. Five new genes induced by luteolin were identified as well as 9 new genes induced in mature N_2-fixing bacteroids. A bacterial (*bacA*; see Introduction) and a plant symbiotic mutant affected in nodule development (*M. truncatula* TE7) were found to be altered in their gene-expression patterns and are, therefore, of particular interest in deciphering gene expression at the intermediate stage of the symbiotic interaction. *S. meliloti* gene expression in both the cultivated legume, *Medicago sativa* (alfalfa), and the model plant, *M. truncatula*, were compared and only a small number of differences was found (Ampe *et al.*, 2003).

Thus in addition to exploring conditions for a genome-wide transcriptome analysis of the model rhizobium, *S. meliloti*, in culture, the second pilot experiment highlighted the differential expression of several classes of genes during symbiosis. These genes are related to invasion, oxidative-stress protection, iron mobilization, and signaling, thus, emphasizing possible common mechanisms between symbiosis and pathogenesis (Ampe *et al.*, 2003).

These pilot experiments that were carried out on a limited number of ORFs also allowed us to define the conditions that can be used for whole genome *S. meliloti* macroarrays to study the mRNA global changes (for details, see Ampe and Batut, 2003). In fact, such whole *S. meliloti* genome macroarrays have been produced by spotting 6063 gene-specific PCR fragments, which represent 97.7% of the genome and were obtained by PCR amplification using the specific primers and re-amplification with the standard primers (Table 1). These macroarrays are presently being used extensively in hybridization experiments using mRNA generated both from cultures grown under varying physical conditions as well as from nodule bacteroids. Because largely the same PCR products are used as target for both macro- and micro-array experiments, it will also be possible to directly compare the results of both types of approaches, including their relative sensitivity.

5. CONCLUSIONS AND PERSPECTIVES

The recent availability of the entire annotated genome sequence of *S. meliloti* (Galibert *et al*, 2001) has opened new perspectives in the study of its biology through the use of DNA arrays. DNA arrays have already been used to study gene expression in several model bacteria, such as *Escherichia coli, Bacillus subtilis*, and several pathogens (*e.g.*, Arfin *et al.*, 2000; Sekowska *et al.*, 2001), but were so far not used for symbiotic model plant-associated bacteria, although Perret *et al.* (1999) pioneered this approach in rhizobia by determining the complete nucleotide sequence of the symbiotic plasmid of *Rhizobium* NGR234 as well as its transcriptome during symbiosis.

Different questions can be addressed by gene-expression profiling using the *S. meliloti* whole-genome arrays. Expression analyses on a global scale can greatly improve our understanding of regulatory networks on the level of transcription. Alterations of expression patterns can be monitored in response to either a specific stimulus or specific environmental conditions or *in planta*. Expression patterns of mutant and wild-type bacteria can be compared. Operons and regulons provide a conceptional biological framework for analysis of comprehensive expression profiling experiments. Comparisons of expression patterns of regulatory mutants and wild-type strains in response to a specific stimulus either in free-living culture and *in planta* contribute to a definition of regulons and a decription of regulatory responses to a perturbation. In addition to exploring conditions for a genome-wide transcriptome analysis of the model rhizobium *S. meliloti*, the present work has highlighted the differential expression of several classes of genes during symbiosis, suggesting possible common mechanisms between symbiosis and pathogenesis.

However, if a large proportion of the genes of a genome change transcript levels, understanding the regulatory networks underlying this response becomes a difficult task. The measurements from array experiments are estimates of transcript levels, but regulatory mechanisms work at the level of gene expression and protein activity. A search of upstream sequences of co-regulated genes for conserved motifs, in some cases, results in the prediction of binding sites of regulatory proteins. Classical approaches are then taken to analyze further the predicted regulons and regulatory mechanisms.

ACKNOWLEDGEMENTS

Construction of *S. meliloti* microarrays was funded by grants 0311752 and 031U213D from Bundesministerium für Forschung und Technologie, Germany, and BIZ 7 from Deutsche Forschungsgemeinschaft "Bioinformatik Initiative". A number of people from both the Institut für Genomforschung (Universität Bielefeld) and the Lehrstuhl für Genetik (Universität Bielefeld) contributed to microarray construction and establishment of protocols. In particular, we are grateful to B. Linke and F. Meyer for automated primer design, to S. Rüberg and E. Schulte-Berndt for performing PCRs, and to E. Krol for her contributions to the development of protocols and microarray hybridizations. Manual design of 143 primer pairs was carried out in the laboratory of J. Batut (INRA-CNRS, Toulouse). F. Ampe, H. Bergès, and E. Lauber carried out the work on macroarrays reviewed here. F. Ampe was supported by the Institut de Recherche pour le Développement (IRD, France). H. Bergès was supported by a fellowship from the Toulouse Génopôle. We thank V. Le Berre and A. Dagkesamanskaya for the spotting of PCR products on membranes. This work was supported by the Toulouse Génopôle, the department "Santé des Plantes et Environnement" of INRA (grant 0441-04), and the CNRS programme "Puces à ADN". We also thank Jacques Batut for directing the studies in Toulouse and for helpful discussions and comments on the manuscript.

REFERENCES

Ampe, F., Kiss, E., Sabourdy, F., and Batut, J. (2003). Transcriptome analysis of *Sinorhizobium meliloti* during symbiosis. *Genome Biol.*, *4*, R15.

Ampe, F., and Batut, J. (2003). Macro-arrays protocols for gene expression studies in bacteria. In A. Akkermans, F. J. de Bruijn, G. Kowalchuk and J. van Elsas (Eds.), *Molecular microbial ecology manual, 2nd Edit.* (in press). Dordrecht, The Netherlands: Kluwer Academic Publishers.

Arfin, S. M., Long, A. D., Ito, E. T., Tolleri, L., Riehle, M. M., Paegle, E. S., *et al.* (2000). Global gene expression profiling in *Escherichia coli* K12. The effects of integration host factor. *J. Biol. Chem.* *275*, 29672-29684.

Bardin, S., Dan, S., Osteras, M., and Finan, T. M. (1996). A phosphate transport system is required for symbiotic nitrogen fixation by *Rhizobium meliloti*. *J. Bacteriol.*, *178*, 4540-4547.

Bardin, S., Voegele, R. T. and Finan, T. M. (1998). Phosphate assimilation in *Rhizobium* (*Sinorhizobium*) *meliloti*: Identification of a *pit*-like gene. *J. Bacteriol.*, *180*, 4219-4226.

Barnett, M. J., Fisher, R. F., Jones, T., Komp, C., Abola, A. P., Barloy-Hubler, F., *et al.* (2001). Nucleotide sequence and predicted functions of the entire *Sinorhizobium meliloti* pSymA megaplasmid. *Proc. Natl. Acad. Sci. USA*, *98*, 9883-9888.

Becker, A. (2003). Design of microarrays for genome-wide expression profiling. In A. Akkermans, F. J. de Bruijn, G. Kowalchuk and J. van Elsas (Eds.), *Molecular microbial ecology manual, 2nd Edit.* (in press). Dordrecht, The Netherlands: Kluwer Academic Publishers.

Bergès, H., Lauber, E., Liebe, C., Batut, J., Kahn, D., de Bruijn, F. J., *et al.* (2003) Development of *Sinorhizobium meliloti* pilot macroarrays for transcriptome analysis. *Appl. Environ. Microbiol.*, *69*, 1214-1219.

Beringer, J. E. (1974). R factor transfer in *Rhizobium leguminosarum*. *J. Gen. Microbiol.*, *84,* 188-198.

Brewin, N. J. (1998). Tissue and cell invasion by *Rhizobium*: The structure and development of infection threads and symbiosomes. In H. P. Spaink, A Kondorosi, and P. J. J. Hooykaas (Eds.). *The rhizobiaceae* (pp.417-429). Dordrecht, The Netherlands: Kluwer Academic Publishers.

Capela, D., Barloy-Hubler, F., Gouzy, J., Bothe, G., Ampe, F., Batut, J., *et al.* (2001). Analysis of the chromosome sequence of the legume symbiont *Sinorhizobium meliloti*. *Proc. Natl. Acad. Sci. USA*, *98*, 9877-9882.

Catoira, R., Galera, C., de Billy, F., Penmetsa, V., Journet, E.-P., Maziiet, F., *et al.* (2000). Four genes of *Medicago truncatula* controlling components of a Nod factor pathway. *Plant Cell*, *12*, 1647-1666.

Cook, D. R., VandenBosch K., de Bruijn F. J., and Huguet T. (1997). Model legumes get the *nod*. *Plant Cell*, *3*, 275-281

Davey, M. E., and de Bruijn, F. J. (2000). A homologue of the tryptophan-rich sensory protein and FixL regulate a novel nutrient-deprivation *Sinorhizobium meliloti* locus. *Appl. Envron. Microbiol.*, *66*, 5353-5359.

David, M., Daveran, M. L., Batut, J., Dedieu, A., Domergue, O., Ghai, J., *et al.* (1988). Cascade regulation of *nif* gene expression in *Rhizobium meliloti*. *Cell*, *54*, 671-683.

Denarie, J., Debelle, F., and Prome, J. C. (1996). *Rhizobium* lipo-chitooligosaccharide nodulation factors: Signalling molecules mediating recognition and morphogenesis. *Annu. Rev. Biochem.*, *65*, 503-535.

Didier, G., Brézellee, P., Remy E., and Henaut, A (2001). GeneAnova-gene expression analysis of variance. *Bioinformatics*, *1,* 490-491.

Ferguson, G. P., Roop, I. I., and Walker, G. C. (2002). Deficiency of *a Sinorhizobium meliloti bacA* mutant in alfalfa symbiosis correlates with alteration of the cell envelope. *J. Bacteriol.*, *184*, 5625-5632.

Finan, T. M., Weidner, S., Chain, P., Buhrmester, J., Wong, K., Vorhölter, F.-J., *et al.* (2001). The complete sequence of the 1,683-kb pSymB megaplasmid from the N_2-fixing endosymbiont *Sinorhizobium meliloti*. *Proc. Natl. Acad. Sci. USA*, *98*, 9889-9894.

Fousard, M., Garnerone, A. M., Ni, F., Soupene, E., Boistard, P., and Batut, F. (1997) Negative autoregulation of the *Rhizobium meliloti fixK* gene is indirect and requires a newly identified regulator, FixT. *Mol. Microbiol.*, *25*, 27-37.

Galibert, F., Finan, T. M., Long, S. R., Pühler, A., Abola, P., Ampe, F., *et al.* (2001). The composite genome of the legume symbiont *Sinorhizobium meliloti*. *Science*, *293*, 668-672.

Glazebrook, J., Ichige, A., and Walker, G. C. (1993). A *Rhizobium meliloti* homolog of the *Escherichia coli* peptide-antibiotic transport protein SbmA is essential for bacteroid development. *Genes Dev.*, *7*, 1485-1497.

Gmuender, H., Kuratli, K., Di Padova, K., Gray, C., Keck, W., and Evers, S. (2001). Gene expression changes triggered by exposure of *Haemophilus influenzae* to novobiocin or ciprofloxacin, combined transcription and translation analysis. *Genome Res.*, *11*, 28-42.

Kaminski, P. A., Batut, J., and Boistard, P. (1998). A survey of nitrogen fixation by rhizobia. In H. P. Spaink, A Kondorosi, and P. J. J. Hooykaas (Eds.). *The rhizobiaceae* (pp. 431-460). Dordrecht, The Netherlands: Kluwer Academic Publishers.

Kane, D. K., Jatkoe, T. A., Stumpf, Lu, J., Thomas, J. D., and Madore, S. J. (2000). Assessment of the sensitivity and specificity of oligonucleotide (50mer) microarrays. *Nucl. Acids Res.*, *28*, 4552-4557.

Lockhard, D. J., and Winzeler, E. A. (2000). Genomics, gene expression and DNA arrays. *Nature*, *405*, 827-836.

Loos, A., Glaneman, C., Willis, L. B., O'Brian, M., Lessard, P. A., Gerstmeir, R., *et al.* (2001). Development and validation of *Corynebacterium* DNA microarrays. *Appl. Environ. Microbiol*, *67*, 2310-2318.

Milcamps , A., Ragatz, D. M., Lim, P., Berger, K. A. and de Bruijn, F. J. (1998). Isolation of carbon- and nitrogen-deprivation induced loci of *Sinorhizobium meliloti* by *Tn5-luxAB* mutagenesis. *Microbiol.*, *144*, 3205-3218.

Pellock, B. J., Cheng, H. P., and Walker, G. C. (2000) Alfalfa root nodule invasion is dependent on *Sinorhizobium meliloti* polysaccharides. *J. Bacteriol.*, *182*, 4310-4318.

Perret, X., Freiberg, C., Rosenthal, A., Broughton, W. J., and Fellay, R. (1999). High-resolution transcriptional analysis of the symbiotic plasmid of *Rhizobium* sp. NGR234. *Mol. Microbiol.*, *32*, 415-425.

Petersohn, A., Brigulla, M., Haas, S., Hoheisel, J., Völker, U., and Hecker, M. (2001). Global analysis of the general stress response of *Bacillus subtilis*. *J. Bacteriol.*, *183*, 5617-5631.

Rhodius, V., Van Dyk, T. K., Gross, C., and LaRossa, R. A. (2002). Impact of genomic technologies on studies of bacterial gene expression. *Ann. Rev. Microbiol.*, *56*, 599-624.

Richmond, C. S., Glasner, J. D., Mau, R., Jin, F. R., and Blattner, F. R. (1999) Genome-wide expression profiling in *Escherichia coli* K-12. *Nucl. Acids Res.*, *27*, 3821-3835.

Ricillo, P. M., Collavino, M. M., Grasso D. H., England, R., de Bruijn, F. J., and Aguilar, O. M. (2000). A *guaB* mutant strain of *Rhizobium tropici* CIAT899 pleiotropically defective in thermal tolerance and symbiosis. *Mol. Plant-Microbe Interact.*, *13*, 1228-1236.

Ricillo, P. M., Muglia, C. I., de Bruijn, F. J., Roe, A. J., Booth, I. R., and Aguilar, O. M. (2001) Glutathione is involved in environmental stress responses in *Rhizobium tropici*, including acid tolerance. *J. Bacteriol.*, *182*, 1748-1753.

Rozen, S., and Skaletsky, H. J. (1996, 1997, 1998). Primer3. Code available at http, //www-genome.wi.mit.edu/genome_software/other/primer3.html

Santos, R., Herouard, D., Sigaud, S., and Puppo, A. (2000). Oxidative burst in alfalfa-*Sinorhizobium meliloti* symbiotic interactions. *Mol. Plant-Microbe Interact.*, *14*, 86-89.

Sekowska, A., Robin, S., Daudin, J. J., Henaut, A. (2001). Extracting biological information from DNA arrays: an unexpected link beteen arginine and methionine metabolism in *Bacillus subtilis*. *Genome Biol.*, *2*, 0019.

Soupene, E., Fousard, M., Boistard, P., Truchet, G., and Batut, J. (1995). Oxygen as a key develomental regulator of *Rhizobium meliloti* N2 fixation gene expression within the alfalfa root nodule. *Proc. Natl. Acad. Sci. USA*, *92*, 3759-3763.

Trebiatowski, J. R., Ragatz, D. M., and de Bruijn, F. J. (2000). Isolation and regulation of *Sinorhizobium 1021* loci induced by oxygen limitation. *Appl. Envir. Micobiol.*, *67*, 3728-3731.

Vasse, J., de Billy, F., Camut, S., and Truchet, G. (1990). Correlation between ultrastructural differentiation of bacteroids and nitrogen fixation in alfalfa nodules. *J. Bacteriol.*, *172*, 4295-4306.

Voegele, R. T., Bardin, S., and Finan, T. M. (1997). Characterization of the *Rhizobium* (*Sinorhizobium*) *meliloti* high- and low-affinity phosphate uptake systems. *J. Bacteriol.*, *179*, 7226-7232.

Watson, R. J., Chan, Y. K., Wheatcroft, R., Yang, A. F., and Han, S. H. (1988). *Rhizobium meliloti* genes rquired for C4-dicarboxylate transport and symbiotic nitrogen fixation are located on a megaplasmid. *J. Bacteriol.*, *170*, 927-934.

Yoshida, K., Kobayashi, K., Miwa, Y., Kang, C., Matsunaga, M., *et al.* (2001). Combined transcriptome and proteome analysis as a powerful approach to study genes under glucose repression in *Bacillus subtilis*. *Nucleic Acids Res.*, *29*, 683-692.

CHAPTER 12

GENOME DYNAMICS IN RHIZOBIAL ORGANISMS

R. PALACIOS AND M. FLORES
Centro de Investigación sobre Fijación de Nitrógeno, UNAM, P.O. Box 565-A, Cuernavaca, Morelos 62170, México

1. INTRODUCTION

Bacterial genomes should be considered as dynamic structures; they are prone to generate rearrangements at relatively high frequencies. The most conspicuous rearrangements are due to homologous recombination between reiterated DNA sequences. According to the location and orientation of the reiterated elements in a genome, different types of rearrangements may be produced. As shown in the scheme presented in Figure 1, recombination between directly oriented repeated sequences in the same replicon can lead to either deletion or amplification of the segment bordered by the repeats. Conversely, recombination between sequences in different orientations can generate an inversion of the region bordered by such sequences. Recombination between reiterated elements located in different replicons may result in the co-integration of the corresponding replicons.

Some rearrangements have interesting biological consequences. In particular, gene amplification has been correlated with the adaptation of microorganisms to survival under either limiting or stress conditions. Deletions may result in the loss of specific non-essential functions. Co-integration between different replicons alters the general architecture of the genome. With the exception of deletions, rearrangements that are generated by homologous recombination are reversible events. Thus, alternative structures, which are generated by either amplifications or inversions or co-integrations, should be considered as transient states that do not alter permanently the structure of a genome.

Rearrangements generated by homologous recombination are found at relatively high frequency, usually from 10^{-2} to 10^{-5}. This means that a bacterial culture contains cells presenting all the different rearrangements that a genome may generate as long as they do not alter the survival of the cell in the particular conditions used. Derived strains containing specific rearrangements may be

R. Palacios and W. E. Newton (eds.), Genomes and Genomics of Nitrogen-fixing Organisms, 183-200.

isolated by different experimental strategies. Some of these strategies do not involve the introduction of exogenous DNA and, thus, the rearranged strains obtained are natural derivatives of wild-type strains. These strains should be ideal for both scientific and applied studies.

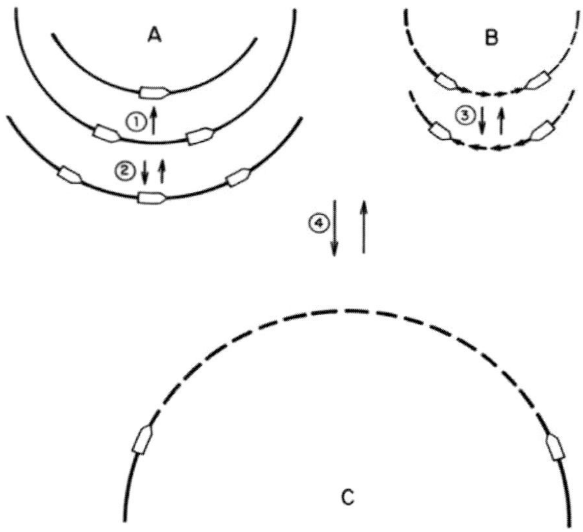

Figure 1. Different types of genomic rearrangements generated by homologous recombination.
The scheme represents regions of two replicons of a genome, A and B. Reiterated DNA elements are represented by open arrow heads. In replicon A, recombination between two directly oriented reiterated sequences can generate either a deletion (1) or a tandem duplication (2). In replicon B, recombination between two reiterated sequences located in different orientation can generate an inversion of the respective region (3). Recombination between reiterated elements present in replicons A and B (4) can generate a new replicon (C), product of the co-integration of the two original replicons.

The availability of the nucleotide sequence of the genome of a large number of microorganisms has open new possibilities for the study of genome dynamics. The reiterated elements of a genome can be localized and the potential rearrangements generated by homologous recombination can be predicted. Pathways of consecutive rearrangements leading to alternative genomic structures can be designed and the desired structures can be isolated. This chapter will discuss different aspects of genome dynamics using as a model the nitrogen-fixing symbiotic bacteria that belong to the genus *Rhizobium* and related genera, hereinafter referred to as rhizobial organisms.

2. REITERATED DNA SEQUENCES

About twenty years ago, at the time of the inauguration of the Nitrogen Fixation Research Center of the National University of México, an experiment aimed at the cloning of the nitrogenase genes in *Rhizobium etli*, which was known at that time as *Rhizobium leguminosarum* bv. *phaseoli*, revealed that the nitrogen-fixation gene sequences were reiterated in this organism (Quinto *et al.*, 1982). Southern blots of total restricted DNA from different strains were hybridized against a probe of the nitrogenase genes from *Klebsiella pneumoniae*. The autoradiography showed several restriction fragments in different isolates of *R. etli*. Further experiments showed that the symbiotic plasmid (pSym) of the model *R. etli* strain CFN42 contained three regions with nitrogenase genes; two of them contained functional nitrogenase *HDK* operons, whereas the third contained a functional *nifH* gene (Quinto *et al.*, 1985). The mapping (Girard *et al.*, 1991) and the recently obtained nucleotide sequence (González *et al.*, 2003) of the pSym of *R. etli* CFN42 have established the structure of the reiterated *nif* regions. Region *nif*-a contains *nifHDK* genes and a truncated *nifE* pseudogene; region *nif*-b contains the *nifHDKENX* genes; and region *nif*-c contains *nifH* and a truncated *nifD* pseudogene. The two largest reiterated regions, of *ca.* 5-kb, are located in direct orientation in the regions *nif*-a and *nif*-b, whereas the reiteration in region *nif*-c is located in an inverted orientation.

The presence of a large number of reiterated DNA sequences in rhizobial genomes was suggested (Flores *et al.*, 1987) and then confirmed by the current availability of the nucleotide sequence either of the complete genomes or of the symbiotic compartments of different strains. Reiterated sequences include complete operons, genes, pseudogenes, regulatory regions, and insertion sequences. Genomic projects have revealed that must of the reiterated sequences correspond to elements related to insertion sequences. Such sequences are particularly abundant in the symbiotic compartments. About 18% of the sequence of the pSym of *Rhizobium* sp NGR234 (Freiberg *et al.*, 1997) corresponds to elements related to insertion sequences. Figure 2 schematizes the position of such elements.

Figure 2. Schematic representation of elements related to insertion sequences in the pSym of R. *sp NGR234.*

According to the annotation of the 536,165 bp long pSym of NGR234 (Freiberg et al.*, 1997), the elements related to insertion sequences are represented as black bands.*

In the pSym of *R. etli* CFN42, 10% of the DNA sequence is related to insertion sequences (González *et al.*, 2003). The genome of *Mesorhizobium loti* strain MAFF303099 consists of a chromosome and two plasmids, pMLa and pMLb (Kaneko *et al.*, 2000). The chromosome contains a symbiotic island of 611 kb. In the symbiotic island, 19.6% of the sequence corresponds to elements related to insertion sequences, which contrasts dramatically with only 0.5% in the chromosome outside of the island, 7.5% in pMLa, and 2.9% in pMLb (Kaneko *et al.*, 2000). The genome of *Bradyrhizobium japonicum* USDA110 is a circular chromosome that contains a presumptive symbiotic island of 681 kb (Kaneko *et al.*, 2002). About 60% of the 167 transposase genes found in the chromosome are present in the symbiotic island (Kaneko *et al.*, 2002). The genome of *Sinorhizobium meliloti* (Galibert *et al.*, 2001; Capela *et al.*, 2001; Finan *et al.*, 2001; Barnett *et al.*, 2001) is composed of a chromosome and two megaplasmids, pSymA (which contains most of the nodulation and nitrogen-fixation genes) and pSymB. The whole genome of *S. meliloti* contains a relatively low amount of elements related to insertion sequences; however, their distribution is asymmetric, being more abundant in pSymA, especially near the symbiotic genes (Barnett *et al.*, 2001).

3. GENOMIC INSTABILITY

It is common knowledge among both research laboratories and industries working with rhizobial organisms that the symbiotic properties of some strains are unstable. Initial experiments performed with *Rhizobium trifolii*, *Bradyrhizobium japonicum*, and *Rhizobium etli* suggested that such instability is due to genomic rearrangements and, in particular, deletions that may affect the symbiotic properties of the strain (Djordjevic *et al.*, 1982; Zurkowski, 1982; Berry and Atherly, 1984; Kaluza *et al.*, 1985; Soberón-Chávez *et al.*, 1986; Hahn and Hennecke, 1987; Flores *et al.*, 1988). The occurrence of genomic rearrangements was observed by analyzing direct descendants of single *Rhizobium* cells with regard both to their DNA-hybridization patterns obtained against different recombinant plasmids (Flores *et al.*, 1988) and to their plasmid profile (Brom *et al.*, 1991).

The mechanisms responsible for the generation of genomic rearrangements in *Rhizobium* were first analyzed by the construction and use of genetic elements known as GDYN elements (Romero *et al.*, 1991). These elements allow the positive selection of different kinds of genomic rearrangements, in particular, amplifications and deletions. The first of these elements used in *Rhizobium* was GDYN1, which is a DNA cassette that carries both kanamycin/gentamycin- and spectinomycin/streptomycin-resistant markers. It also contains a region with the *SacR-SacB* genes of *Bacillus subtilis* that confer sucrose sensitivity on several Gram-negative bacteria, including *Rhizobium*. Upon introduction into *Rhizobium,* this element confers a high level of resistance to spectinomycin (100μg ml^{-1}), but a low level of resistance to kanamycin (15μg.ml^{-1}). Higher levels of resistance to kanamycin can be obtained by an increase in gene dosage, thus, providing a way to isolate amplified variants. On the other hand, either deletions or loss of plasmids can be isolated by selecting for sucrose-resistant derivatives (Romero *et al.*, 1991).

The GDYN1 element was first introduced in the *nifH* gene of region *nif*-c (see above) of the pSym of *R. etli* CFN42 (Romero *et al.*, 1991). An increase in the kanamycin concentration in the culture medium allowed selection of cells containing amplifications of a region of *ca.* 120 kb. The mechanism responsible for such amplification was the homologous recombination between the reiterated *nif* operons in regions *nif*-a and *nif*-b (see above). Duplications of the region were found in about 10^{-3} cells, whereas a higher level of amplification, up to 8-fold, was found in about 10^{-5} cells. The addition of sucrose to the culture medium selected for cells that had lost the GDYN1 element. Analysis of the DNA of such cells indicated that the 120-kb segment that extended from region *nif*-a to region *nif*-b was deleted as a consequence of the homologous recombination between the two reiterated nitrogenase operons located in these regions. Such deletions occurred at a frequency of about 10^{-4} (Romero *et al.*, 1991). Figure 3 shows the dynamics of amplification and deletion of the *nod-nif* region bordered by the reiterated nitrogenase operons of the pSym of *R. etli* CFN42.

The introduction of GDYN elements in different regions of the pSym indicated the existence of other end points for amplification and deletion (Romero *et al.*, 1995). Moreover, the use of a GDYN element as a transposon led to the conclusion that gene amplification is a general feature in the genome of *R. etli*, occurring at high frequency in both the chromosome and in the different plasmids (Flores *et al.*, 1993).

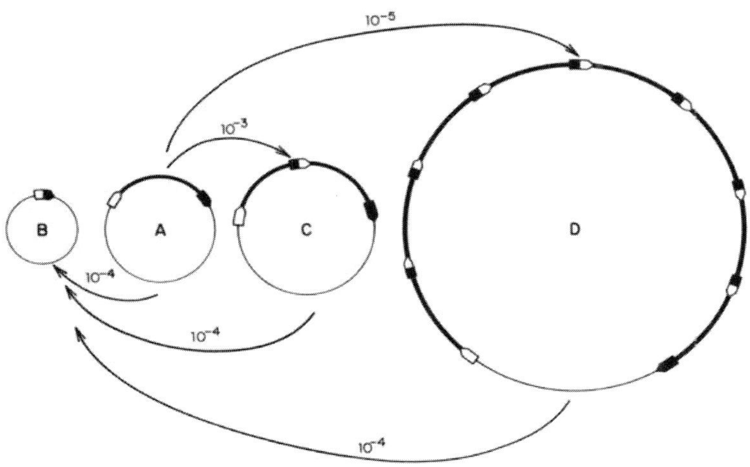

Figure 3. Amplification and deletion of a nod-nif *amplicon in the pSym of* R. etli.
The original structure of the pSym (A) presents an amplicon bordered by the two reiterated nif *operons (see text) represented by white and black arrow heads. This amplicon can generate deletion (B), tandem duplication (C), or amplification (D). The relative frequencies at which the different structures are found are shown in the respective connecting arrows.*

4. NATURAL GENE AMPLIFICATION

The analysis of the genome of different *Rhizobium* strains using GDYN elements (Romero *et al.*, 1991; Flores *et al.*, 1993; Romero *et al.*, 1995; Mavingui *et al.*, 1998) (see above), as well as more recent approaches based on the prediction of genomic rearrangements from the DNA sequence (see below), led to the conclusion that gene amplification occurs at high frequency in different regions of the rhizobial genome. Actually, gene amplification is probably ubiquitous in the genomes of prokaryotic organisms. The foundations for current interpretation of gene amplification were established by the genetic work of Roth, Hill and their associates in *Escherichia coli* and *Salmonella typhimurium* (Anderson and Roth, 1977; Petes and Hill, 1988). These pioneering studies were followed by an expansion of knowledge of natural gene amplification in a variety of prokaryotes (reviewed by Romero and Palacios, 1997).

Gene amplification may affect almost any region on the bacterial genome. The frequency of duplication for specific loci varies, usually from 10^{-2} to 10^{-5}. A common characteristic of amplifiable regions is the presence of long repeated sequences, located in direct orientation, flanking the region that undergoes tandem duplication and amplification. We have referred to those structures as "amplicons" (Flores *et al.*, 1993; Romero *et al.*, 1995; Romero and Palacios, 1997). Figure 4 schematizes the structure and dynamics of an amplicon. It can be defined as a region of the genome bordered by two repeated sequences present in direct orientation (Figure 4A). Homologous recombination between the repeated sequences may lead either to a tandem duplication (Figure 4C) or to a deletion of the amplicon sequence (Figure 4B). Once a tandem duplication is formed, recombination may lead to either further amplification (Figure 4D) or to return to the basal, non-amplified state.

The rate-limiting step for amplification is the initial duplication. An amplified region of the genome is a highly dynamic structure. Recombination is continually increasing or decreasing the level of amplification and, eventually, unless selective pressure is applied, the structure returns to the basal state. Amplification is, therefore, a transient phenomenon that does not compromise the general structure of the genome. It is interesting to point out that recombination in amplified regions results in the excision of complete amplicon sequences, either as monomeric or as polymeric closed circular structures (Figure 4E). We have developed an experimental strategy for cloning such amplicon structures *in vivo* from *Rhizobium* into *E. coli* (Flores *et al.*, 1993). The bacterial genome can be considered as a structure formed by overlapping amplicons. The location of the different families of reiterated elements determines the "amplicon structure" of the genome. Such structure may be predicted from the nucleotide sequence (see below).

Several lines of evidence suggest that the general biological role of gene amplification in prokaryotes is related to adaptation to extreme conditions that cannot be handled by the regulatory systems of the cell. Among such conditions are: antibiotic resistance (Perlman and Stickgold, 1977; Yagi and Clewel, 1980; Nichols and Guay, 1989; Matheus and Stewart, 1988); resistance to heavy metals (Kondratyeva *et al.*, 1995); growth under conditions of nutrient scarcity (Sonti and

Roth, 1989; Tlsty *et al*., 1984); and growth on exotic nutrient sources (Neuberger and Hartley, 1981; Ghosal and You, 1988; McBeth and Shapiro, 1984).

Other examples of the role of gene amplification in adaptive processes have been derived from both pathogenic and symbiotic interactions. Increased pathogenicity, due to amplification of specific DNA regions, has been shown in *Vibrio cholerae* (Goldberg and Mekalanos, 1986), *Haemophilus influenzae* (Hoiseth *et al*., 1986) and *Pseudomonas aeruginosa* (Deretic *et al*., 1986). In regard to symbiotic interactions, amplification of a region in the pSym of *Rhizobium tropici* leads to enhanced production of nodulation factors, which are necessary for nodule development (Mavingui *et al*., 1988).

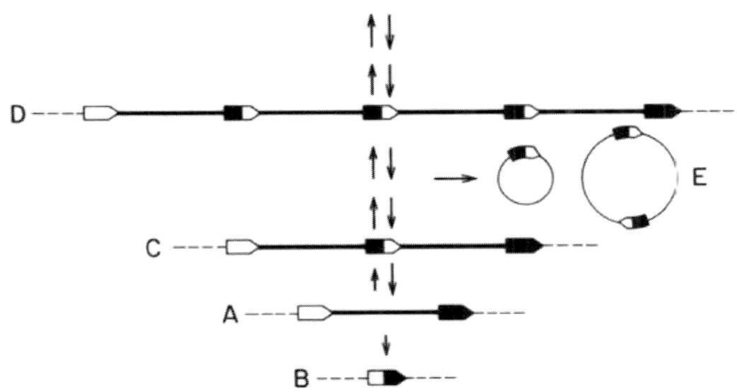

Fig. 4. Structure and dynamics of an amplicon.
An amplicon structure (A) is a region of the genome (solid line) bordered by two reiterated sequences present in direct orientation (white and black arrow heads). Recombination between the reiterated sequences can generate either a deletion (B) or a tandem duplication (C). The duplicated region can recombine generating either further amplification (D) or a return to the basal non amplified structure (A). The amplified state (D) is a highly dynamic structure; recombination can increase or decrease the level of amplification, and generates closed circular structures containing monomers or multimers of the whole amplicon structure (E). The rate limiting step in the amplification is the first tandem duplication (see text).

5. ARTIFICIAL GENE AMPLIFICATION

As discussed above, the location of reiterated sequences in a genome determines which regions may be amplified at high frequency. This is an inherent property of any particular genome. In addition to this "natural gene amplification", the genome can be manipulated to generate new amplicons, thus allowing the amplification of any desirable region. We will refer to such manipulations as "artificial gene amplification".

We have used different rhizobial strains to develop experimental approaches for artificial gene amplification. Two general strategies have been followed. In the first one, different segments of a genome are amplified at random. In the second strategy, a defined region of the genome is specifically amplified. Both strategies, random (RDA) (Mavingui *et al.*, 1997) and specific (SDA) DNA amplification (Castillo *et al.*, 1999), have resulted in the generation of derivative rhizobial strains with improved symbiotic properties (see below).

The RDA approach is based in the generation of amplicon-type structures by the co-integration of recombinant plasmids, which harbor randomly cloned DNA sequences, into the homologous region of the genome (Mavingui *et al.*, 1997). This strategy is schematized in Figure 5 as applied to the pSym of *Rhizobium tropici*.

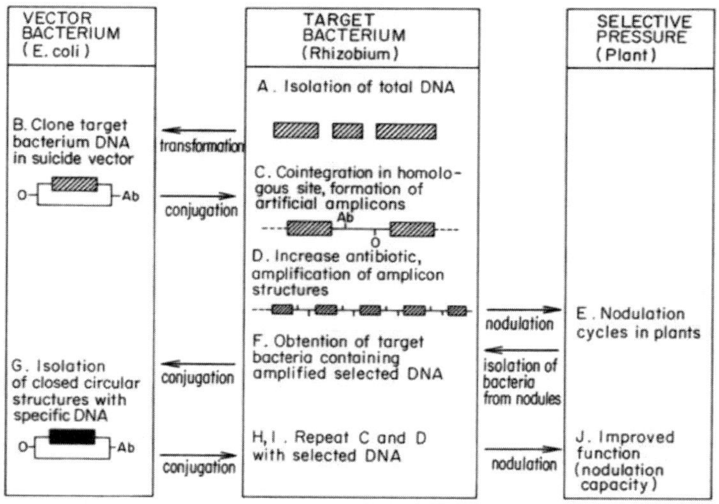

*Figure 5. Schematic representation of the random DNA amplification strategy. This strategy produces DNA amplifications in random regions of a genome of a target bacterium (*Rhizobium*). For the purpose of DNA transfer, a vector bacterium is used (*E. coli*). The amplified derivative strains of the target bacterium are subjected to a selective pressure (nodulation in plants). The different steps of the strategy (A-J) are explained in the text.*

Total DNA from the target bacterial strain is isolated (Figure 5A) and cloned in an appropriate plasmid in the vector bacterium, *E. coli* (Figure 5B). In this case, the plasmid is a suicide vector that is able to replicate in *E. coli* but not in *R. tropici*. The recombinant suicide vector is mobilized by conjugation from the vector bacterium to the target bacterium, by selecting for antibiotic resistance. In the target bacterium, the recombinant plasmids co-integrate into the homologous sites, so forming artificial amplicon structures (Figure 5C). An increase in antibiotic concentration selects for cells containing random amplifications of DNA (Figure 5D). The amplified target bacteria are then challenged with a selective pressure, in this case, nodulation cycles in plants (Figure 5E). In these conditions, the cells, which contain amplifications that better fit the particular conditions used, are selected and isolated (Figure 5F). Selected bacteria are conjugated with the vector bacteria to obtain plasmids containing the selected amplicon-type structures (Flores *et al.*, 1993) (Figure 5G). Such plasmids, which are in fact suicide vectors, are mobilized again to the target bacterium for co-integration and amplification of the corresponding selected DNA regions (Figures 5H and 5I). Finally, these derivative strains are tested for improved function (Figure 5J).

The potential success of the RDA approach relies on imposing a strong selective pressure on the amplified derivative strains generated in the procedure. If amplification of a particular region of the genome confers an advantage under the specific conditions used, then the better fit strains may be selected. When this strategy was applied to the pSym of *R. tropici*, strains with an increase in competitiveness for nodule formation were obtained (Mavingui *et al.*, 1997).

In the SDA approach (Castillo *et al.*, 1999), the target for amplification is a defined DNA region of the genome. The desired region is first cloned in a suicide vector in *E. coli*. The corresponding recombinant plasmid is mobilized by conjugation into the target bacterium, by selecting for an antibiotic-resistance marker of the vector. In the target bacterium, the recombinant plasmid co-integrates into the homologous site, forming an amplicon-type structure. Different concentrations of an antibiotic marker in the vector will select for cells containing different copies of the amplicon structure. Cells with different gene dosage of the amplified region are then tested for the particular desired function.

We applied the SDA approach to a region of the *nod* regulon of *Sinorhizobium meliloti* (Castillo *et al.*, 1999). This region contains the regulatory gene *nodD1*; the nodulation genes, *nodA*, *nodB*, and *nodC*, which encode the enzymes responsible for the synthesis of the core structure of Nod factors; and an operon, which is essential for nitrogen fixation. Derivatives of *S. meliloti* containing this fragment with an average copy number of 2.5 to 3 showed a significant increase in nodulation and nitrogen fixation and also promoted alfalfa growth.

6. DYNAMICS OF GENOME ARCHITECTURE

The genomes of rhizobial organisms are compartmentalized. In particular, the region that contains most of the nitrogen-fixation and nodulation genes is commonly present as either a plasmid (pSym) or as a symbiotic island in the chromosome. Genome architecture differs according to the species but is usually constant among different strains of the same species. The genome of *S. meliloti* 1021 (Galibert *et al.*, 2001) is composed of a chromosome (3,654,135 bp) (Capela *et al.*, 2001) and two megaplasmids, pSymA (1,354,325 bp) (Barnett *et al.*, 2001) and pSymB (1,683,333 bp) (Finan *et al.*, 2001). *Mesorhizobium loti* strain MAFF303099 consists of a chromosome (7,036,072 bp) and two plasmids, pMLa (351,911 bp) and pMLb (208,315 bp) (Kaneko *et al.*, 2000). *Bradyrhizobium japonicum* USDA110 has only a single replicon, a circular chromosome of 9,105,828 bp (Kaneko *et al.*, 2002). *Rhizobium* sp NGR234 contains a chromosome, a megaplasmid (larger than 2MB) and a symbiotic plasmid of 563,165 bp (Freiberg *et al.*, 1997; Flores *et al.*, 1998). Finally, the genome of *R. etli* CFN42 contains a chromosome and 6 plasmids, one of them (371,255 bp) corresponds to the pSym (González *et al.*, 2003). Figure 6 schematizes the genomic architecture of different rhizobial strains.

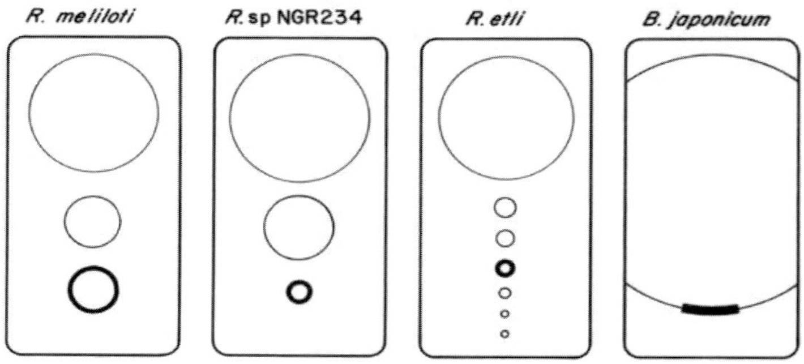

Figure 6. Genome architecture of rhizobial organisms.
The replicons that constitute the genome of different rhizobial model strains are represented by circles. The diameter is proportional to the size of the replicon; The size in bp of some of the replicons are given in the text. In all cases, the largest replicon is the chromosome. In R. *sp NGR234 and in* R. etli, *the size of the chromosome is tentative. In each of the genomes, the symbiotic compartment is represented by either a thick black circle (pSym A in* R. meliloti, *pSym in* R. *sp NGR234 and* R. etli*) or by an arc of a circle (symbiotic region in the chromosome of* B. japonicum*).*

As mentioned above (see Figure 1), homologous recombination between reiterated sequences located in different replicons may lead to co-integration of such replicons, thus altering the genomic architecture. Genomic projects have actually revealed the presence of repeated elements, mainly insertion sequences, which are shared by more than one replicon, in different genomes analyzed. We searched for changes in genomic architecture by analyzing the replicon profiles of individual colonies derived from the same bacterial culture. These studies were first performed in *R. sp* NGR234 (Mavingui *et al.*, 2002) and, more recently, in *S. meliloti* (Guo *et al.*, 2003).

As mentioned above, strain NGR234 presents three replicons, a chromosome, a megaplasmid, and the pSym, whereas *S. meliloti* stain 1021 presents a chromosome and two megaplasmids, pSymA and pSymB. The analysis of the replicon profiles of individual colonies led to the retrieval of alternative genomic architectures. From both strains, we could isolate derivatives containing the genetic information in either two replicons or in a single replicon, which are products of the co-integration of the three original structures. It is interesting to point out that different architectures do not alter significantly the symbiotic properties of the strains, suggesting that genome architecture can be manipulated without impairing important biological functions (Mavingui *et al.*, 2002; Guo *et al.*, 2003).

7. PREDICTION OF GENOMIC REARRANGEMENTS

The knowledge of the DNA sequence of a genome indicates the location of the different reiterated DNA families that may participate in homologous recombination events. From this knowledge, the potential rearrangements that a genome may generate can be predicted and, thus a dynamic map of the genome can be produced. A schematic dynamic map is presented as Figure 7, showing different types of potential genomic rearrangements.

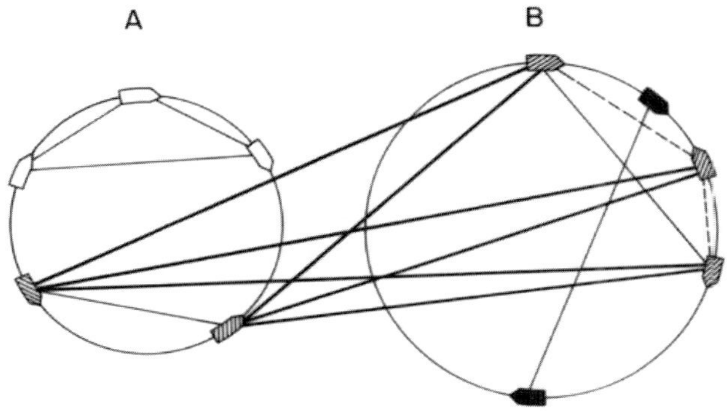

Figure 7. Prediction of genomic rearrangements based on the nucleotide sequence. A simplified hypothetical genome formed by two replicons (A and B) is shown. According to the nucleotide sequence, the reiterated DNA families may be located. White arrow heads represent a family of 3 direct repeated elements in replicon A; black arrow heads represent a family of 2 direct repeated elements in replicon B; dashed arrow heads represent a family of 5 elements shared by both replicons, two direct elements in replicon A and two direct and one inverted elements in replicon B. Potential rearrangements are indicated by lines joining the two elements involved in the recombination events. Solid thin lines, amplifications or deletions; broken lines, inversions; solid thick lines, replicon cointegrations.

We first predicted the potential rearrangements that could result from recombination of direct repeated DNA sequences in the pSym of R. sp NGR234 (Flores *et al.*, 2000). A dynamic map of this replicon is schematized in Figure 8. For the construction of this map, we took into account only those reiterated elements longer than one kilobase, had 100% sequence homology, and were present in direct orientation. There are 13 such elements in the plasmid. As discussed above, DNA regions bordered by two reiterated sequences in direct orientation (amplicons) can generate either deletions or amplifications. The pSym of NGR234 contains eight overlapping amplicons (Figure 8). We have shown that both amplification and deletion of each of these amplicons occur at high frequency in a culture of the wild-type strain (Flores *et al.*, 2000).

Based on the reported genomic sequence (Galibert *et al.*, 2001; Capela *et al.*, 2001; Barnett *et al.*, 2001; Finan *et al.*, 2001), we have recently generated a dynamic map of the whole genome of *S. meliloti* strain 1021, indicating potential sites for the generation of amplifications, deletions, inversions and replicon cointegrations (Guo *et al.*, 2003).

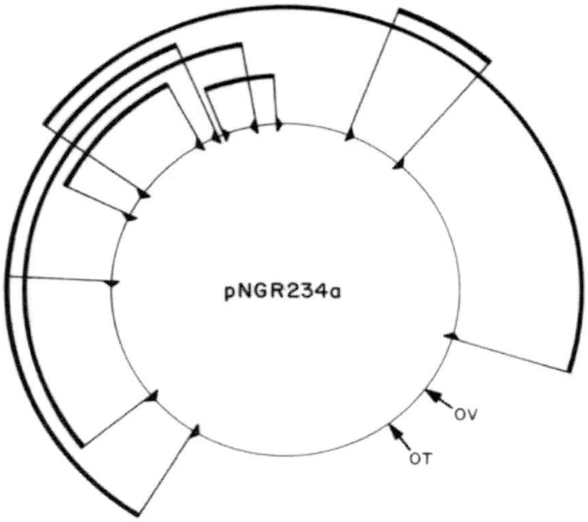

Figure 8. Prediction of genomic rearrangements in the pSym of R. *sp NGR234.*
Based on the nucleotide sequence of the pSym of R. *sp NGR234 (Freiberg* et al., *1997), the different families of reiterated elements present in direct orientation were localized (arrow heads). Such elements form 8 overlapping amplicon structures (thick solid lines joining the corresponding pair of reiterated elements). Each amplicon can be either deleted or amplified (see text). The origins of replication (OV) and transference (OT) are shown.*

8. IDENTIFICATION OF GENOMIC REARRANGEMENTS

We have developed a general experimental approach, based on the polymerase chain reaction (PCR), to detect different types of genomic rearrangements in a bacterial culture (Flores *et al.*, 2000). In accord with the DNA sequence, PCR primers, which border the reiterated sequences that may generate rearrangements, are synthesized. Such primers should be located near, but outside of, the two ends (5' and 3') of each reiterated sequence. The use of different combinations of primers to drive the PCR allows the detection of different types of rearrangements. This strategy is exemplified in Figure 9. The use of the 5' primer of the first repeat and the 3' primer of the second repeat detect a deletion; the 5' primer of the second repeat and the 3' primer of the first detect an amplification; either the 5' primer of the first repeat plus the 5' primer of the second or the 3' primer of the first repeat plus the 3' primer of the second detect an inversion; the 5' primer of a repeat located in one replicon and the 3' primer of a repeat located in another replicon detect a co-integration between the two replicons.

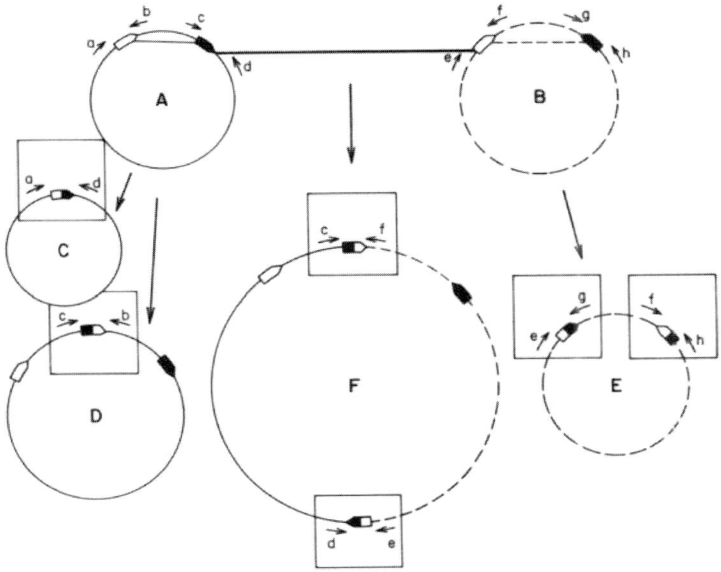

Figure 9. Identification of genomic rearrangements by a PCR strategy.
A simplified hypothetical genome containing two replicons (A and B) is shown. Reiterated elements and their orientation are represented by arrow heads. The relative positions of the oligonucleotides used to prime the PCR reaction for identification of the rearrangement are indicated (a-h). The structures generated by different types of rearrangements are presented: C, deletion; D, amplification; E, inversion; F, replicon cointegration. Squares indicate the joint point structures characteristic of each rearrangement, showing the position of the corresponding oligonucleotides. Details of the strategy are discussed in the text.

We have used successfully this approach to detect different types of rearrangements (Flores *et al.*, 2000; Mavingui *et al.*, 2002; Guo *et al.*, 2003). In our experience, rearrangements present in 10^{-5} cells in a bacterial culture can be clearly detected by this approach.

9. ARTIFICIAL SELECTION OF GENOMIC REARRANGEMENTS

Once a desirable genome rearrangement is identified in a bacterial culture, it is possible to obtain a cell sub-population that is "pure" for such a rearrangement. Actually, a sub-population can only be pure in the case of non-reversible rearrangements, such as deletions. In the case of reversible rearrangements (amplifications, inversions, and replicon co-integrations or excisions), some cells will return to the basal state. However, sub-populations more than 99% pure for a particular rearrangement can be obtained.

We have developed a strategy to purify specific rearrangements from wild-type bacterial cultures. In the first step, the goal is to obtain an enriched sub-population (see Figure 10). The culture is plated and colonies derived from individual cells are isolated. When a single cell divides to form a colony, at certain stage a particular rearrangement will appear. If such a rearrangement is generated at an early stage of the formation of the colony, a large proportion of cells of the derived culture will contain the rearrangement. Conversely, if the rearrangement first occurs at the later stages of colony formation, a relatively low proportion of the cells will present the rearrangement. If the cell from which the derived culture is generated contains the rearrangement, the derived culture should be "pure" for the particular rearrangement. Thus, cultures derived from individual cells should contain different relative proportions of specific rearrangements. We have experimentally confirmed these predictions (Flores *et al.*, 2000).

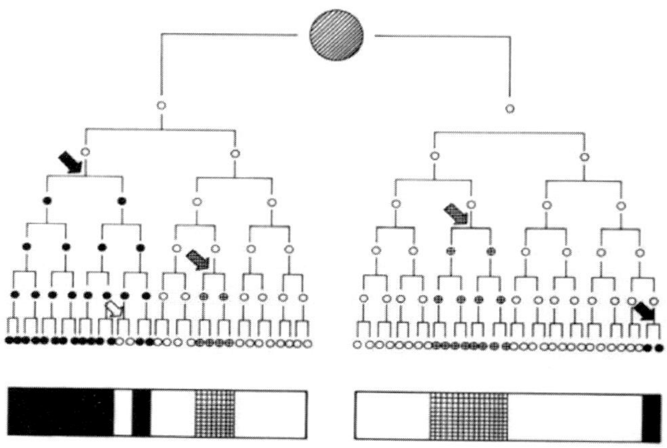

Figure 10. Relative proportion of specific rearrangements in colonies derived from single cells of a bacterial culture.
A bacterial culture (dashed circle in the top) is plated and colonies are derived from single cells. The scheme represents the development of two of such colonies. Arrows indicate the occurrence of specific rearrangements. The presence of cells harboring such rearrangements is indicated with the same graphic motive as the corresponding arrow. The white arrow indicates a reversion to the original non-rearranged structure. At the bottom, the areas with the corresponding graphic motifs indicate the relative proportion of the different rearrangements in the resulting cultures.
Details are presented in the text.

The cultures derived from single cells are then analyzed for the specific rearrangement. For this purpose, a quantitative PCR method can be used, priming the reaction with the appropriate oligonucleotides to identify the join-point structure characteristic of the rearrangement. Usually, the analysis of 100-1000 cultures derived from single cells is sufficient to obtain an enriched sub-population in which 10^{-1}-10^{-3} cells contain the desired rearrangement.

In the second step, the enriched sub-population is plated again and colonies derived from single cells are analyzed, searching for a sub-population "pure" for the desired rearrangement. To confirm the purity, the analysis of most of the cultures derived from single cells of the potential "pure" sub-population must reveal the presence of the rearrangement at high concentration. An optimal concentration of product after the corresponding PCR is a good criterion for different types of rearrangements. In the case of rearrangements that alter the architecture of the genome, such as replicon cointegrations or excisions, the analysis of the replicon profile by gel electrophoresis is a method of choice.

Using this strategy, we have obtained sub-populations "pure" for specific rearrangements both in *R.* sp NGR234 (Flores *et al.*, 2000; Mavingui *et al.*, 2002) and in *S. meliloti* (Guo *et al.*, 2003). It is interesting to point out that this strategy does not use the introduction of exogenous DNA and that no selective pressure is applied. The sub-populations "pure" for specific rearrangements are actually natural derivatives of wild-type strains. These sub-populations are thus ideal strains for either scientific or applied purposes.

10. NATURAL GENOMIC DESIGN

The capacity to predict, identify, and purify specific genomic rearrangements allows a novel type of genome manipulation that we have proposed as "natural genomic design" (Flores *et al.*, 2000). In the first step, the nucleotide sequence of a genome is analyzed and the different potential genomic rearrangements are predicted. In the second step, a pathway of consecutive rearrangements leading to a desired alternative genome structure is designed. In the third step, a sub-population "pure" for the first rearrangement of the design is obtained by an artificial selection procedure (see above). Then, in further steps, sub-populations pure for each of the remaining rearrangements of the design are consecutively isolated until the final desired structure is obtained. This procedure is schematized in Figure 11. We have successfully used this strategy to obtain novel genomic architectures in *S. meliloti* (Guo *et al.*, 2003).

11. CONCLUDING REMARKS

The knowledge accumulated in regard to the mechanisms responsible for the generation of genomic rearrangements and their biological consequences, together with the availability of the complete nucleotide sequence of a continuously growing number of prokaryotic genomes, is resulting in novel strategies for the generation of alternative genome structures. Some of the alternative structures obtained could be of great value for either scientific or applied purposes. In a broad sense, this type of manipulation could be considered as part of a man-guided prospective evolution of prokaryotic organisms.

ACKNOWLEDGEMENTS

The work from our laboratory was supported in part by Grant ER025/028 from CONACyT-México. The authors wish to thank Virginia Quinto, Rosa Ma. Ocampo, Angeles Moreno, and Marisa Rodríguez for technical assistance.

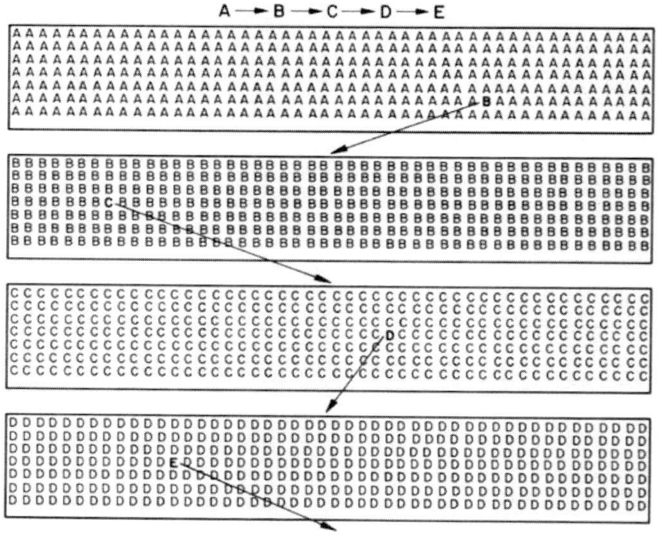

Figure 11. Strategy for a natural genomic design.

A wild-type strain has a particular genomic structure (A). Based on the nucleotide sequence, a series of consecutive rearrangements are designed to obtain a desired alternative structure of the genome (E). In the original culture (A),some cells contain the first rearrangement (B). A culture "pure" for the structure B is obtained (see text). Such culture contains some cells presenting the next rearrangement of the design (C). A culture pure for structure C is then obtained. The strategy continues until a culture "pure" for structure E is obtained (see text).

REFERENCES

Anderson, R. P., and Roth, J. R. (1977). Tandem genetic duplications in phage and bacteria. *Annu. Rev. Microbiol., 31,* 473-505.

Barnet, M. J., Fisher, R. F., Jones, T., Komp, C., Abola, A. P., Barloy-Hubler, F., *et al.* (2001). Nucleotide sequence and predicted functions of the entire *Sinorhizobium meliloti* pSymA megaplasmid. *Proc. Natl. Acad. Sci. USA, 98*, 9883-9888.

Berry, J. O., and Atherly, A. G. (1984). Induced plasmid-genome rearrangements in *Rhizobium japonicum*. *J. Bacteriol., 157*, 218-224.

Brom, S., García de los Santos, A., Girard, L., Dávila, G., Palacios, R., and Romero, D. (1991). High frequency rearrangements in *Rhizobium leguminosarum* bv *phaseoli* plasmids. *J. Bacteriol., 173*, 1344-1346.

Capela, D., Barloy-Hubler, F., Gouzy, J., Bothe, G., Ampe, F., Batut, J., *et al.* (2001). Analysis of the chromosome sequence of the legume symbiont *Sinorhizobium meliloti* strain 1021. *Proc. Natl. Acad. Sci. USA, 98*, 9877-9882.

Castillo, M., Flores, M., Mavingui, P., Martínez-Romero, E., Palacios, R., and Hernández, G. (1999). Increase in alfalfa nodulation, nitrogen fixation and plant growth by specific DNA amplification (SDA) in *Sinorhizobium meliloti*. *Appl. Enrivon. Microbiol., 65*, 2716-2722.

Deretic, V. P., Darzins, T. A., and Chakrabarty, A. M. (1986). Gene amplification induces mucoid phenotype in *rec-2 Pseudomonas aeruginosa* exposed to kanamycin. *J. Bacteriol., 165*, 510-516.

Djordjevic, M. A., Zurkowski, W., and Rolfe, B. G. (1982). Plasmids and stability of symbiotic properties of *Rhizobium trifolii*. *J. Bacteriol., 161*, 560-568.

Finan, T. M., Weidner, S., Wong, K., Buhrmester, J., Chain, P., Vorhölter, F. J., *et al.* (2001). The complete sequence of the 1,683-kb pSymB megaplasmid from the N_2-fixing endosymbiont *Sinorhizobium meliloti*. *Proc. Natl. Acad. Sci. USA, 98*, 9889-9894.

Flores, M., González, V., Brom, S., Martínez, E., Piñero, D., Romero, *et al.* (1987). Reiterated sequences in *Rhizobium* and *Agrobacterium* spp. *J. Bacteriol., 169*, 5782-5788.

Flores, M., González, V., Pardo, M. A., Leija, A., Martínez, E., Romero, *et al.* (1988). Genomic instability in *Rhizobium phaseoli*. *J. Bacteriol., 170*, 1191-1196.

Flores, M., Brom, S., Stepkowski, T., Girard, M. L., Dávila, G., and Palacios, R. (1993). Gene amplification in *Rhizobium:* Identification and *in vivo* cloning of discrete amplifiable DNA regions (amplicons) from *Rhizobium leguminosarum* bv. *phaseoli*. *Proc. Natl. Acad. Sci. USA, 90*, 4932-4936.

Flores, M., Mavingui, P., Girard, L., Perret, X., Broughton, W. J., Martínez-Romero, E., *et al.* (1998). Three replicons of *Rhizobium* sp. strain NGR234 harbor symbiotic gene sequences. *J. Bacteriol., 180*, 6051-6053.

Flores, M., Mavingui, P., Perret, X., Broughton, W. J., Hernández, G., Dávila. G., *et al.* (2000). Prediction, identification and artificial selection of DNA rearrangements in *Rhizobium*: Toward a natural genomic design. *Proc. Natl. Acad. Sci. USA, 97*, 9138-9143.

Freiberg, C., Fellay, R., Bairoch, A., Broughton, W. J., Rosenthal, A., and Perret, X. (1997). Molecular basis of symbiosis between *Rhizobium* and legumes. *Nature, 387*, 394-401.

Galibert, F., Finan, T. M., Long, S. R., Puhler, A., Abola, P., Ampe, F., *et al.* (2001) The composite genome of the legume symbiont *Sinorhizobium meliloti*. *Science, 293*, 668-672.

Girard, M. L., Flores, M., Brom, S., Romero, D., Palacios, R., and Dávila, G. (1991). Structural complexity of the symbiotic plasmid of *Rhizobium leguminosarum* bv. *phaseoli*. *J. Bacteriol., 173*, 2411-2419.

Ghosal, D., and You, I. S. (1988). Gene duplication in haloaromatic degradative plasmids pJP4 and pJP2. *Can. J. Microbiol., 34*, 709-715.

Goldberg, I., and Mekalanos. J. J. (1986). Effect of a *recA* mutation on cholera toxin gene amplification and deletion events. *J. Bacteriol., 165*, 723-731.

González, V., Bustos, P., Ramírez-Moreno, M. A., Medrano-Soto, A., Salgado, H., Hernández-González, I., *et al.* (2003). The mosaic structure of the symbiotic plasmid of *Rhizobium etli* CFN42 and its relation with other symbiotic genome compartments. *Genome Biol., 4*, R36.

Guo, X., Flores, M., Mavingui, P., Fuentes, S., Hernández, G., Dávila, G., *et al.* (2003). Natural genomic design in *Sinorhizobium meliloti*: Novel genomic architecture. *Genome Res., 13*, 1810-1817.

Hahn, M., and Hennecke, H. (1987). Mapping of a *Bradyrhizobium japonicum* DNA region carrying genes for symbiosis and an asymmetric accumulation of reiterated sequences. *Appl. Environ. Microbiol., 53*, 2247-2252.

Hoiseth, S. K., Moxon, E. R., and Silver, R. P. (1986). Genes involved in *Haemophilus influenzae* type b capsule expression are part of an 18-kilobase tandem duplication. *Proc. Natl. Acad. Sci. USA, 83*, 1106-1110.

Kaluza, K., Hahn, M., and Hennecke, H., (1985). Repeated sequences similar to insertion elements clustered around the *nif* region of the *Rhizobium japonicum* genome. *J. Bacteriol., 162*, 535-542.

Kaneko, T., Nakamura, Y., Sato, S., Asamizu, E., Kato, T., Sasamoto, S., *et al.* (2000). Complete genome structure of the nitrogen-fixing symbiotic bacterium *Mesorhizobium loti*. *DNA Res., 7*, 331-338.

Kaneko, T., Nakamura, Y., Sato, S., Minamisawa, K., Uchiumi, T., Sasamoto, S., *et al.* (2002). Complete genome sequence of nitrogen-fixing symbiotic bacterium *Bradyrhizobium japonicum* USDA 110. *DNA Res., 9*, 189-197.

Kondratyeva, T. F., Muntyan, L. N., and Karavaiko, G. I. (1995). Zinc- and arsenic-resistant strains of *Thiobacillus ferrooxidans* have increased copy numbers of chromosomal resistance genes. *Microbiol., 141*, 1157-1162.

Mathews, P. R., and Stewart, P.R. (1988). Amplification of a section of chromosomal DNA in methicillin-resistant *Staphylococcus aureus* following growth in high concentrations of methicillin. *J. Gen. Microbiol., 134*, 1455-1464.

Mavingui, P., Flores, M., Romero, D., Martínez-Romero, E., and Palacios, R. (1997). Generation of *Rhizobium* strains with improved symbiotic properties by random DNA amplification (RDA). *Nature Biotechnol., 15*, 564-569.

Mavingui, P., Laeremans, T., Flores, M., Romero, D., Martínez-Romero, E., and Palacios, R. (1998). Genes essential for nod factor production and nodulation are located on a symbiotic amplicon (AMP Rtrpc60) in *Rhizobium tropici*. *J. Bacteriol., 180*, 2866-2984.

Mavingui, P., Flores, M., Guo, X., Dávila, G., Perret, X., Broughton, W. J., *et al.* (2002). Dynamics of genome architecture in *Rhizobium* sp. strain NGR234. *J. Bacteriol., 184*, 171-176.

McBeth, D. L., and Shapiro, J. A. (1984). Reversal by DNA amplifications of an unusual mutation blocking alkane and alcohol utilization in *Pseudomonas putida*. *Mol. Gen. Genet., 197*, 384-391.

Neuberger, M. S., and Hartley, B. S. (1981). Structure of an experimentally evolved gene duplication encoding ribitol dehydrogenase in a mutant of *Klebsiella aerogenes*. *J. Gen. Microbiol., 122*, 181-191.

Nichols, B. P., and Guay, G. G. (1989). Gene amplification contributes to sulfonamide resistance in *Escherichia coli*. *Antimicrob. Agents Chemother., 12*, 2042-2048.

Perlman, D., and Stickgold, R. (1977). Selective amplification of genes on the R plasmid NR1, in *Proteus mirabilis*: An example of the induction of selective gene amplification. *Proc. Natl. Acad. Sci. USA, 74*, 2518-2822.

Petes, T. D., and Hill, C. W. (1988). Recombination between repeated genes in microorganisms. *Annu. Rev. Genet., 22*, 147-168.

Quinto, C., De la Vega, H., Flores, M., Fernández, L., Ballado, T., Soberón, G., *et al.* (1982). Reiteration of nitrogen fixation gene sequences in *Rhizobium phaseoli*. *Nature, 299*, 724-726.

Quinto, C., De la Vega, H., Flores, M., Leemans, J., Cevallos, M. A., Pardo, M. A., *et al.* (1985). Nitrogenase reductase: A functional multigene family in *Rhizobium phaseoli*. *Proc. Natl. Acad. Sci. USA, 82*, 1170-1174.

Romero, D., Brom, S., Martínez-Salazar, J., Girard, M. L., Palacios, R., and Dávila, G. (1991). Amplification and deletion of a non-*nif* region in the symbiotic plasmid of *Rhizobium phaseoli*. *J. Bacteriol., 173*, 2435-2441.

Romero, D., Martínez-Salazar, J., Girard, L., Brom, S., Dávila, G., Palacios, R., *et al.* (1995). Discrete amplifiable regions (amplicons) in the symbiotic plasmid of *Rhizobium etli* CFN42. *J. Bacteriol., 177*, 973-980.

Romero, D., and Palacios, R. (1997). Gene amplification and genomic plasticity in prokaryotes. *Annu. Rev. Genet., 31*, 91-111.

Soberón-Chavez G., Nájera, R., Olivera, H., and Segovia, L. (1986). Genetic rearrangements of a *Rhizobium phaseoli* symbiotic plasmid. *J. Bacteriol., 167*, 487-491.

Sonti, R. V., and Roth, J. R. (1989). Role of gene duplications in the adaptation of *Salmonella typhimurium* to growth on limiting carbon sources. *Genetics, 123*, 19-28.

Tlsty, T. D., Albertini, A. M., and Miller, J. H. (1984). Gene amplifiction in the *lac* region of *E. coli*. *Cell, 37*, 217-224.

Yagi, Y., and Clewell, D. B. (1980). Amplification of the tetracycline resistance determinant of plasmid pAMa1 in *Streptococcus faecalis*: Dependence on host recombination machinery. *J. Bacteriol., 143*, 1070-1072.

Zurkowski, W. (1982). Molecular mechanism for loss of nodulation properties of *Rhizobium trifolii*. *J. Bacteriol., 150*, 999-1007.

Chapter 13

IMPACT OF GENOMICS ON THE RECONSTRUCTION OF EVOLUTIONARY RELATIONSHIPS OF NITROGEN-FIXING BACTERIA AND IMPLICATIONS FOR TAXONOMY

P. VAN BERKUM[1] AND B. D. EARDLY[2]
[1]*Soybean Genomics and Improvement Laboratory, Agricultural Research Service, U. S. Department of Agriculture, 10300 Baltimore Blvd., Beltsville, MD 20705, and* [2]*Pennsylvania State University, Berks Campus, Tulpehocken Road, P.O. Box 7009, Reading, PA 19610, USA*

1. SYSTEMATICS

Swedish botanist Carolus Linnaeus (Karl von Linné) introduced a system of classification of plants based on overall morphological similarities among their sexual organs. He established a workable classification of living organisms in his Systema Naturae published in 1735 that resulted in the birth of modern taxonomy. The Linnaean System of classification is a hierarchy that includes the categories Kingdom, Phylum, Class, Order and Suborder, Family, Genus, and Species, and still remains the basic framework for taxonomy in the biological sciences. The primary category in his classification scheme is the species for which a "type specimen" is chosen to reflect the belief that all earthly items are projections of their perfect heavenly forms.

There are several ways in which organisms are currently classified. Evolutionary classification uses any pertinent information available to construct a dendrogram that illustrates phylogenetic relatedness. Evidence may be gathered from a variety of sources, including the fossil record, comparative morphological and biochemical studies, and through comparisons of any other available characteristic, such as ecology and lifecycle. The problem with this approach is that there are many gaps in the fossil record and assumptions have to be made. Also, there is no standard definition of each taxonomic group and the classification method is not measurable. Phenetic classification, or numerical taxonomy, is based

R. Palacios and W. E. Newton (eds.), Genomes and Genomics of Nitrogen-fixing Organisms, 201-219.

solely on the shared similarity of traits between two organisms and is calculated using a large number of morphological characteristics. Because there is the problem of convergent evolution, many different characteristics need to be examined. One disadvantage of this method is that it is necessary to decide whether two traits are either variations within the same character or are characters in their own right. Cladistic classification (sometimes called phylogenetics) is a scheme that uses variation in characters reflected by differences in the genetic composition of organisms. The similarity of characters and their derivation are represented in a cladogram that is intended to portray the ancestry of extant species. Unlike phenetic classification methods, which are often used to infer the evolutionary time span (or distance) that separates two species, cladistic analysis is intended to infer branch points in a tree that represents the evolutionary relationships among the organisms. There is a trend, especially when classifying organisms in the Kingdom Monera, to rely heavily on cladistic classification to both estimate evolutionary relationships and guide decisions of taxonomy and nomenclature. However, with the dawn of genome sequencing and comparative genomics, it is becoming increasingly evident that genome plasticity should have appreciable influence on interpretations of phylogenetic relationships.

Nitrogen fixation is a biological process confined to the Super Kingdom Prokarya (or the Prokaryotes) and the Kingdom Monera, which by some has been suggested to consist of the Subkingdoms, Archaea and Eubacteria. They are distinguished from life forms in all other kingdoms in that they do not have a membrane-bound nucleus containing the genetic material of the cell. The study of microbes was initiated within the fields of hygiene, medicine, fermentation, and food technology. Bacteriology as a science originated from botany, which supplied the basis for microbial systematics. Classifying bacteria on the basis of their appearance or morphology is extremely difficult because they are generally quite small and have simple shapes. However, there are some bacteria with nitrogen-fixation capability, notably the cyanobacteria and actinomycetes that have sufficiently complex morphology to permit some classification by shape. In addition to shape, bacteria have traditionally been identified and classified on the basis of their biochemistry and the conditions under which they grow. Using the methods of numerical taxonomy, morphological variation and cultural characteristics have frequently been used to produce phenetic classification schemes for bacteria. More recently, phenetic classification of bacteria has been combined with cladistic classification in an attempt to increase the number of characters for taxonomic purposes. The combination of these two schemes has been called the Polyphasic Approach.

Cladistic classification of bacteria originally relied on rudimentary determinations of traits, such as either genome G+C content or DNA homology, to ascertain differences or similarities among entire genomes. However, these two characters failed to provide sufficient detail about the evolutionary relationships among the genomes. Therefore, the goal became the reconstruction of evolutionary history through the study of patterns of the genetic diversity, and from these results to predict the ancestral types. There is some disagreement among bacterial systematists about defining a bacterial species but, until the advent of genomics, there was the general consensus that a bacterial species should be monophyletic or at least approximately so. This meant that most of the DNA of the members of a given species should be descended from a single common ancestral genome.

2. CURRENT REFLECTIONS FOR EVOLUTION OF DIAZOTROPHY

Newton (2000) has outlined the possible circumstances under which Earth's environment may have provided the necessary conditions for the evolution of nitrogen fixation, first as an abiological process ultimately incorporated by living cells. A case was made that initially the Fe-only nitrogenase came into existence (Chisnell *et al.*, 1988) and that this was the precursor for the three nitrogenases that have been described among extant microbial species (Newton, 2000). However, two alternative hypotheses have been suggested to explain the development of the nitrogen-fixing process and the sequence with which the three different nitrogenases evolved (Newton, 1993; Newton, 2000; Postgate, 1974; Postgate and Eady, 1988). Whatever the view relative to development of diazotrophy, it is evident that the process is widely distributed among highly divergent members of the prokaryotes. This has fostered suggestions that maybe either diazotrophy initially was a common character that, with evolutionary time, was randomly lost (Newton, 2000) or that evolution of extant *nif* clusters was more recent and was followed by their spread by lateral gene transfer (Postgate, 1974).

3. RECONSTRUCTION OF EVOLUTIONARY RELATIONSHIPS AMONG MEMBERS OF THE KINGDOM MONERA.

From evidence based on the reconstruction of evolutionary relationships, it would appear that there are two distinct groups of prokaryotes (Figure 1), the Bacteria (or Eubacteria) and the Archaea (or Archaebacteria). It has been suggested that these two groups diverged near the time of the origin of life and that they each belong to one of the three domains of life (the Bacteria, the Archaea and the Eucarya). Because each of these domains has two or more kingdoms, it is possible that the Kingdom Monera eventually will be replaced by these two domains.

The reconstruction of microbial phylogeny has undergone dramatic progress with the advent of sequencing analysis of the ribosomal genes. Sequencing the 16S rRNA gene, in particular, has profoundly affected the description of the relationships among the bacteria (Maidak *et al.*, 1994; Olsen *et al.*, 1994). The 16S rRNA gene sequence has been declared useful for this purpose because it is slowly evolving and the gene product is both universally essential and functionally conserved. Basing bacterial phylogeny on 16S rRNA gene sequence variation not only presupposes that evolution throughout the genome progresses at a constant rate by mutation and Darwinian selection, but it also assumes that the evolution of the genome and of the 16S rRNA gene, in particular, is strictly hierarchical. In a hierarchical evolutionary pattern, genes are passed from generation-to-generation vertically by descent and are not shared between existing cells as a result of horizontal transfer. Furthermore, the assumption is that the 16S rRNA gene can map evolutionary paths of entire genomes. From a practical standpoint, this approach requires either that each genome harbors a single copy of the 16S rRNA gene or that multiple alleles within single genomes have identical sequences.

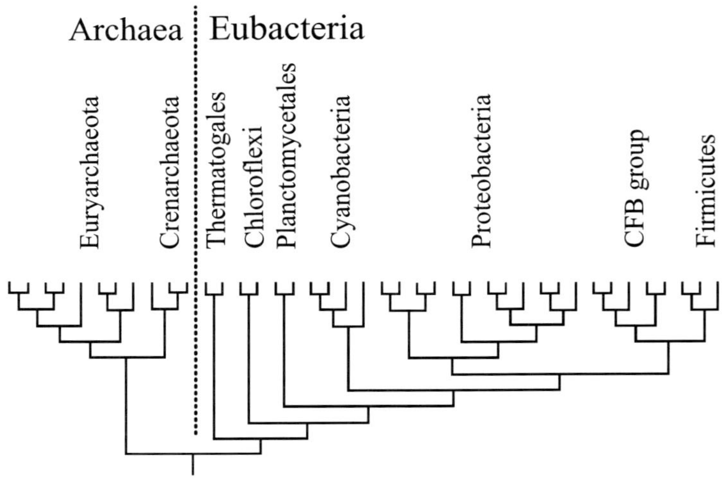

Figure 1. Phylogenetic relationships among members of the Kingdom Monera reconstructed from 16S rRNA gene sequence divergence.

Although the results from some multi-locus sequencing studies give the impression of congruence between 16S rRNA gene trees and corresponding trees based on other loci (Gaunt *et al.*, 2001), usually the discordance is more readily apparent (Feil *et al.*, 2001). Other lines of evidence also may be cited to suggest that some of the assumptions mentioned above may not always be justified. For example, linear discontinuities within 16S rRNA gene sequences were suggested as evidence for intragenic recombination among divergent alleles of this locus (Eardly, *et al.*, 1996; Smith *et al.*, 1999). Furthermore, the species *Thermomonospora chromogena* harbors two different sets of functional expressed 16S rRNA genes, one of which was acquired by a horizontal-transfer event (Yap *et al.*, 1999). This observation is supported by evidence indicating that the presence of heterologous rRNA molecules within the same cell may not be as problematic as once believed (Asai *et al.*, 1999).

Besides the uncertainties regarding the assumptions made when interpreting evolutionary relationships from divergent 16S rRNA gene sequences, there also are problems with the data when drawing conclusions for taxonomic purposes. Because the 16S rRNA gene sequences are highly conserved, it has been suggested that, perhaps for reasons described above, the scale of variation in the data is not sensitive enough to differentiate between closely related species (Fox *et al.*, 1992; Stackebrandt and Goebel, 1994). Although Stackebrandt and Goebel (1994) concluded that a 16S rRNA gene sequence similarity of less than 97% may be interpreted as representing strains of different species, this suggestion may be limited by the possible presence of divergent 16S rRNA alleles within single cells of magnitudes that may reach up to 6.4% sequence variation (Wang *et al.*, 1997).

4. THE RAPID SPREAD OF ANTIBIOTIC RESISTANCE: IMPLICATIONS OF RETICULATE MICROBIAL EVOLUTION

The discovery of penicillin, which became widely available in the 1940s, paved the way for development of many different antibiotics. Antibiotic use in medicine transformed the treatment of many infectious diseases that did not respond to other forms of medication. Besides medical applications, *ca.* 50% of the antibiotics manufactured are used in agriculture and aquaculture both to stimulate growth in livestock and to reduce disease caused by overcrowded animal husbandry practices (Teuber, 1999). The disadvantage associated with overuse of antibiotics, both in medicine and agriculture, is that it provides selective pressure on microbial communities for the survival of resistant mutants and elimination of those susceptible. The result is a dramatic increase in the incidence of resistance to antibiotics among many pathogenic or clinical microorganisms (Mazel and Davies, 1999).

Although the origin of antibiotic resistance genes is unclear, plausible sources that have been proposed include housekeeping genes, self-protection genes in antibiotic-producing microbes, and naturally evolved genes in soil communities (Bushman, 2002; Davies, 1997). Mutation as a mechanism by which resistance genes emerged among clinically important microbes was considered less likely than acquisition of those genes by horizontal transfer (Maiden *et al.*, 1998; Mazel and Davies, 1999). Horizontal gene transfer is defined as the movement of DNA between bacteria other than by descent in which genetic information travels through the generations as the cell divides. It is more similar to a sexual process that requires a mechanism for the mobilization of chromosomal DNA among bacterial cells. Therefore, horizontal gene transfer is a reticulate rather than a hierarchical evolutionary process that ensures survival of recombinants with genes that provide selective advantage over other members of the microbial community. In this manner, it was the clinical and agricultural overuse of antibiotics that provided selective pressure for the rapid spread of resistance genes by horizontal transfer among microbial communities (Amabile-Cuevas *et al.*, 1995; Nwosu, 2001; Ochman *et al.*, 2000; Teuber, 1999).

5. MECHANISMS OF HORIZONTAL GENE TRANSFER IN MICROBES

Genetic exchange in prokaryotes differs from genetic exchange in higher organisms (Cohan, 2001). There are three processes by which microbes are able to share genetic information through horizontal gene transfer. These processes are transformation, conjugation, and transduction (Droge *et al.*, 1999; Ochman *et al.*, 2000; Zgur-Bertok, 1999).

Transformation is the uptake of naked DNA that can mediate the exchange of any part of a chromosome and is most common in bacteria that are naturally transformable. Competence for transformation develops during the life cycle of both Gram-positive and Gram-negative bacteria. Typically, only short DNA fragments are exchanged by transformation. Mechanisms of transformation in Gram-positive and Gram-negative bacteria have been reviewed (Dubnau, 1999).

Conjugation is the transfer of DNA mediated by conjugal plasmids that may carry transposons. The process requires cell-to-cell contact and can occur between distantly related bacteria or even between bacterial and eukaryotic cells. An example of conjugation between a bacterial species and eukaryotic cells is the transfer of either the entire Ti plasmid or a 20-kb fragment (T-DNA) from *Agrobacterium tumefaciens* to plant cells, which is followed by the expression of agrobacterial genes within the transformed cells. Conjugation can involve the transfer of long fragments of DNA.

Transduction is the transfer of DNA by phage requiring the donor and recipient cells to share surface receptors for phage binding, which limits the mechanism to closely related bacteria. However, this process can be divided into two types, generalized and specialized transduction. In generalized transduction, host DNA is randomly packaged into the phage head. A second cycle of infection may transfer this DNA to a recipient, followed by its incorporation into the genome. Specialized transduction includes the integration of phage DNA into the genome called prophages. Phage head packaging in the infected cell can only include regions of host DNA directly adjacent to the insertion sites. This is a common mechanism by which diphtheria toxin genes (Groman, 1984) and the cholera toxin locus (Waldor and Mekalanos, 1996) are transferred. The length of DNA transferred is limited by the size of the phage head.

Processes and genetic determinants for homologous recombination in prokaryotes have been reviewed (Smith, 1988).

6. SIGNIFICANCE OF HORIZONTAL GENE TRANSFER IN NATURE.

Horizontal gene transfer among bacteria allows the rapid, effective, and competitive exploration of new ecological niches. The three processes of horizontal gene transfer in microbes are well understood and their utility in molecular-biology research is well documented. What is less clear is the extent by which these processes occur outside the laboratory in nature. A particular concern is that genes, which are the products of genetic engineering either in novel microbes or in modified crop plants, upon their release may transfer by one or more of these processes to the general microbial population in the environment (Davison, 1999; Droge *et al.*, 1998). The concern for taxonomy is that such genetic exchange would challenge the key assumption that evolutionary relationships among members of the Kingdom Monera are strictly hierarchical.

6.1. Transformation

DNA is present in most environments as the result of excretion, cell death, and autolysis (Davison, 1999; Droge *et al.*, 1999). Because the DNA in the environment originated from a multitude of sources, it is highly variable containing an abundance of different gene sequences and it is present mostly in a form that is available to bacteria, which are naturally transformable (Day, 1998; Lorenz and Wackernagel, 1994). Naturally transformable bacteria include diazotrophs; for example, the heterotroph, *Azotobacter vinelandii*, the cyanobacterium, *Nostoc muscorum*, and the legume symbiont, *Sinorhizobium meliloti* (Day, 1998; Lorenz and Wackernagel, 1994). DNA is very sensitive to hydrolysis (Blum *et al.*, 1997; Fibi *et al.*, 1991; Maeda and Taga, 1974; Paul *et al.*, 1987; Phillips *et al.*, 1989; Romanowski *et al.*, 1992; Romanowski *et al.*, 1993) by DNAases produced by the micro flora (Greaves and Wilson, 1970; Paul *et al.*, 1988), but it can persist for up to 2 months in soil (Romanowski *et al.*, 1992; Romanowski *et al.*, 1993). Although DNA hydrolysis would appear to be more rapid in aquatic than in terrestrial environments (Droge *et al.*, 1999), to our knowledge, the only known reported example of transformation in nature is in a freshwater habitat (Williams *et al.*, 1996). Natural transformation in aquatic environments and in soil has been reviewed (Davison, 1999; Day, 1998; Droge *et al.*, 1999; Wackernagel *et al.*, 1998).

6.2. Transduction

The potential for transduction as a mechanism for genetic recombination of bacteria in the environment has been described and several different environmental factors have been postulated to affect this process in nature (Droge *et al.*, 1999). Phage particles may be found in high numbers in soils (Campbell *et al.*, 1995; Germida and Casida, 1983; Lanning and Williams, 1982; Reanney and Marsh, 1973) and in aquatic environments (Berg *et al.*, 1989; Bratbak *et al.*, 1990; Ewert and Paynter, 1980; Hara *et al.*, 1991; Paul *et al.*, 1991; Paul *et al.*, 1993; Proctor and Fuhrman, 1990; Wommack *et al.*, 1992). Because the isolation of phage particles that infect nitrogen-fixing bacteria has been described (Abdel Basit *et al.*, 1991; Abebe *et al.*, 1992; Ahmad and Morgan, 1994; Ali *et al.*,1998; Bancroft and Smith, 1989; Barnet, 1972; Bishop *et al.*, 1977; Dhar and Ramkrishna, 1987; Dhar *et al.*, 1993; Elmerich *et al.*, 1982; Germida, 1986; Hashem *et al.*, 1986; Hegazi and Jensen, 1973; Hegazi and Leitgeb, 1976; Hu *et al.*, 1981; Kankila and Lindstrom, 1994; Kowalski *et al.*, 1974; Lajudie and Bogusz, 1984; Lawson and Barnet, 1984; Lindstrom and Kaijalainen, 1991; Patel and Craig, 1984; Patel *et al.*, 1985; Seldin *et al.*, 1984; Wdowiak *et al.*, 2000; Werquin *et al.*, 1988), there also is the potential for transduction as a mechanism for genetic exchange in diazotrophs in the environment. Although from laboratory studies it can be inferred that transduction is an important mechanism for the exchange of genetic information among bacteria, only some definitive results have been obtained from analyses in aquatic environments (Morrison *et al.*, 1978; Ripp and Miller, 1995; Saye *et al.*, 1987, 1990). Nevertheless, transduction is thought to be important also in terrestrial environments because many bacterial genomes harbor phage sequences (Blattner *et al*., 1997; Bolotin *et al*., 2001; Ferretti *et al*., 2001; Kunst *et al*., 1997; Miller and Sayler, 1992; Ogunseitan *et al*., 1992; Perna *et al*., 2001; Simpson *et al*., 2000).

6.3. Conjugation

Conjugation involves physical contact between cells for the transfer of distinct plasmids from donor to recipient. Often the plasmids are transferred across highly divergent genetic backgrounds. Plasmids are circular, autonomously replicating DNA elements that bear a variety of different genes and may be vehicles by which other non-plasmid-borne genes are transferred. The *tra* gene is required for the conjugal transfer of self-transmissible plasmids and also provides in *trans* the conjugal mechanism for mobilizable plasmids rendered by products of the *mob* gene. However, self-transmissible plasmids also may provide the mechanism for conjugal transfer in *cis* by combining with another plasmid (co-integration) before transfer. In some cases, chromosomally-located genes may be mobilized by plasmids that have the ability to co-integrate into the chromosome, as for example plasmid R68.45 (Currier and Morgan, 1982; Kinkle *et al*., 1993). The majority of the reports describing bacterial gene transfer in the natural environment are related to conjugation (Davison, 1999). Conjugation has been documented in animal ecosystems, the rhizosphere, on plant leaves, in non-polluted water and soil, and in polluted environments (Davison, 1999). Particularly relevant are the transfer of: symbiotic determinants between *Sinorhizobium fredii* and *Rhizobium leguminosarum* (Kinkle and Schmidt, 1991); numerous genes, including *nod* and *nif*, between bacterial genomes within the genus *Mesorhizobium* (Sullivan *et al*., 1995; Sullivan *et al*., 1996); and genes for antibiotic resistance, mercury resistance, and the degradation of 2,4-dichlorophenoxyacetic acid among serogroups of *Bradyrhizobium japonicum* (Kinkle *et al*., 1993).

7. EVIDENCE FOR LATERAL GENE TRANSFER AND RECOMBINATION IN MICROBIAL GENOMES

With some significant exceptions (*e.g*., Sullivan and Ronson, 1998), it has not been possible to directly monitor the transfer of DNA from one bacterial genome to another in natural populations. Consequently, the evidence for the acquisition of genes usually relies on unusual features of the transferred regions when compared to the characteristics of the DNA of the remaining genome. Because the evidence is indirect, it is possible that the extent of gene transfer and recombination among microbial genomes may be underestimated. Evidence for lateral gene transfer may be inferred from phylogenetic analyses when comparing evolutionary relationships from results of different loci and subsequently identifying incongruence (Eisen, 2000; Herrick *et al*., 1997; Koonin *et al*., 2001; Ragan, 2001). Although there are many possible explanations for incongruence besides horizontal gene transfer, the phylogenetic reconstruction method is probably the only approach to infer historical events from gene sequences (Koonin *et al*., 2001). As an alternative, lateral gene transfer may also be inferred from a best sequence match for a locus harbored by highly divergent genomes (Koonin *et al.*, 2001; Logsdon and Faguy, 1999; Markham *et al.*, 1999). The identification of regions within the genome that have either unusual G+C contents or unusual codon usage may also be indicative of gene transfer and recombination (Delorme *et al.*, 1994; Eisen, 2000; Garcia *et al.*, 2000; Kuroda *et al.*, 2001; Lawrence and Ochman, 1998; Martin, 1999). Finally, from either partial or complete genome sequencing projects, insertion-sequence (IS) elements, phage remnants, and many other patches of unusual composition indicate genome plasticity through horizontal transfer among divergent bacterial genomes (Blattner *et al.*, 1997; Bolotin *et al.*, 2001; Bourgoin *et al.*, 1996; Edwards *et al.*, 2002; Ferretti *et al.*, 2001; Kunst *et al.*, 1997; Kuroda *et al.*, 2001; Nelson *et al.*, 1999; Nolling *et al.*, 2001; Perna *et al.*, 2001; Ruepp *et al.*, 2000; Salanoubat *et al.*, 2002; Simpson *et al.*, 2000; Takami *et al.*, 2000).

8. GENOMIC ISLANDS.

"Genomic islands" is a collective term used to describe mobile elements that occur as distinct units on the core chromosomes of bacteria. Although they can vary in size from 1 to 500 kb or more, they exhibit several unifying characteristics. It is particularly important to understand that they may be present in the genomes of closely related strains and that their presence may confer a significant change in bacterial phenotype. Frequently, they differ in both G+C content and codon usage from the remaining genome, and often they are flanked by specific repeat sequences that are generated following integration into the host genome *via* recombination. Usually, they are associated with tRNA loci that have 3' ends identical to attachment sites of bacteriophages, indicating that these are regions for the insertion of foreign DNA. Genomic islands often possess either genes or cryptic pseudo genes coding for genetic mobility and origins for replication and may be, but are not necessarily, unstable (Hentschel and Hacker, 2001).

Genomic islands were first defined following a study of the chromosomally-located hemolysin determinants in *E. coli* (Hacker *et al.*, 1990; Hacker *et al.*, 1983) and they were termed pathogenicity islands or PAIs (Hacker *et al.*, 1990) because they conferred the bacterial strain with a pathogenic phenotype. Hacker and Kaper (2000) have presented a simplified model diagram of a bacterial pathogenicity island, showing its presence integrated into a tRNA locus in a pathogenic strain and its absence from the tRNA locus in a closely related non-pathogenic variant. Since the original description, it has become evident that pathogenicity islands occur in the genomes of various human, animal, and plant pathogens and may carry determinants for adherence, toxins, iron uptake, invasion, and type III or type IV secretion. However, not all genomic islands are associated with pathogenicity but may encode different functions.

Although genomic islands that are associated with resistance to either antibiotics (Ito *et al.*, 1999) or sucrose metabolism (Hochhut *et al.*, 1997) have been described, the symbiosis island of *Mesorhizobium loti* (Sullivan and Ronson, 1998) is the most relevant in a description of evolution of nitrogen-fixing bacteria. Although the evidence for the presence of a symbiosis island in the genome of *M. loti* appears conclusive, to our knowledge, no other example within the genus

Mesorhizobium or any other rhizobial genus has been described. Similarly, there is no other evidence for a genomic island that carries a full complement of *nif* genes. Certainly, if evidence for the existence of a "*nif* island" is obtained, it would in part offer some explanation for the presence of diazotrophy among bacteria with widely divergent genomic backgrounds. It may be relevant in the search for genomic islands to be aware that they need not necessarily be located in chromosomes, but also may be plasmid-borne or be part of phages (Hacker and Kaper, 1999). Also of relevance is the realization that only *ca.* 75% of genomic islands are associated with tRNA loci (Hacker and Kaper, 1999).

9. MICROBIAL EVOLUTION AND GENETIC RECOMBINATION

Genetic variability is fundamental to Darwinian evolution. Point mutations, genomic rearrangements, and horizontal gene transfer are important driving forces of microbial evolution, creating genetic variants for the selection and survival of the most successful genomic combinations (Arber, 1993). By their nature, point mutations result in small changes, which require long time frames for their cumulative effect to result in the emergence of descendants by a hierarchical evolutionary process. In contrast, genomic rearrangements and lateral gene transfer, which are mediated by plasmids, phages, and genomic islands, result in almost instantaneous reticulate microbial evolutionary histories. However, evolution of the core microbial genome is complemented by evolution of the genetic elements that transfer between different genomes (Hacker and Kaper, 2000). Therefore, microbial evolution is a complex process permitting genomes to quickly adapt to rapidly changing environmental pressures.

The biomass of prokaryotic protoplasm has been suggested to exceed half that of all life on earth with a total number of cells of approximately 1.0×10^{30} that have an annual cellular production rate of about 1.7×10^{30} cells per year (Whitman *et al.*, 1998). There is a vast potential for genetic diversity among the prokaryotes because of this large population size and production rate. Diversity resulting from point mutations has been estimated at five such events occurring in every gene shared by the population every 60 years (Whitman et al. 1998). However, estimates for the influence of recombination on diversity may be much higher. Feil *et al.* (1999) concluded from an analysis of genetic variation among housekeeping genes of *Neisseria meningitides*, *Escherichia coli*, or *Streptococcus pneumoniae* that a single nucleotide was from 10-fold to 80-fold more likely to change as a result of recombination than as a result of mutation. Therefore, by combining the conclusion of Feil *et al.* (1999) with the estimated rate of mutation (Whitman et al. 1998), there would seem to be the potential for approximately 5 simultaneous recombination events in every gene shared by prokaryotes every year.

10. ESTABLISHED SPECIES CONCEPTS APPLIED TO BACTERIA

Considerable attention has been given to the various species concepts that have been proposed for bacteria because the species is the most fundamental unit in both taxonomy and systematics. Kluyver and van Neil (1936) stated: "that the only truly scientific foundation of classification is in appreciating the available facts from a phylogenetic view". In the most recent version of the Bergey's Manual of Systematic Bacteriology, Brenner *et al.* (2001) indicated that most new bacterial species are defined using a pragmatic polyphasic (consensus) approach that integrates all available data, but places considerable emphasis on DNA similarity between organisms from measurements of DNA homology. Although there has been little disagreement as to the relative merits of a polyphasic approach for defining taxonomic species limits, the suggested emphasis on DNA-DNA hybridization as the primary criterion has been criticized. The focus of the criticisms have been on: (i) the labor-intensiveness of the method and non-transportability of the results, which in practical terms does not allow for sufficient sampling of genetic diversity in natural populations; (ii) an arbitrary cutoff (70% similarity) to define species limits; and (iii) that a species definition using these criteria are not real entities and, thus, may be of only limited value in understanding evolutionary dynamics in natural populations. A related criticism, which is particularly relevant to genera within the *Rhizobiaceae*, is the potential ambiguity that may arise as the result of acquisition (or loss) of large, horizontally transferred genomic segments (*e.g.*, a symbiotic island; Sullivan *et al.*, 1996).

Over the years, population biologists have proposed two species concepts that may be used to address some of the problems associated with the standard taxonomic (or typological) species concept. They are the Biological Species Concept (BSC) and the Ecological Species Concept (ESC).

Genetic exchange is frequently cited as being a cohesive force in plant and animal populations (Meglitch, 1954). The BSC, as proposed by Mayr (1982), defines a species as a population of organisms that share a common gene pool by genetic exchange. The application of this concept to bacteria was first proposed by Dykhuizen and Green (1991) on the basis of observations from extensive bacterial population surveys. From these surveys, it was evident that intraspecific recombination among strains of *Escherichia coli* and among strains of its sister species, *Salmonella enterica*, occurred much more often within, rather than between, these lineages. Dykhuizen and Green (1991) suggested that a species could be recognized as a group of bacteria whose individual gene trees would be incongruent because of allele scrambling during recombination. The subsequent realization that gene transfer can, and occasionally does, occur across vast genetic distances, necessitated an alternative definition of the concept of a "common gene pool" for bacteria. Lawrence (2001) suggested that the BSC could be revised to describe a species as a group of organisms that mutually exchange genes substantially more often with each other than with other organisms. The BSC also has been criticized because rates of intraspecific recombination for certain bacteria may be very different than for others (Maynard Smith *et al.*, 2000). Maynard Smith *et al.* (2000) also questioned the practicality of the BSC because recombinational frequencies decrease as the lineages diverge, leaving no obvious discontinuities by which to circumscribe the groups. Such considerations have led to the suggestion that additional criteria are needed to define bacterial species, particularly in a population's context (Lawrence, 2001).

The ESC defines a species as a group of organisms that exploit the same ecological niche (van Valen, 1973). The assumption behind this concept is that if two sets of organisms are attempting to occupy the same ecological niche, at the same place and at the same time, then stochastic processes will inevitably result in one of the groups displacing the other. The fine-scale dynamics that mediate the development and/or demise of such incipient ecological species have been described by Cohan (2001). He refers to them as ecotypes and describes them as being populations that can be represented genetically as discrete DNA sequence clusters. Theoretically, the longevity of a particular sequence cluster will be limited because, over time, a given cluster will likely be out-competed by an adaptive mutant from within the population. This type of periodic selection is sometimes referred to as a "selective sweep" (Guttman and Dykhuizen, 1994). The net effect of these selective sweeps would be to purge the population of nearly all its genetic diversity, conferring an appearance of genetic cohesion among the survivors. Cohan (2001) pointed-out that the resulting ecotypes that are successful in occupying a particular niche, share many properties with eukaryotic species. He also indicated that the role of genetic exchange among bacterial species is markedly different than its role among eukaryotic species, in that bacterial genetic exchange does not significantly hinder adaptive divergence. Consequently, he suggested that there is no justification for classifying bacteria solely on the basis of patterns of recombination.

11. A PROPOSED UNIFIED SPECIES CONCEPT FOR BACTERIA

Lawrence (2001) has proposed a Unified Species Concept (USC) for bacteria that is based principally on the BSC (as outlined above), but which modifies that concept by invoking the ESC as the primary arbiter of natural selection. According to the USC, a bacterial species is defined as: "a group of organisms that occupy a common niche and, as a result, exhibit effective rates of recombination that are greater among members within the group than with other organisms outside the group", *i.e.*, although interspecific recombination may occur, the resulting recombinants are, on average, less successful because the hybrids do not exploit the parental environment as effectively. Over time, the divergence of nucleotide sequences among members of the populations will reduce the likelihood of homologous DNA exchange between lineages, effectively imposing premating genetic isolation. Although the USC appears to address many of the criticisms directed toward the two earlier species concepts as applied to bacteria, it remains to be seen whether species limits in highly diverse natural populations (*e.g.*, in the soil) can be resolved using this approach.

12. RELEVANT INSIGHTS FROM RECENT GENOMIC COMPARISONS

Because the field of genomics is young and because the number of available genomic sequences for nitrogen-fixing species is small, the amount of taxonomic information that can be derived is somewhat limited. Nevertheless, from the genomic sequences that are currently available, it is evident that there are some interesting differences (and similarities) among the genomic architectures of symbiotic nitrogen-fixing bacteria and their pathogenic relatives. For instance, the symbiotic genes of some of these species have been acquired through a variety of evolutionary routes. Wood *et al.* (2001) concluded that, because the *nod* genes of *Sinorhizobium meliloti* and the *vir* genes of *Agrobacterium tumefaciens* contain G+C contents and codon usage patterns that are distinctly different from that of the genomes in which they reside, they probably were acquired relatively recently in evolutionary time. This observation affirms previous conclusions that genes involved in specialized symbiotic associations cannot be expected to provide an evolutionarily coherent basis for taxonomic classification (Piñero *et al.*, 1988, Eardly *et al.*, 1990).

Because the vast majority of genes that encode essential cellular functions are located on the bacterial chromosome (Capela *et al.*, 2002), it is reasonable to suggest that comparative analyses of these replicons might form a sound basis for a phylogenetically coherent taxonomic classification system. This suggestion is supported by an analysis of 2,573 chromosomally encoded orthologs in *M. loti* and *Sinorhizobium meliloti* (Morton, 2002), which indicated that the genes that appear to have been inherited vertically from the common ancestor of these species are those that are required for a free-living lifestyle. In a comparison between the chromosomes of *S. meliloti, M. loti*, and *A. tumefaciens*, Wood *et al.*, (2001) observed more substantive nucleotide co-linearity and gene-order conservation between the circular chromosomes of *A. tumefaciens* and *S. meliloti* than between the chromosomes of *S. meliloti* and *M. loti*. This observation is supported by proteomic and multi-locus DNA sequence comparisons, from which it may be inferred that *A. tumefaciens* shares a more recent common ancestor with *S. meliloti* than with *M. loti*. Galibert (2001) speculated that the *S. meliloti* chromosome was probably present in a progenitor of both *S. meliloti* and *M. loti* and that progeny of this ancestral lineage later acquired the genomic elements that have since evolved into the symbiotic megaplasmids, pSymA and pSymB. Building on this scenario, Wood *et al.* (2001) suggested that, because both the linear chromosome and the smaller plasmids of *A. tumefaciens* contain orthologs that are distributed across the three replicons of *S. meliloti*, *A. tumefaciens* probably diverged from its common ancestor with *S. meliloti* after *S. meliloti* had acquired its pSymA and pSymB megaplasmids.

13. IMPLICATIONS AND FUTURE STRATEGIES.

Advances in genomic characterization methods, such as genomic sequencing and multi-locus sequence typing, have made it possible to envisage a global DNA sequence-based framework upon which to base future taxonomic decisions (Palys *et al.*, 1997). In the implementation of this type of approach, however, important issues will need to be addressed, *e.g.*, which genomes should be sequenced and which species concept would be most relevant? Cot curve analyses, which have been used to examine the diversity of bacterial species in soils (Torsvik *et al.*, 1990), have led to estimates that more than one billion bacterial species may inhabit the biosphere (Dykhuizen, 1998). Even if this figure is a gross overestimate and the efficiency of DNA sequencing continues to improve, it is still likely that, within the foreseeable future, it will only be possible to sequence a tiny fraction of the genomes present on the planet. Consequently, if the breadth of bacterial species diversity is to be catalogued, it will require a much more efficient means of estimating genomic diversity. The problem of how to best define a bacterial species also will probably persist into the foreseeable future. Although it is encouraging that both population geneticists and taxonomists agree that phylogenetic relatedness should be a central guiding principle, the present lack of a method for reconstructing complex networks of relationships among recombining bacterial lineages presents a significant and perhaps even intractable problem for all except for the most clonal of species (Spratt, 1999). For this reason, existing methods for estimating phylogenetic relationships, such as those based on comparative 16S rRNA gene sequence analyses, will probably continue to enjoy popularity among taxonomists, in spite of the inherent risks of inferring the evolutionary history of an entire genome from the analysis of a single, perhaps unrepresentative, gene.

ACKNOWLEDGEMENTS

We thank Susan Wilzer, Current Awareness Literature Specialist, National Agricultural Library, USDA, Beltsville, for her outstanding assistance with the literature search for this chapter.

REFERENCES

Abdel Basit, H., Angle, J. S., Salem, S., Gewaily, E. M., Kotob, S. I., and van Berkum, P. (1991). Phenotypic diversity among strains of *Bradyrhizobium japonicum* belonging to serogroup 110. *Appl. Environ. Microbiol., 57*, 1570-1572.

Abebe, H. M., Sadowsky, M. J., Kinkle, B. K., and Schmidt, E. L. (1992). Lysogeny in *Bradyrhizobium japonicum* and its effect on soybean nodulation. *Appl. Environ. Microbiol., 58*, 3360-3366.

Ahmad, M. H., and Morgan, V. (1994). Characterization of a cowpea (*Vigna unguiculata*) rhizobiophage and its effect on cowpea nodulation and growth. *Biol. Fertil. Soils, 18*, 297-301.

Ali, F. S., Loynachan, T. E., Hammad, A. M. M., and Anarchi, Y. (1998). Polyvirulent rhizobiophage from a soybean rhizosphere soil. *Soil Biol. Biochem., 30*, 2171-2175.

Amabile-Cuevas, C. F., Cardenas-Garcia, M., and Ludgar, M. (1995). Antibiotic Resistance. *American Scientist, 83*, 320-329.

Arber, W. (1993). Evolution of prokaryotic genomes. *Gene, 135*, 49-56.

Asai, T., Zaprojets, D., Squires, C., and Squires, C. L. (1999). An *Escherichia coli* strain with all chromosomal rRNA operons inactivated: complete exchange of rRNA genes between bacteria. *Proc. Natl. Acad. Sci. USA, 96*, 1971-1976.

Bancroft, I., and Smith, R. J. (1989). Restriction mapping of genomic DNA from five cyanophages infecting the heterocystous cyanobacteria *Nostoc* and *Anabaena*. *New Phytol., 113*, 161-166.

Barnet, Y. M. (1972). Bacteriophages of *Rhizobium trifolii* I. Morphology and Host Range. *J. gen. Virol., 15*, 1-15.

Bergh, Ø., Børsheim, K. Y., Bratbak, G., and Heldal, M. (1989). High abundance of viruses found in aquatic environments. *Nature (London), 340*, 467-468.

Bishop, P. E., Supiano, M. A., and Brill, W. J. (1977). Technique for isolating phage for *Azotobacter vinelandii*. *Appl. Environ. Microbiol., 33*, 1007-1008.

Blattner, F. R., Plunketti, G. I., Bloch, C. A., Perna, N. T., Burland, V., Riley, M., *et al.* (1997). The complete genome sequence of *Escherichia coli* K-12. *Science, 277*, 1453-1462.

Blum, S. A. E., Lorenz, M. G., and Wackernagel, W. (1997). Mechanism of retarded DNA degradation and prokaryotic origin of DNases in nonsterile soils. *Syst. Appl. Microbiol., 20*, 513-521.

Bolotin, A., Wincker, P., Mauger, S., Jaillon, O., Malarme, K., Weissenbach, J., *et al.* (2001). The complete genome sequence of the lactic acid bacterium *Lactococcus lactis* ssp. lactis IL1403. *Genome Res., 11*, 731-753.

Bourgoin, F., Guedon, G., Pebay, M., Roussel, Y., Panis, C., and Decaris, B. (1996). Characterization of a mosaic ISS1 element and evidence for the recent horizontal transfer of two different types of ISS1 between *Streptococcus thermophilus* and *Lactococcus lactis*. *Gene*, 178, 15-23.

Bratbak, G., Heldal, M., Norland, S., and Thingstad, T. F. (1990). Viruses as partners in spring bloom microbial trophodynamics. *Appl. Environ. Microbiol., 56*, 1400-1405.

Bushman, F. (2002). *Lateral DNA transfer. Mechanisms and consequences*. Cold Spring Harbor, NY: Cold Spring Harbor Laboratory Press.

Campbell, J. I. A., Albrechtsen, M., and Sorensen, J. (1995). Large *Pseudomonas* phages isolated from barley rhizosphere. *FEMS Microbiol. Ecol., 18*, 63-74.

Chisnell, J. R., Premakumar, R., and Bishop, P. E. (1988). Purification of a second alternative nitrogenase from a *nif*HDK deletion strain of *Azotobacter vinelandii*. *J Bacteriol., 170*, 27-33.

Cohan, F. M. (2001). Bacterial species and speciation. *Syst. Biol., 50*, 513-524.

Currier, T. C., and Morgan, M. K. (1982). Direct DNA repeat in plasmid R68.45 is associated with deletion formation and concomitant loss of chromosome mobilization ability. *J. Bacteriol., 150*, 251-259.

Davies, J. E. (1997). Origins, acquisition and dissemination of antibiotic resistance determinants. In Ciba Foundation Symposium No. 207, *Antibiotic resistance: Origins, evolution, selection and spread.* (pp. 15-35). New York, NY: John Wiley and Sons.

Davison, J. (1999). Genetic exchange between bacteria in the environment. *Plasmid, 42*, 73-91.

Day, M. (1998). Transformation in aquatic environments. In C. I. Kado (Ed.), *Horizontal gene transfer* (pp. 144-167). London: Chapman and Hall.

Delorme, C., Godon, J. J., Ehrlich, S. D., and Renault, P. (1994). Mosaic structure of large regions of the *Lactococcus lactis* subsp. cremoris chromosome. *Microbiology Reading, 140*, 3053-3060.

Dhar, B., and Ramkrishna, K. (1987). Morphology and general characteristics of phages of chickpea rhizobia. *Arch. Microbiol., 147*, 121-125.

Dhar, B., Upadhyay, K. K., and Singh, R. M. (1993). Isolation and characterization of bacteriophages specific for *Rhizobium leguminosarum* biovar phaseoli. *Can. J. Microbiol., 39*, 775-779.

Droge, M., Puhler, A., and Selbitschka, W. (1998). Horizontal gene transfer as a biosafety issue: A natural phenomenon of public concern. *Journal of Biotechnology, 64*, 75-90.

Droge, M., Puhler, A., and Selbitschka, W. (1999). Horizontal gene transfer among bacteria in terrestrial and aquatic habitats as assessed by microcosm and field studies. *Biology and Fertility of Soils, 29*, 221-245.

Dubnau, D. (1999). DNA uptake in bacteria. *Ann. Rev. Microbiol., 53*, 217-244.

Eardly, B. D., Wang, F. S., and van Berkum, P. (1996). Corresponding 16S rRNA gene segments in *Rhizobiaceae* and *Aeromonas* yield discordant phylogenies. *Plant and Soil, 186*, 69-74.

Edwards, R. A., Olsen, G. J., and Maloy, S. R. (2002). Comparative genomics of closely related salmonellae. *Trends in Microbiology, 10*, 94-99.

Eisen, J. A. (2000). Horizontal gene transfer among microbial genomes: New insights from complete genome analysis. *Curr. Opin. Gen. Dev., 10*, 606-611.

Elmerich, C., Quiviger, B., Rosenberg, C., Franche, C., Laurent, P., and Dobereiner, J. (1982). Characterization of a temperate bacteriophage for *Azospirillum*. *Virology, 122*, 29-37.

Ewert, D. L., and Paynter, M. J. B. (1980). Enumeration of bacteriophages and host bacteria in sewage and the activated sludge treatment process. *Appl. Environ. Microbiol., 39*, 576-583.

Feil, E. J., Holmes, E. C., Bessen, D. E., Chan, M. S., Day, N. P., Enright, M. C., *et al.* (2001). Recombination within natural populations of pathogenic bacteria: short-term empirical estimates and long-term phylogenetic consequences. *Proc. Natl. Acad. Sci. USA, 98*, 182-187.

Feil, E. J., Maiden, M. C. J., Achtman, M., and Spratt, B. G. (1999). The relative contributions of recombination and mutation to the divergence of clones of *Neisseria meningitidis*. *Mol. Biol. Evol., 16*, 1496-1502.

Ferretti, J. J., McShan, W. M., Ajdic, D., Savic, D. J., Savic, G., Lyon, K., *et al.* (2001). Complete genome sequence of an M1 strain of *Streptococcus pyogenes*. *Proc. Natl. Acad. Sci. USA, 98*, 4558-4663.

Fibi, M. R., Broker, M., Schulz, R., Johannsen, R., and Zettlmeissl, G. (1991). Inactivation of recombinant plasmid DNA from a human erythroprotein-producing cell mouse line grown on a large scale. *Appl. Microbiol. Biotechnol., 35*, 622-630.

Fox, G. E., Wisotzkey, J. D., and Jurtshuk, P. (1992). How close is close: 16S rRNA sequence identity may not be sufficient to guarantee species identity? *Int. J. Syst. Bacteriol., 42*, 166-170.

Garcia, V. S., Romeu, A., and Palau, J. (2000). Horizontal gene transfer in bacterial and archaeal complete genomes. *Genome Res., 10*, 1719-1725.

Gaunt, M. W., Turner, S. L., Rigottier-Gois, L., Lloyd-Macgilp, S. A., and Young, J. P. W. (2001). Phylogenies of *atp*D and *rec*A support the small subunit rRNA-based classification of rhizobia. *Int. J. Syst. Evol. Microbiol., 51*, 2037-2048.

Germida, J. J. (1986). Population dynamics of *Azospirillum brasilense* and its bacteriophage in soil. *Plant and Soil, 90*, 117-128.

Germida, J. J., and Casida, L. E. J. (1983). *Ensifer adhaerens* predatory activity against other bacteria in soil, as monitored by indirect phage analysis. *Appl. Environ. Microbiol., 45*, 1380-1388.

Greaves, M. P., and Wilson, M. J. (1970). The degradation of nucleic acids and montmorillonite-nucleic-acid complexes by soil microorganisms. *Soil Biol. Biochem., 2*, 257-268.

Groman, N. B. (1984). Conversion of corynephages and its role in the natural history of diptheria. *J. Hyg., 93*, 405-417.

Hacker, J., Bender, L., Ott, M., Wingender, J., Lund, B., Marre, R., *et al.* (1990). Deletions of chromosomal regions coding for fimbriae and hemolysins occur *in vitro* and *in vivo* in various extraintestinal *Escherichia coli* isolates. *Microbial Pathogenesis, 8*, 213-225.

Hacker, J., and Kaper, J. B. (1999). The concept of pathogenicity islands. In J. Hacker (Ed.), *Pathogenicity islands and other mobile virulence elements.* (pp. 1-11). Washington, DC: American Society of Microbiology.

Hacker, J., and Kaper, J. B. (2000). Pathogenicity islands and the evolution of microbes. *Ann. Rev. Microbiol., 54*, 641-679.

Hacker, J., Knapp, S., and Goebel, W. (1983). Spontaneous deletions and flanking regions of the chromosomally inherited hemolysin determinants of an *Escherichia coli* O6 strain. *J. Bacteriol., 154*, 1145-1152.

Hara, S., Terauchi, K., and Koike, I. (1991). Abundance of viruses in marine waters assessment by epifluorescence and transmission electron microscopy. *Appl. Environ. Microbiol., 57*, 2731-2734.

Hashem, F. M., Angle, J. S., and Ristiano, P. A. (1986). Isolation and characterization of rhizobiophages specific for *Bradyrhizobium japonicum* USDA 117. *Can. J. Microbiol., 32*, 326-329.

Hegazi, N. A., and Jensen, V. (1973). Studies of *Azotobacter* bacteriophages in Egyptian soils. *Soil Biol. Biochem., 5*, 231-243.

Hegazi, N. A., and Leitgeb, S. (1976). *Azotobacter* bacteriophages in Czeckoslovakian soils. *Plant and Soil, 45*, 379-395.

Hentschel, U., and Hacker, J. (2001). Pathogenicity islands: The tip of the iceberg. *Microbes and Infection, 3*, 545-548.

Herrick, J. B., Stuart, K. K. G., Ghiorse, W. C., and Madsen, E. L. (1997). Natural horizontal transfer of a naphthalane dioxygenase gene between bacteria native to a coal tar-contaminated field site. *Appl. Environ. Microbiol., 63*, 2330-2337.

Hochhut, B., Jahreis, K., Lengeler, J. W., and Schmid, K. (1997). CTnscr94, a conjugative transposon found in enterobacteria. *J. Bacteriol., 179*, 2097–2102.

Hu, N., Thiel, T., Giddings Jr., T. H., and Wolk, C. P. (1981). New *Anabaena* and *Nostoc* cyanophages from sewage settling ponds. *Virology, 114*, 236-246.

Ito, T., Katayama, Y., and Hiramatsu, K. (1999). Cloning and nucleotide sequence determination of the entire *mec* DNA of pre-methicillin-resistant *Staphylococcus aureus* N315. *Antimicrob. Agents Chemother., 43*, 1449-1458.

Kankila, J., and Lindstrom, K. (1994). Host range, morphology and DNA restriction patterns of bacteriophage isolates infecting *Rhizobium leguminosarum* bv. trifolii. *Soil Biol. Biochem., 26*, 429-437.

Kinkle, B. K., Sadowsky, M. J., Schmidt, E. L., and Koskinene, W. C. (1993). Plasmids pJP4 and R68.45 can be transferred between populations of bradyrhizobia in nonsterile soil. *Appl. Environ. Microbiol., 59*, 1762-1766.

Kinkle, B. K., and Schmidt, E. L. (1991). Transfer of the pea symbiotic plasmid pJB5JI in nonsterile soil. *Appl. Environ. Microbiol., 57*, 3264-3269.

Koonin, E. V., Makarova, K. S., and Aravind, L. (2001). Horizontal gene transfer in prokaryotes: Quantification and classification. In E. Gottesman-Susan (Ed.), *Annual review of microbiology* (Vol. 55, pp. 709-742) Palo Alto, CA: Annual Reviews.

Kowalski, M., Ham, G. E., Frederick, L. R., and Anderson, I. C. (1974). Relationship between strains of *Rhizobium japonicum* and their bacteriophages from soil and nodules from field grown soybeans. *Soil Sci., 118*, 221-228.

Kunst, F., Ogasawara, N., Moszer, I., Albertini, A. M., Alloni, G., Azevedo, V., *et al.* (1997). The complete genome sequence of the Gram-positive bacterium *Bacillus subtilis*. *Nature (London), 390*, 249-256.

Kuroda, M., Ohta, T., Uchiyama, I., Baba, T., Yuzawa, H., Kobayashi, I., *et al.* (2001) Whole genome sequencing of meticillin-resistant *Staphylococcus aureus*. *Lancet (North American Edition), 357*, 1225-1240.

Lajudie, P. d., and Bogusz, D. (1984). Isolation and characterization of two bacteriophages of a stem-nodulating *Rhizobium* strain from *Sesbania rostrata* [Nitrogen fixation]. *Can. J. Microbiol., 30*, 521-525.

Lanning, S., and Williams, S. T. (1982). methods for direct isolation and enumeration of actinophages in soil. *J. Gen. Microbiol., 128*, 2063-2071.

Lawrence, J. G., and Ochman, H. (1998). Molecular archaeology of the *Escherichia coli* genome. *Proc. Natl. Acad. Sci. USA, 95*, 9413-9417.

Lawson, K. A., and Barnet, Y. M. (1984). Factors influencing growth of *Rhizobium* and its bacteriophage in soil. *Adv. Agric. Biotechnol., 4*, 347.

Lindstrom, K., and Kaijalainen, S. (1991). Genetic relatedness of bacteriophage infecting *Rhizobium galegae* strains. *FEMS Microbiol. Lett., 82*, 241-246.

Logsdon, J. M., Jr., and Faguy, D. M. (1999). Evolutionary genomics: *Thermotoga* heats up lateral gene transfer. *Current Biology, 9*, R747-R751.

Lorenz, M. G., and Wackernagel, W. (1994). Bacterial gene transfer by natural genetic transformation in the environment. *Microbiol. Rev., 58*, 563-602.

Maeda, M., and Taga, N. (1974). Occurrence and distribution of deoxyribonucleic acid-hydrolyzing bacteria in sea water. *J. Exp. Mar. Biol. Ecol., 14*, 157-169.

Maidak, B. L., Larsen, N., McCaughey, M. J., Overbeek, R., Olsen, G. J., Fogel, K., *et al.* (1994). The ribosomal database project. *Nucleic Acids Res., 22*, 3485-3487.

Maiden, M. C. J., Bygraves, J. A., Feil, E., Morelli, G., Russell, J. E., Urwin, R., *et al.* (1998). Multilocus sequence typing: A portable approach to the identification of clones within populations of pathogenic microorganisms. *Proc. Natl. Acad. Sci. USA, 95*, 3140-3145.

Markham, P. F., Duffy, M. F., Glew, M. D., and Browning, G. F. (1999). A gene family in *Mycoplasma imitans* closely related to the PMGA family of *Mycoplasma gallisepticum*. *Microbiology, 145*, 2095-2103.

Martin, W. (1999). Mosaic bacterial genomes: A challenge en route to a tree of genomes. *Bioessays, 21*, 99-104.

Mazel, D., and Davies, J. (1999). Antibiotic resistance in microbes. *CMLS Cell. Mol. Life Sci., 56*, 742-754.

Miller, R. V., and Sayler, G. S. (1992). Bacteriophage/host interaction in aquatic systems. In J. D. van Elsas (Ed.), *Genetic interactions among microorganisms in the natural environment* (pp. 176-193). Oxford, UK: Pergamon Press.

Morrison, W. D., Miller, R. V., and Saylar, G. S. (1978). Frequency of F116-mediated transduction of *Pseudomonas aeruginosa* in a freshwater environment. *Appl. Environ. Microbiol., 36*, 724-730.

Nelson, K. E., Clayton, R. A., Gill, S. R., Gwinn, M. L., Dodson, R. J., Haft, D. H., *et al.* (1999). Evidence for lateral gene transfer between Archaea and bacteria from genome sequence of *Thermotoga maritima*. *Nature (London), 399*, 323-329.

Newton, W. E. (1993). Nitrogenases: Distribution, composition, structure and function. In R. Palacios, J. Mora and W. E. Newton (Eds.), *New horizons in nitrogen fixation* (pp. 5-18). Dordrecht. The Netherlands: Kluwer Academic Publishers.

Newton, W. E. (2000). Nitrogen fixation in perspective. In F. O. Pedrosa, M. Hungria, M. G. Yates, and W. E. Newton (Eds.), *Nitrogen fixation: From molecules to crop productivity* (pp. 3-8). Dordrecht, the Netherlands: Kluwer Academic Publishers.

Nolling, J., Breton, G., Omelchenko, M. V., Makarova, K. S., Zeng, Q., Gibson, R., *et al.* (2001). Genome sequence and comparative analysis of the solvent-producing bacterium *Clostridium acetobutylicum*. *J. Bacteriol., 183*, 4823-4838.

Nwosu, V. C. (2001). Antibiotic resistance with particular reference to soil microorganisms. *Res. Microbiol., 152*, 421-430.

Ochman, H., Lawrence, J. G., and Groisman, E. A. (2000). Lateral gene transfer and the nature of bacterial innovation. *Nature (London), 405*, 299-304.

Ogunseitan, O. A., Sayler, G. S., and Miller, R. V. (1992). Application of DNA probes to analysis of bacteriophage distribution patterns in the environment. *Appl. Environ. Microbiol., 58*, 2046-2052.

Olsen, G. J., Woese, C. R., and Overbeek, R. (1994). The winds of (evolutionary) change: breathing new life into microbiology. *J. Bacteriol., 176*, 1-6.

Patel, J. J., and Craig, A. S. (1984). Isolation and characterisation of bacteriophages active against strains of *Rhizobium trifolii* used in legume inoculants in New Zealand. *N. Z. J. Sci., 27*, 81-86.

Patel, J. J., Craig, A. S., and Chandra, P. (1985). Characterisation of bacteriophages virulent to field isolates of *Rhizobium trifolii*. *N. Z. J. Agric. Res., 28*, 283-288.

Paul, J. H., DeFlaun, M. F., and Jeffrey, W. H. (1988). Mechanisms of DNA utilization by estuarine bacterial populations. *Appl. Environ. Microbiol., 54*, 1682-1688.

Paul, J. H., Jeffrey, W. H., and DeFlaun, M. F. (1987). Dynamics of extracellular DNA in the marine environment. *Appl. Environ. Microbiol., 53*, 170-179.

Paul, J. H., Jiang, S. C., and Rose, J. B. (1991). Concentrations of viruses and dissolved DNA from aquatic environments by vortex flow filtration. *Appl. Environ. Microbiol., 57*, 2197-2204.

Paul, J. H., Rose, J. B., Jiang, S. C., Kellog, C. A., and Dickson, L. (1993). Distribution of viral abundance in the reef environment of Key Largo, Florida. *Appl. Environ. Microbiol., 59*, 718-724.

Perna, N. T., Plunkett, G., III, Burland, V., Mau, B., Glasner, J. D., Rose, D. J., *et al.* (2001). Genome sequence of enterohaemorrhagic *Escherichia coli* O157:H7. *Nature (London), 409*, 529-533.

Phillips, S. J., Dalgarn, D. S., and Young, S. K. (1989). Recombinant DNA in wastewater, pBR322 degradation kinetics. *J. Water Pollut. Control Fed., 61*, 1588-1595.

Postgate, J. R. (1974). Evolution within nitrogen-fixing systems. *Symp. Soc. Gen. Microbiol., 24*, 265-292.

Postgate, J. R., and Eady, R. R. (1988). The evolution of biological nitrogen fixation. In H. Bothe, F. J. DeBruijn, and W. E. Newton (Ed.), *Nitrogen fixation: Hundred years after*. Stuttgart, Germany: Gustav Fischer.

Proctor, L. M., and Fuhrman, J. A. (1990). Viral mortality of marine bacteria and cyanobacteria. *Nature (London), 343*, 60-62.

Ragan, M. A. (2001). Detection of lateral gene transfer among microbial genomes. *Curr. Op. Genet. Develop., 11*, 620-626.

Reanney, D. C., and Marsh, S. C. N. (1973). The ecology of viruses attacking *Bacillus stearothermophilus* in soil. *Soil Biol. Biochem., 5*, 399-408.

Ripp, S., and Miller, R. V. (1995). Effects of suspended particulates on the frequency of transduction among *Pseudomonas aeruginosa* in a fresh water environment. *Appl. Environ. Microbiol., 61*, 1214-1219.

Romanowski, G., Lorenz, M., Saylar, G., and Wackernagel, W. (1992). Persistence of free plasmid DNA in soil monitored by various methods. *Appl. Environ. Microbiol., 58*, 3012-3019.

Romanowski, G., Lorenz, M. G., and Wackernagel, W. (1993). Use of polymerase chain reaction and electroporation of *Escherichia coli* to monitor the persistence of extracellular plasmid DNA introduced into natural soils. *Appl. Environ. Microbiol., 59*, 3438-3446.

Ruepp, A., Graml, W., Santos, M. M. L., Koretke, K. K., Volker, C., Mewes, H. W., *et al.* (2000). The genome sequence of the thermoacidophilic scavenger *Thermoplasma acidophilum*. *Nature (London), 407*, 508-513.

Salanoubat, M., Genin, S., Artiguenave, F., Gouzy, J., Mangenot, S., Arlat, M., *et al.* (2002). Genome sequence of the plant pathogen *Ralstonia solanacearum*. *Nature (London), 415*, 497-502.

Saye, D. J., Ogunseitan, O. A., Sayler, G. S., and Miller, R. V. (1987). Potential for transduction of plasmids in a natural freshwater environment, effect of plasmid donor concentration and a natural microbial community on transduction of *Pseudomonas aeruginosa*. *Appl. Environ. Microbiol., 53*, 987-995.

Saye, D. J., Ogunseitan, O. A., Sayler, G. S., and Miller, R. V. (1990). Transduction of linked chromosomal genes between *Pseudomonas aeruginosa* strains during incubation in situ in a freshwater habitat. *Appl. Environ. Microbiol., 56*, 140-145.

Seldin, L., van Elsas, J. D., and Penido, E. G. C. (1984). *Bacillus polymyxa* bacteriophages from Brazilian soils. *Antoine van Leeuwenhoek, 50*, 39-51.

Simpson, A. J. G., Reinach, F. C., Arruda, P., Abreu, F. A., Acencio, M., Alvarenga, R., *et al.* (2000). The genome sequence of the plant pathogen *Xylella fastidiosa*. *Nature (London), 406*, 151-157.

Smith, G. R. (1988). Homologous Recombination in Procaryotes. *Microbiol. Rev., 52*, 1-28.

Smith, N. H., Holmes, E. C., Donovan, G. M., Carpenter, G. A., and Spratt, B. G. (1999). Networks and groups within the genus *Neisseria*: analysis of *arg*F, *rec*A, *rho*, and 16S rRNA sequences from human *Neisseria* species. *Mol. Biol. Evol., 16*, 773-783.

Stackebrandt, E., and Goebel, B. M. (1994). Taxonomic note: A place for DNA-DNA reassociation and 16S rRNA sequence analysis in the present species definition in bacteriology. *Int. J. Syst. Bacteriol., 44*, 846-849.

Sullivan, J. T., Eardly, B. D., van Berkum, P., and Ronsom, C. W. (1996). Four unnamed species of nonsymbiotic rhizobia isolated from the rhizosphere of *Lotus corniculatus*. *Appl. Environ. Microbiol., 62*, 2818-2825.

Sullivan, J. T., Patrick, H. N., Lowther, W. L., Scott, D., and Ronsom, C. W. (1995). Nodulating strains of *Rhizobium loti* arise through chromosomal symbiotic gene transfer in the environment. *Proc. Natl. Acad. Sci. USA, 92*, 8985-8989.

Sullivan, J. T., and Ronson, C. W. (1998). Evolution of rhizobia by acquisition of a 500-kb symbiosis island that integrates into a phe-tRNA gene. *Proc. Natl. Acad. Sci. USA, 95*, 5145-5149.

Takami, H., Nakasone, K., Takaki, Y., Maeno, G., Sasaki, R., Masui, N., *et al.* (2000). Complete genome sequence of the alkaliphilic bacterium *Bacillus halodurans* and genomic sequence comparison with *Bacillus subtilis*. *Nucleic Acids Res., 28*, 4317-4331.

Teuber, M. (1999). Spread of antibiotic resistance with food-borne pathogens. *Cell. Mol. Life Sci., 56*, 755-763.

Wackernagel, W., Sikorski, J., Blum, S. A. E., Lorenz, M. G., and Graupner, S. (1998). Natural genetic transformation of bacteria in soil. In C. I. Kado (Ed.), *Horizontal gene transfer* (pp. 168-178). New York, NY: Chapman and Hall.

Waldor, M. K., and Mekalanos, J. J. (1996). Lysogenic conversion by a filamentous phage encoding cholera toxin. *Science, 272*, 1910-1914.

Wang, Y., Zhang, Z., and Ramanan, N. (1997). The actinomycete *Thermobispora bispora* contains two distinct types of transcriptionally active 16S rRNA genes. *J. Bacteriol., 179*, 3270-3276.

Wdowiak, S., Malek, W., and Grzadka, M. (2000). Morphology and general characteristics of phages specific for *Astragalus cicer* rhizobia. *Curr. Microbiol., 40*, 110-113.

Werquin, M., Ackermann, H. W., and Levesque, R. C. (1988). A study of 33 bacteriophages of *Rhizobium meliloti*. *Appl. Environ. Microbiol., 54*, 188-196.

Whitman, W. B., Coleman, D. C., and Wiebe, W. J. (1998). Prokaryotes: The unseen majority. *Proc. Natl. Acad. Sci. USA, 95,* 6578-6583.

Williams, H. G., Day, M. J., Fry, J. C., and Stewart, G. J. (1996). Natural transformation in river epilithon. *Appl. Environ. Microbiol., 58*, 2965-2970.

Wommack, K. E., Hill, R. T., Kessel, M., Russek-Cohen, E., and Colwell, R. R. (1992). Distribution of viruses in the Chesapeake Bay. *Appl. Environ. Microbiol.*, 58, 2965-2970.

Yap, W. H., Zhang, Z., and Wang, Y. (1999). Distinct types of rRNA operons exist in the genome of the actinomycete *Thermomonospora chromogena* and evidence for horizontal gene transfer of an entire rRNA operon. *J. Bacteriol., 181*, 5201-5209.

Zgur-Bertok, D. (1999). Mechanisms of horizontal gene transfer (Review). *Folia Biologica Prague, 45*, 91-96.

CHAPTER 14

THE PHYLOGENY AND EVOLUTION OF NITROGENASES

J. P. W. YOUNG
Dept. of Biology, University of York, York, Y0105YW, UK

1. INTRODUCTION

The ability to fix N_2 has a very wide taxonomic distribution, but a patchy one. It is found in most major groups of bacteria and in the methanogenic archaea, but only in a minority of species. In a comprehensive survey published over a decade ago, reports of nitrogen fixation were found for almost a hundred genera (Young, 1992). The number has increased since then through the discovery of nitrogen fixation in new organisms, but also because advances in microbial taxonomy have led to the recognition of more genera. Strikingly though, nitrogen fixation has never been reported in eukaryotes.

The nitrogenase enzymes of all these nitrogen-fixing organisms are so similar (with one exception) that they are clearly derived from a common ancestor. We can, therefore, rule out multiple independent inventions of the process as a possible explanation for its patchy distribution. The remaining possibilities are either horizontal gene transfer or independent loss in many lineages. We can hope to distinguish between these by examining the phylogeny of the nitrogenases themselves in relation to that of the organisms that carry them. If the trees match, this would be persuasive evidence that horizontal gene transfer had not played a significant role. In fact, the trees are not the same, but they do reveal another complication, namely that many organisms have more than one nitrogenase system and these can be highly divergent. Once the possibility of gene duplication is included, discordance between phylogenies could be explained by the inclusion of genes that were not orthologs (derived from the same copy in the common ancestor) but paralogs (derived from different members of a gene family).

These issues have been discussed repeatedly in the literature. Postgate (1974; 1982) challenged the earlier view that nitrogen fixation was present in the common ancestor of life and had frequently been lost since. Instead, he suggested that

R. Palacios and W. E. Newton (eds.), Genomes and Genomics of Nitrogen-fixing Organisms, 221-241.

diazotrophy arose relatively recently, less than 1,500 million years ago, and was transferred laterally to achieve its present distribution. Molecular sequences, as they became available, were interpreted by some to support a phylogeny for nitrogenase that matched that of the ribosomal genes and so were consistent with an ancestral origin (Hennecke *et al.*, 1985, Postgate and Eady 1988), whereas others saw evidence for lateral transfer (Normand and Bousquet 1989). Once the existence of deep paralogous branches was taken into account, the data could be seen as consistent with multiple losses, with multiple transfers, or with some combination of both processes (Young, 1992). As DNA sequencing proceeds apace and the complete genomes of many diazotrophs have now been determined, new groups of nitrogenase homologs are discovered and the web of relationships grows more tangled. Although horizontal gene transfer has probably played a role, there are few clear-cut examples and our understanding of the evolutionary history of nitrogenases is far from complete.

Within the nitrogenases, there are a number of distinct clades, but the overall subunit organisation is almost invariant, and there are enough conserved features to guide an alignment of the sequences of each of the polypeptide subunits. The only known exception is a pathway in *Streptomyces thermoautotrophicus* in which N_2 reduction is coupled to the oxidation of carbon monoxide by enzymes that are unrelated to other nitrogenases (Ribbe *et al.*, 1997). It appears to be a true "independent invention" but, as it has not so far been reported in other organisms, it will not be considered further here.

2. THE GENETIC ORGANISATION OF NITROGENASE GENES

The widespread molybdenum-containing forms of the nitrogenase enzyme are composed of two alpha and two beta subunits (together forming component I, dinitrogenase, MoFe protein), encoded by the *nifD* and *nifK* genes, respectively, in association with a dimer encoded by *nifH* (component II, dinitrogenase reductase, Fe protein). It is the relationships of these catalytic proteins that I will discuss in this chapter, together with the products of the *nifE* and *nifN* genes, which are homologs of *nifD* and *nifK*. In addition, I will include the nitrogenases that either have vanadium rather than molybdenum or that have only iron and which have similar subunit components plus additional delta subunits encoded by either *vnfG* or *anfG*. There are many other gene products necessary to support the nitrogen-fixation process, but the work needed to piece together the history of this whole network lies in the future.

Surprisingly, there is a "standard arrangement" of the nitrogenase genes that can be discerned despite the large evolutionary distances that separate them. An operon arranged as either *nifHDKEN* or *vnfHDGKEN* or *anfHDGKEN* could be seen as the archetypal form. However, this arrangement is clearly not critical for function because many variants are found. For example, either *nifH* or *nifEN* may be separated from *nifDK* and individual genes may be duplicated (especially *nifH* or *nifK*). Some examples of variant arrangements are illustrated by Dean and Jacobson (1992) and Kaminski *et al.* (1998) and others are listed in Table 1.

Table 1. The arrangement of nitrogenase gene clusters in diazotroph genomes

Organism	B-type	C-type	A-type Vnf	A-type Anf	D-type
PROTEOBACTERIA					
Bradyrhizobium japonicum	**H, DKEN**[1] N768409, 383-6[2]				
Mesorhizobium loti	**HDKEN** N106489-93				
Sinorhizobium meliloti	**HDKE, N** N435696-8, 724				
Rhodopseudomonas palustris	**HDKEN** Z09392-88		**NE--H-DGK** Z10950-58	**HDGK** Z11017-14	**HEN** Z12151-3 **HEN** Z12171-3 **H----EN** Z11901-895
Rhodospirillum rubrum	**HDK** Z14415-3 **EN** Z15686-7			**HDGK** Z15420-18	**HEN** Z14613-5 **EN** Z14636-7
Rhodobacter sphaeroides	**HDKEN** Z07624-28				
Burkholderia fungorum	**HDK--EN** Z34963-57				
Azotobacter vinelandii	**HDK----EN** Z90773-65		**EN------H-DGK** Z89155-63	**HDGK** Z92169-72	
Magnetococcus sp. MC-1	**HDK-----EN** Z44347-56				
CYANOBACTERIA					
Nostoc sp. PCC7120	**H, DKEN** N485497, 84-80				
Nostoc punctiforme	**HD, KEN** Z112319-41 **HEN** Z111243-45				
Trichodesmium erythraeum	**HDK[EN]** Z73548-45				
CLOSTRIDIA					
Desulfitobacterium hafniense	**HDKEN** Z99588-92				**HNE** Z98133-5
Clostridium acetobutylicum		**H—DKEN** N346894-900			
CHLOROBI					
Chlorobium tepidum		**H—DKEN** N662417-23			
ARCHAEA					
Methanosarcina acetivorans		**H—DKEN** N618766-72	**H--DGKEN** N616152-9	**DGK--H** N616149-4	**ENH** N616561-3 **NH** N618503-2 **[HE]N** N616955-6
Methanosarcina barkeri		**H—DKEN** Z78739-33	**H--DGK-EN** Z78063-71	**DGK** Z77947-8	**NH** Z78936-7 **[HE]N?** Z76604-5
Methanosarcina mazei		**H—DKEN** N632743-49			**NH** N632538-9
Methanothermobacter thermautotrophicus		**H—DKEN** N276673-79			

[1]– intervening gene, <30 intervening genes, [ ] fused genes, _ reverse orientation.
[2]database accession numbers, NP_ abbreviated to N, ZP_000 to Z.

As gene sequences became available, it was realised that *nifE* is a homolog of *nifD* and *nifN* is a homolog of *nifK* (Brigle *et al.*, 1987; Aguilar *et al.*, 1987). Hence, it seems certain that *nifDK* and *nifEN* are the products of ancient gene duplication. Furthermore, at a deeper level, *nifD* and *nifK* are also homologous (Holland *et al.*, 1987), so we can explain the *nifDKEN* genes, and their order, as a result of two successive duplications. The arguments for this have been discussed in detail by Fani *et al.* (2000).

3. NITROGENASE GENES FROM GENOME SEQUENCING PROJECTS

There are now hundreds of published nitrogenase gene sequences from many different species, especially for *nifH*, because this gene has been widely used for diversity studies. In most cases, these were obtained by cloning or amplifying sequences from organisms in which nitrogen fixation had been demonstrated experimentally. One advantage of the current availability of complete genome sequences is that one can make a complete list of the relevant genes in an organism, including those that might be cryptic because they are either seldom expressed or defective. This approach might reveal unsuspected abilities in some organisms, but it also allows us, for the first time, to state categorically that some organisms lack nitrogenase genes and are, therefore, incapable of nitrogen fixation under any circumstances (unless by a radically novel process that we know nothing about).

To provide a consistent set of sequences, the phylogenies presented here are based primarily on the genomes that can be searched in the database available at NCBI (http://www.ncbi.nlm.nih.gov/sutils/genom_table.cgi). A comprehensive search of the complete genomes was made using PSI-Blast (Altschul *et al.*, 1997) to find all homologs of NifH, NifDKEN, and NifG. The predicted peptide sequences were aligned using ClustalX (Thompson *et al.*, 1997) and phylogenies constructed using Mega 2.1 (Kumar *et al.*, 2001). All trees shown were based on the neighbor-joining method with rate variation corrected by a gamma distribution with a parameter value of 2 (approximating Dayhoff distances).

At the time of writing, the complement of proteins had been predicted for 143 bacterial and 18 archaeal genomes. Of these, 16 bacterial and 4 archaeal genomes had identifiable sets of *nif* genes (including at least *H*, *D*, and *K*). However, *Magnetospirillum magnetotacticum* was not analysed further because the assembly was too fragmentary to give full-length genes and *Rhodospirillum rubrum* could only be included after careful scrutiny. These genomes are listed in Table 1 and form the main basis for this review. Nine are fully annotated and published (Deppenmeier *et al.*, 2002; Eisen *et al.*, 2002; Galagan *et al.*, 2002; Galibert *et al.*, 2001; Kaneko *et al.*, 2000; 2001; 2002; Nölling *et al.*, 2001; Smith *et al.*, 1997). Incomplete DNA sequences of an additional 87 genomes were available, but without gene predictions. Six of these had close matches to *nifH*, *D*, and *K* genes; they were from *Methylococcus capsulatus*, *Acidithiobacillus ferrooxidans*, *Rhizobium leguminosarum*, *Erwinia chrysanthemi*, *E. carotovora*, and *Geobacter sulfurreducens*, all of which are proteobacteria.

Although this survey does not cover a completely representative sample of prokaryotes, it does emphasise that the potential for nitrogen fixation is confined to a minority of species. The only sequenced organism that has *nif* genes but was not previously known to be capable of fixing N_2 (or closely related to a fixer) is *Desulfitobacterium hafniense.* A preliminary report, prompted by the discovery of *nif* genes in the genome, has now demonstrated that this organism can use N_2 gas as sole nitrogen source (Tiedje and Davis, 2002). There is no evidence for cryptic nitrogenase systems in organisms that are not known to fix N_2. It is clear that there is a patchy distribution of the genes, not just of the phenotype.

4. ORGANISATION OF THE NITROGENASE GENES

Genes encoding the different subunits of nitrogenase are clustered together in the genome, usually in a characteristic order (see Table 1). There are three types of nitrogenase, A, B, and C, that will be described in detail in the next section. In the case of the common B type of enzyme, the *nifHDKEN* genes are typically adjacent, without intervening genes, as in *Mesorhizobium loti*, for example. However, this arrangement is clearly not essential for function because this group can be fragmented in many different ways. In fact, of the thirteen complete examples in Table 1, only five are unbroken, and all possible breaks between the genes are found in the remaining examples. The intervening genes are sometimes directly relevant to nitrogenase, *e.g.*, *nifT* and *nifY* following *nifK* in *Azotobacter vinelandii*, but in other cases, the gap may be longer and include seemingly unrelated genes.

In contrast, the C-type systems have a very consistent organisation, even though they are found in both bacteria and archaea. Between the *nifH* and *nifD* genes lie two small genes (either *nifI* or *glnB*) that encode P_{II} proteins, which are implicated in regulation of nitrogenase by ammonia in *Methanococcus maripaludis* (Kessler and Leigh 1999). In all six examples, the order is *nifHIIDKEN.*

In the Vnf systems, there is much more diversity of gene arrangement, although there is always a *vnfG* gene between *vnfD* and *vnfK*. Similarly, the Anf systems have an *anfG* gene, although this was not always detected by the automated gene-finding software that is used to deduce the protein complement from the genome sequence, for example, in *Rhodospirillum rubrum*. *Methanosarcina barkeri* has *anfDGK* genes, but there is no *nifH* either preceding or immediately following them. However, the genome is incompletely assembled and these genes are at the end of a contig, so it is possible that a *nifH* will eventually be located a few genes beyond *anfK* as seen in *Methanosarcina acetivorans*. The Nif and Vnf systems of *M. barkeri* have been described and studied in some detail (Chien *et al.*, 2000), but there is as yet no experimental evidence for a functional Anf system.

5. EVOLUTIONARY RELATIONSHIPS OF THE NITROGENASE GENES

5.1. Phylogeny of NifH

Figure 1 shows a phylogeny based on NifH proteins in the completely sequenced nitrogen-fixers and their recognisable homologs in the same species. The first thing to note is that the true NifH sequences form a clearly defined clade, but there are

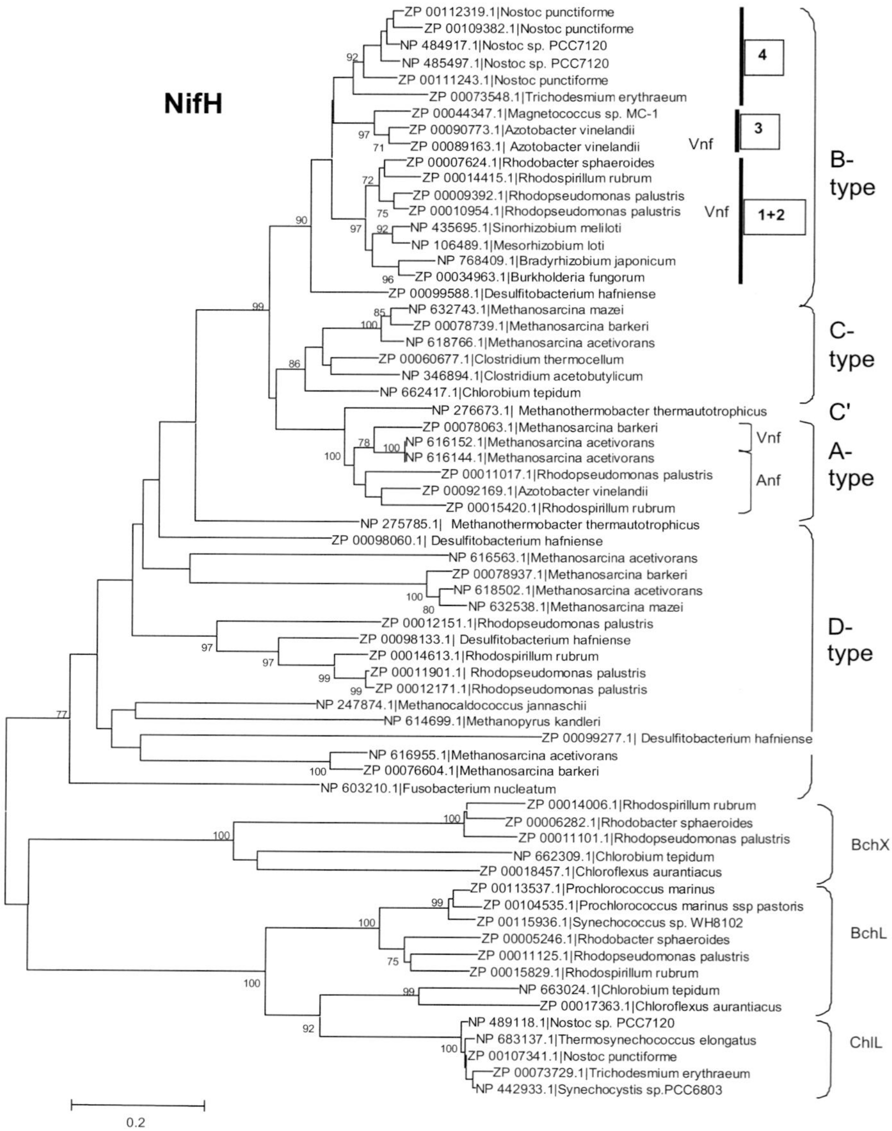

Figure 1. Phylogeny of NifH and homologous proteins from diazotroph genome sequences.

also a large number of more distant relatives. Some of these are well-known proteins involved in the synthesis of photosynthetic pigments, namely protochlorophyllide reductase (either BchL or ChlL) and chlorin reductase (BchX). The similarity between these proteins and NifH was analysed and discussed by Burke *et al.* (1993), who argued that nitrogen fixation probably originated before photosynthesis, so the photosynthesis enzymes would have been derived from NifH rather than the other way round. Nevertheless, modern NifH sequences are more tightly conserved, suggesting that this protein is subject to more rigorous structural constraints.

The true NifH proteins can be divided into three types (Young, 1990; 2000). Type B ("bacterial") is the best represented and includes enzymes from the proteobacteria, cyanobacteria, and firmicutes (the subclusters 1-4 within B are discussed later). Type C ("clostridial") is found in the firmicute bacterium *Clostridium*, the green sulfur bacterium *Chlorobium*, and also in the archaeon *Methanosarcina*.

Type A is associated with the "alternative" nitrogenases, which do not contain molybdenum, and is found in both archaea and proteobacteria. There is a group that is phylogenetically related to A-type, but is associated with Mo-containing enzymes in some archaea. It is represented among the sequenced genomes only by *Methanothermobacter thermautotrophicus*, but is also found in some other methanogens, including *Methanococcus maripaludis*, in which the Mo-dependence of the enzyme has been demonstrated (Kessler *et al.*, 1997). The organisation of the genes in these systems is similar to that in the C-type found in other methanogens (*Methanosarcina*) and clostridia, even though they are phylogenetically remote, so this class is identified here as C'-type.

A more distant and diverse group was designated type D (Young 1990), but it is now clear that these proteins are not part of nitrogenase enzymes as their genes are not associated with *nifD* or *nifK*, although some are adjacent to rather distant homologs of the *nifDKEN* family.

5.2. Phylogeny of the NifDKEN family

The overall relationships among the NifD, K, E, and N proteins are shown in Figure 2. As with NifH, there are homologs involved in photosynthetic-pigment synthesis. These are not related to any particular Nif protein, but all the *bch* and *chl* genes cluster together. This suggests that the photosynthesis and nitrogen-fixation systems separated very early on, and that the two rounds of duplication that gave rise to the NifK, D, E, and N proteins was mirrored by a similar, but separate, succession of duplications to create the photosynthetic-pigment synthesis system. The NifD, NifK, NifE, and NifN proteins each form clearly distinct groups, as expected (Fig. 2). However, there are some exceptions, such as the putative *nifN* in the C-type systems, which is discussed in detail later. The archaeal VnfE and VnfN proteins are so diverged that they do not cluster with anything else on the tree.

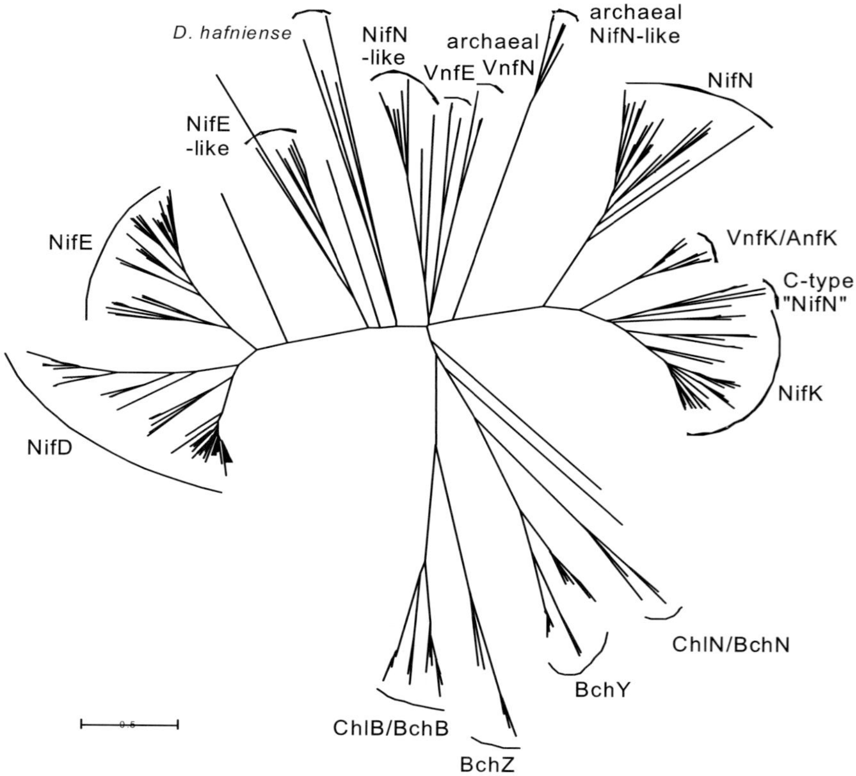

Figure 2. Overall phylogeny of NifD, K, E, N and homologous proteins from diazotroph genome sequences.

The diazotroph genomes contain a number of other homologs of the NifDKEN family. In both the photosynthetic proteobacteria and in *Desulfitobacterium*, there are adjacent pairs of genes that encode vaguely NifE-like and NifN-like proteins, mostly accompanied by a D-type NifH homolog. Their function is at present completely unknown and may have nothing to do with nitrogen fixation. Related genes are found in *Clostridium thermocellum*, which does not have nitrogenase. The archaeal genomes also have genes encoding NifN-like proteins adjacent to D-type NifH homologs, which in two cases are fused to NifE homologs. Again, these may not be relevant to nitrogen fixation. The non-fixers, *Methanopyrus kandleri* and *Methanococcus jannaschii,* also have them.

These very distant homologs are omitted from the phylogenies of the individual NifD, K, E and N proteins (Figures 3-6). Each of these phylogenies can be partitioned into types corresponding to those identified for NifH.

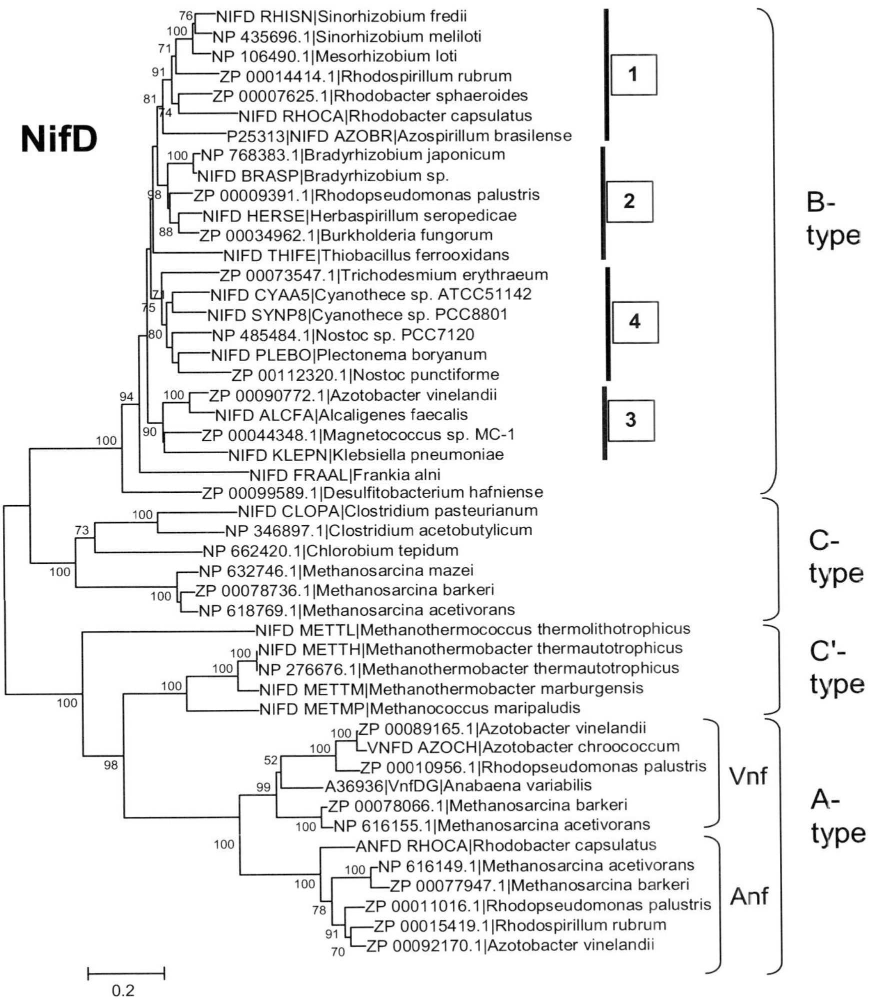

Figure 3. Phylogeny of NifD and homologous proteins from diazotroph genome sequences.

In most cases, a consistent phylogenetic clustering is seen for all the linked genes encoding a system, but there are exceptions. The VnfD and VnfK proteins of *Rhodopseudomonas* and *Azotobacter* are A-type, related to their archaeal counterparts, but their VnfH, VnfE, and VnfN are B-type. On the other hand, their Anf systems are consistently A-type. The C-type systems have a gene following *nifE*, where one would expect to find *nifN*, which encodes a protein that is quite

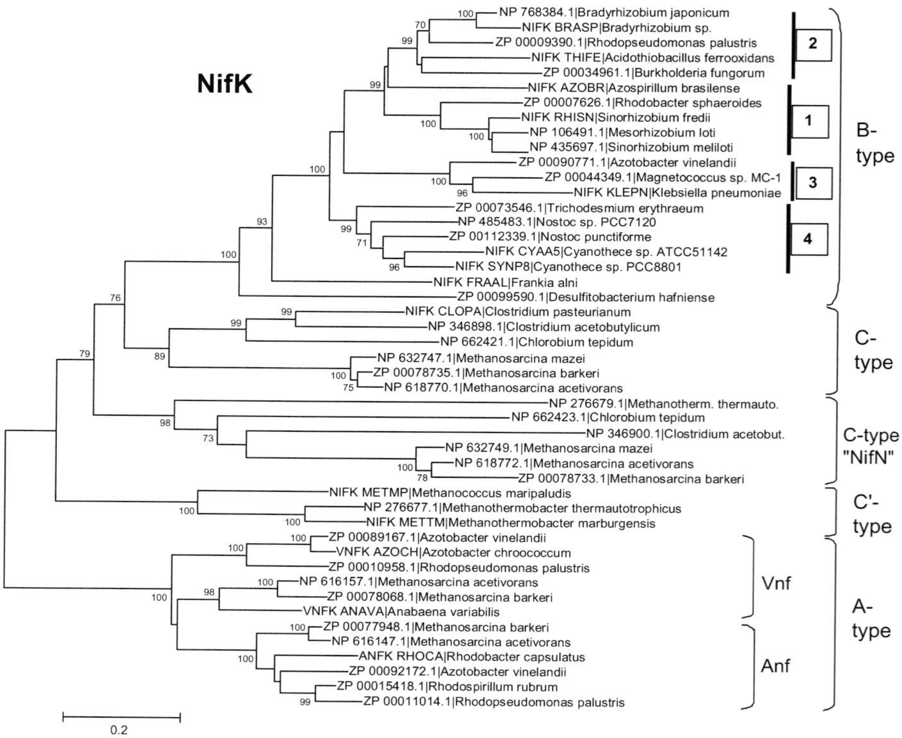

Figure 4. Phylogeny of NifK and homologous proteins from diazotroph genome sequences.

unrelated to other NifN proteins but is much closer to NifK. In fact, it is closer than both the VnfK and AnfK proteins. These putative NifN genes are shown on the NifK phylogeny (Figure 4). A somewhat related NifN is found in the C'-type system of *Methanothermococcus*, even though its other genes are quite unrelated to those of C-type systems.

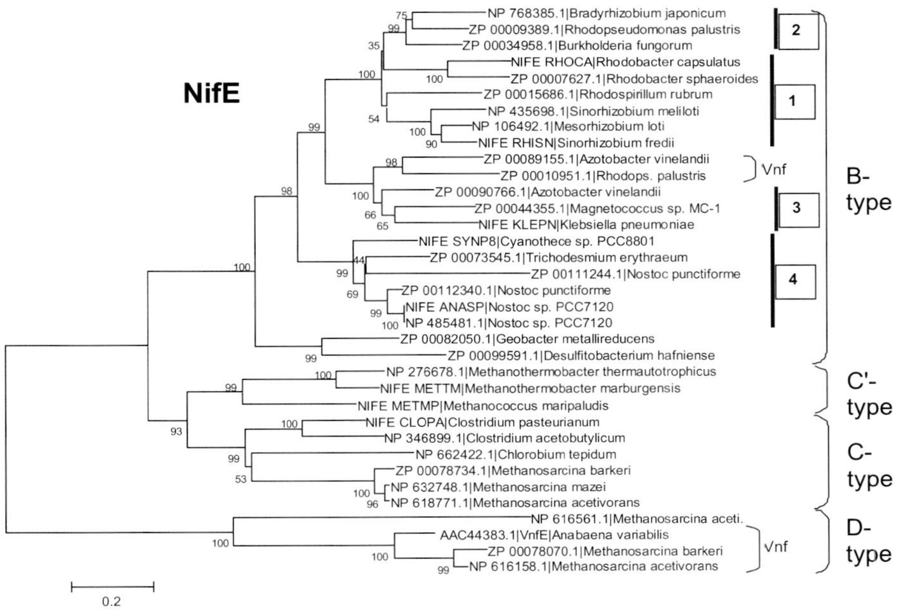

Figure 5. Phylogeny of NifE and homologous proteins from diazotroph genome sequences.

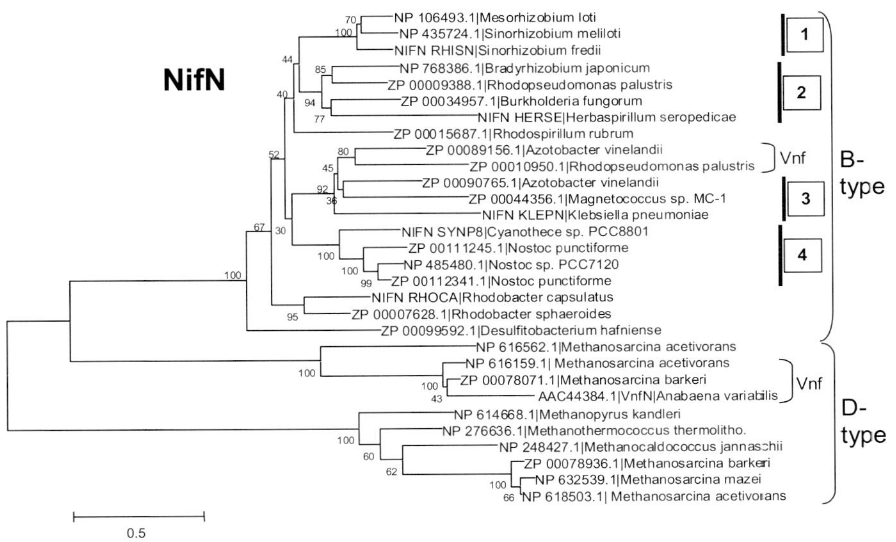

Figure 6. Phylogeny of NifN and homologous proteins from diazotroph genome sequences.

The Mo-independent nitrogenases have an extra delta subunit. This small protein is encoded by either *vnfG* or *anfG* and is located between the *D* and *K* genes. The VnfG proteins of the vanadium enzymes and the AnfG proteins of the iron-only enzymes form two separate groups (Figure 7). They are related to each other but show no significant homology to any other proteins.

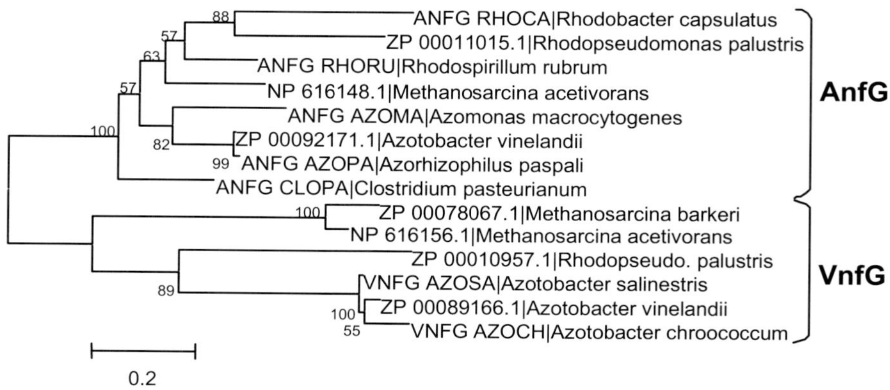

Figure 7. Phylogeny of VnfG and AnfG proteins from diazotroph genome sequences.

6. NITROGENASE PHYLOGENY *VERSUS* ORGANISM PHYLOGENY

Phylogenetic relationships among distantly related organisms are most often described using sequences of their small-subunit ribosomal RNA (SSU rRNA) genes. Such a phylogeny is shown in Figure 8 for the 19 diazotrophs whose nitrogenase proteins we have just discussed. Despite being based on very few taxa, this tree is largely consistent with more comprehensive SSU trees (Cole *et al.*, 2003), although a specific relationship is seen between cyanobacteria and Gram-positive bacteria, such as clostridia, when more species are included and *Rhodobacter* moves deeper in the alpha-proteobacteria branch than it appears in Figure 8. There has been considerable discussion as to whether phylogenies based on rRNA really reflect the relationships of the genomes as a whole, but a recent analysis indicates that there is a large core of hundreds of genes, including rRNA, that does show a consistent phylogeny in most comparisons of bacteria (Daubin *et al.*, 2003). It is, therefore, reasonable to compare the phylogenies of nitrogenase sequences with that of SSU rRNA and to interpret discrepancies as resulting from events in the evolution of nitrogenase. Each of the major types of nitrogenase will be discussed first, before considering the relationship between the types.

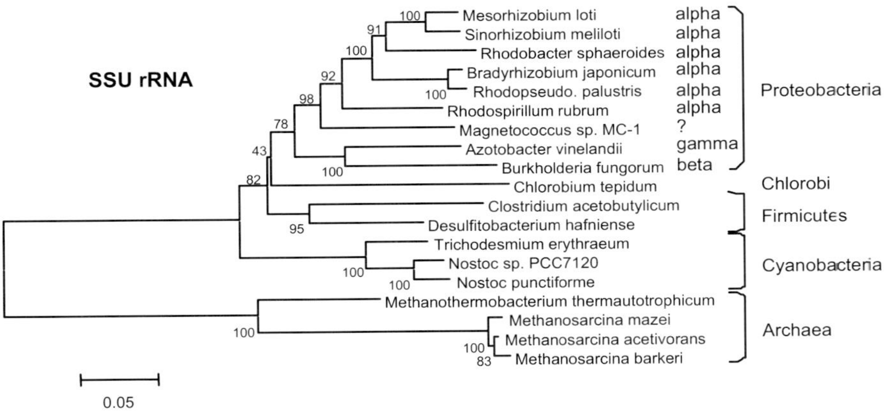

Figure 8. Phylogeny of small subunit ribosomal RNA from diazotroph genome sequences.

6.1. The history of B-type nitrogenases

B-type enzymes are widely distributed in bacteria but have not been found in archaea. Among those included in this survey, the enzyme from *Desulfitobacterium hafniense*, a firmicute or low-GC Gram-positive bacterium related to clostridia, is consistently the most divergent in every subunit (Figures 1, 3-6). The genome of *Frankia*, an actinobacterium or high-GC Gram-positive bacterium, has not been sequenced yet, but its Nif proteins are also outliers within the B-type clade (Figures 3 and 4). This situation makes some phylogenetic sense because the remainder of the B-type enzymes are from two other bacterial phyla, the proteobacteria and the cyanobacteria.

For each protein, the cyanobacterial sequences form a single clade, labelled "4" in the figures. The anomaly is that the cyanobacterial sequences are close to, and in some cases embedded within, the groups of proteobacterial sequences. This location is certainly inconsistent with the phylogenetic position of cyanobacteria which, based on SSU and other criteria, are closer to the firmicutes and certainly not particularly close to the proteobacteria. When this anomaly for NifH, D, and K was last discussed (Young, 2000), it was commented that either gene duplication or gene transfer were both plausible explanations. For example, the unexpectedly distant position of the γ-proteobacteria might reflect paralogy: an ancient duplication created two families of *nif* genes, one of which was retained in the γ-proteobacteria and the other in the remaining proteobacteria and in the cyanobacteria. This explanation is less plausible now that we have complete genome sequences. None of the relevant genomes have traces of an ancient duplication of the B-type nitrogenase system. It now seems more likely that there has been a transfer of nitrogenase genes between cyanobacteria and proteobacteria, although the direction is not immediately obvious. The cyanobacteria have a special control system involving an excision element in *nifD* (Golden *et al.*, 1985), but this is presumably a fairly recent acquisition since the transfer (in whichever direction).

The groups labelled "1", "2" and "3" correspond approximately to the α, β and γ proteobacteria, respectively, and are fairly consistently maintained across the different proteins. However, *Bradyrhizobium japonicum* and *Rhodopseudomonas palustris*, two closely related α-proteobacteria, have nitrogenases that group with those of β-proteobacteria (group 2). On the whole, the different Nif proteins tell the same story, with varying degrees of certainty, but *B. japonicum* and *R. palustris*, closest neighbours for all other Nif sequences and for SSU, have NifH proteins that cluster with different proteobacterial groups (Figure 1). It is tempting to connect this situation with the fact that the *nifH* gene of *B. japonicum* is distant from the rest of the operon (Table 1), suggesting that it might have been acquired by horizontal gene transfer. It has to be said, though, that it is the *R. palustris* sequence, which is further from its expected place. Hurek *et al.* (1997) postulated lateral gene transfer when they found that some *Azoarcus* species had NifH sequences in group 2 and others in group 3. Interestingly, when Moulin *et al.* (2001) described the first example of a β-proteobacterium that nodulated legumes, the partial NifH sequence they determined was extremely similar to that of *Burkholderia fungorum* and, therefore, close to that of *Bradyrhizobium*. It seems that, while there is an overall phylogenetic pattern in the nitrogenases of proteobacteria, some horizontal transfers within and between the α, β and γ orders have occurred. A detailed reconstruction is beyond our scope here, but mismatches between *nifH* and SSU phylogenies have been described previously, particularly among the rhizobia (Haukka *et al.*, 1998).

6.2. The history of C-type nitrogenases

The second major type of Mo-dependent nitrogenase has a quite distinct distribution. There is, so far, no report of an organism with both B-type and C-type nitrogenase. The type-C nitrogenases are found in archaea and in the bacteria, *Clostridium* and *Chlorobium*. The sequences are, in general, more divergent than those of B-type systems and it could be argued that they were consistent with direct inheritance from the common ancestor of bacteria and archaea. This suggestion would require that the C-type genes were completely lost from other bacteria; there is certainly no sign of them now. The alternative idea, that genes have been transferred between archaea and bacteria, is much more acceptable now than it would have been a few years ago. The *Methanosarcina* genomes are the largest yet determined for any archaea and Deppenmeier *et al.* (2002) point out that 30% of the genes in *M. mazei* are closer to bacterial than to archaeal homologs (in fact, half of these have no significant match in any other archaeon). They ascribe this situation to massive horizontal gene transfer from the Bacteria to the Archaea, even emphasising this transfer in the title of their article.

This theme of extensive gene transfer between distant organisms (or, at least, organisms with distant rRNA sequences) has been taken up by many authors recently, including notably Ford Doolittle (*e.g.*, Doolittle *et al.*, 2002), who argue that there may in fact be almost no "core" genome that is immune to horizontal transfer. An example of the kind of analysis that can lead to such a conclusion is provided by Raymond *et al.* (2002), who found that different sets of genes supported different topologies for photosynthetic bacteria. In an interesting parallel with our discussion of nitrogen fixation, they concluded that photosynthesis genes had been subject to repeated horizontal transfer during the evolution of prokaryotes. The extreme view that there is no substantial core of genes with a consistent phylogeny is countered by studies such as that of Daubin *et al.* (2003) but, even if half of the genome were "core", that would leave plenty of scope for transfer. As far as nitrogenase genes are concerned, they do not appear to be part of the core within the proteobacteria because there is evidence for transfer (see previous section), so wider-scale transfers are also plausible.

Although Deppenmeier *et al.* (2002) characterise the *M. mazei* genome as the recipient of massive transfer from bacteria, it is by no means certain that the nitrogenase genes moved in this direction. In fact, Eisen *et al.* (2002) point out that 12% of *Chlorobium tepidum* genes are closer to archaeal than to bacterial homologs, and argue that, in certain cases, the transfer has been from archaea because their phylogenetic origin is within archaea. Eisen *et al.* specifically suggest that the nitrogenase genes were transferred between kingdoms, but without defining the direction. Nölling *et al.* (2001) make similar comments about the nitrogenase genes in *Clostridium acetobutylicum*.

The direction of transfer is not obvious because C-type nitrogenases have a restricted taxonomic distribution in both Bacteria and Archaea. In Bacteria, they are known from the Firmicutes (several species of *Clostridium*) and the unrelated Chlorobi or green sulfur bacteria (*Chlorobium tepidum*). In Archaea, they are restricted to two groups of methanogens, *Methanosarcina* and *Methanothermobacter*. The bacterial proteins are related to those in *Methanosarcina*, but the *Methanothermobacter* genes are much more divergent (the C'-type) and might be thought unrelated except that they share both the gene organisation (two P_{II} proteins encoded after *nifH*) and the NifK-like "NifN". Did these systems originate early in the Archaea as C and C'-types, with C-type later spreading to bacteria (perhaps separately to clostridia and chlorobi because their genes are not closely related)? Or is C'-type the true archaeal version, with *Methanosarcina* having recently adopted a bacterial C-type version?

6.3. The history of Vnf nitrogenases

The vanadium-dependent nitrogenases are also found in both the Archaea and the Bacteria and, in this case, there is much clearer evidence for horizontal gene transfer as well as for reassortment of the genes within the operon(s). The two archaeal examples, from *Methanosarcina acetivorans* and *M. barkeri*, are closely similar in all respects, which is not surprising because these bacteria are close relatives. Their genes are consistently A-type, related to Anf genes. The two bacterial examples from sequenced genomes are both proteobacteria, but a Vnf system is also described in the cyanobacterium, *Anabaena variabilis* (Thiel, 1993; 1996) and this has been included in the phylogenies (Figures 3-6). Each bacterial system shows peculiarities that suggest a mixed origin.

In *Azotobacter vinelandii*, the VnfD, G, and K sequences (Joerger *et al.*, 1990) are A-type, related to the corresponding archaeal Vnf genes. However, VnfH is of B-type and very similar to the NifH in the same organism. It has clearly been recruited by a recent duplication of the NifH. The *vnfE* and *vnfN* genes (Wolfinger and Bishop, 1991) are not after *vnfK*, instead they are few genes upstream of *vnfH* and, like it, they are B-type and related to their *nif* equivalents in the same species. Vnf (and sometimes also Anf) was found in several other γ-proteobacteria by Loveless *et al.* (1999). They used primers in *vnfD* and *vnfK* to amplify *vnfG*, thus, verifying the gene order *vnfDGK*, but did not investigate the other *vnf* genes.

The gene organisation in *R. palustris* is even more disrupted, with *vnfH* inverted relative to *vnfDGK*. Again, VnfD, G, and K are A-type, VnfN and E are B-type and specifically related to the *A. vinelandii* equivalents, *i.e.*, they are in the γ-proteobacterial cluster 3 rather than cluster 2 with the Nif proteins of *R. palustris*. This situation suggests that the α-proteobacterium, *R. palustris*, has acquired the Vnf system from a γ-proteobacterium. The *R. palustris* VnfH, on the other hand, is extremely similar to NifH in the same organism and has clearly been recruited locally in a similar fashion to the VnfH of *A. vinelandii*, but independently.

There is as yet no genome sequence for *Anabaena variabilis* ATCC 29413 as distinct from *Anabaena* sp. PCC1720, which is called *Nostoc* at NCBI and herein. However, the Vnf system of *A. variabilis* is sufficiently distinctive to be worth mentioning here. The arrangement of the genes is *vnf[DG]KEN*, where *D* and *G* are fused (Thiel, 1993; 1996). No *vnfH* has been found. The VnfD part of the VnfDG fused protein is A-type and roughly equidistant between the homologous bacterial (*A. vinelandii* and *R. palustris*) and archaeal (*Methanosarcina*) sequences (Figure 3) and the same is true of the VnfG-like part of this protein (not shown). The *Anabaena* VnfK is unambiguously related to the archaeal rather than the other bacterial sequences (Figure 4). Even more surprisingly, the VnfE and VnfN are very close to the *Methanosarcina* sequences (Figures 5 and 6) and, therefore, completely unlike the B-type proteins found in the other bacterial systems. In other words, all the known components of the *A. variabilis* Vnf system are similar to those in archaea and, especially in the case of VnfE and N, much too similar to be explained by descent from the common ancestor of bacteria and archaea.

There has clearly been horizontal transfer of the *vnf* genes between archaea and bacteria, but can we be sure of the direction? Vnf systems have been found in several taxonomic groups of bacteria, and their organisation is diverse and rather loose, whereas we only know of Vnf in two closely related archaea, which have a compact, self-contained operon that might have been delivered as a package from some bacterium. On the other hand, the bacterial examples of Vnf look like *bricolage*, rather amateur attempts to piece together a working system around a chance acquisition of the core *vnfDGK* genes. Further support for the idea that Vnf comes from archaea is provided by the organisation of the *Methanosarcina acetivorans vnf* operon, which is back-to-back with its *anf* operon in a manner that suggests the possibility of joint control. The arrangement of the genes is

anfH - - *K G D I2 I1* *vnfH I1 I2 D G K E N*

where underlining indicates that the genes are in reverse orientation. The *I1* and *I2* genes encode GlnB-like P_{II} proteins, the *vnf* and *anf* equivalents are related but distinct. On the other hand, the gene labelled *anfH* is 97% identical to the *vnfH* gene and encodes an identical polypeptide; however, there is no evidence about its activity in either the Anf or Vnf systems. The two ORFs that separate it from the *anf* operon have no obvious connection with nitrogen fixation.

6.4. The history of Anf nitrogenases

The distribution of Anf systems among our sequenced genomes is correlated with that of Vnf. *Rhodopseudomonas palustris*, *Azotobacter vinelandii* and *Methanosarcina acetivorans* have both and the incomplete *M. barkeri* genome certainly has part of an Anf system (*anfDGK*, but as yet no *anfI1I2* or *H*). In addition, *Rhodospirillum rubrum* has Anf, although it does not have Vnf. Anf has also been found in *Rhodobacter capsulatus* (Masepohl and Klipp 1996) and in *Clostridium pasteurianum* (Zinoni *et al.*, 1993), but the sequenced *C. acetobutylicum* does not have it. Loveless and Bishop (1999) reported an *anfD*-like sequence in *Heliobacterium gestii*, another firmicute.

Phylogenetically, the Anf systems are much more coherent than Vnf. In all cases, the components are A-type and most closely related to other Anf systems. Only *anfD* and *anfG* have been sequenced from *C. pasteurianum*, but both fall towards the edge of the Anf clade as might be expected (for *anfG* see Figure 7; *anfD* not shown). For each gene, the variation among the proteobacteria is roughly comparable with that seen for the equivalent B-type gene, while the distance to the archaeal sequences is not much greater than this. Unless there are extraordinary functional constraints preventing further divergence of the Anf genes, the conclusion must be that the Anf genes in *Methanosarcina* shared a common ancestor with the bacterial examples early in the history of the proteobacteria, and certainly long after the common ancestor of archaea and bacteria. Until more systems are characterised, it is hard to say whether this represents an old transfer in either direction or a much more recent transfer to *Methanosarcina* from a bacterium that has not yet been examined.

6.5. The origin of nitrogenases

It is clear that there are three major types of nitrogenase, each with phylogenetically distinct set of H, D, K, E and N proteins. The A-type systems are Mo-independent (either Vnf or Anf) and are widespread among bacteria and known from the archaeon, *Methanosarcina*. B-type is common among the proteobacteria and cyanobacteria and is found in *Desulfitobacterium*, a relative of *Clostridium*, but not so far in *Clostridium* itself, nor in any archaeon. C-type is found in various archaea (*Methanosarcina* and *Methanothermobacter*) and in *Clostridium* and *Chlorobium*, which represent two related bacterial groups. All these systems share related components and even a conserved gene order, implying that they must have a common ancestry. They are, however, very divergent, which is consistent with the widely held view that nitrogenase is a very ancient system that was already present in the last universal common ancestor of all modern life. Given that there is evidence for interdomain transfer of both A-type and C-type, it not easy to determine whether the early development of these systems occurred in bacteria or in archaea.

7. NITROGENASE GENES IN THEIR GENOMIC CONTEXT

In addition to the sequence and gene organisation information that has already been discussed, complete genome sequences provide other potential clues about the origin of the nitrogenase genes in them. The location of the genes may indicate whether they are either long-term residents or recent intruders, as may their base composition relative to the surrounding genome.

In the rhizobia, the *nif* genes are on either plasmids or genomic islands, which are, in both cases, generally lower in G+C content than the basic chromosomal genome (Galibert *et al.*, 2001; Kaneko *et al.*, 2000; 2002). These parts of the genome are prone to transfer, at least between related bacteria, and include many genes that are specific to particular isolates or species. It can be said, therefore, that in these bacteria the *nif* genes form part of the accessory genome. Although their context is a relatively low G+C part of the genome, the *nifH*, *D*, *K*, and *E* genes have extremely high G+C (Young, unpublished analysis). The reason is unknown and it is not true in general of other *nif*-related genes in these organisms.

In the other sequenced organisms, the nitrogenase genes are chromosomally located, even when, as in *Nostoc* sp. PCC 7120, the genome also includes large plasmids. However, these genomes have not been systematically examined for genomic islands, so it is not clear whether the context of the nitrogenase genes is truly in the basic chromosome or in the accessory genome. It is possible that a careful analysis would reveal that each of the nitrogenase types was truly "at home" in certain organisms. On the other hand, we know of no major group of organisms in which nitrogen fixation is universal, so perhaps nitrogenase lives a perpetually nomadic existence as a life-long member of the horizontal gene pool.

8. CONCLUSIONS AND PROSPECTS

It is too soon for conclusions. Our access to genome information is still very limited, but some things are already becoming clearer. Now that there are hundreds of microbial genome sequences, we can see that nitrogenase genes are only present in a minority of them (and, incidentally, absent from eukaryotes). These data confirm the observation that most organisms do not fix N_2 and remove the nagging doubt that we just had not found the right conditions. The availability of more complete sets of nitrogenase sequences has reinforced the distinctness of the A, B and C types; the distinction is recognisable in each of the components. Phylogenetic evidence makes it virtually certain that nitrogenase genes have been transferred horizontally, even between bacteria and archaea, but it is difficult to be certain of the direction.

Analysis of the genomic context of the genes is still in its infancy, but promises to provide new insights into the distribution of nitrogenase genes and their integration into the metabolism of their host organisms. Of course, genome-gazing is no substitute for experimental evidence, but it can tie together a mass of observations and can also suggest interesting questions for the future.

REFERENCES

Aguilar, O. M., Reilander, H., Arnold, W., and Pühler, A. (1987). *Rhizobium meliloti nifN* (*fixF*) gene is part of an operon regulated by a NifA-dependent promoter and codes for a polypeptide homologous to the *nifK* gene product. *J. Bacteriol., 169*, 5393-5400.

Altschul, S. F., Madden, T. L., Schaffer, A. A., Zhang, J. H., Zhang, Z., Miller, W., *et al.* (1997). Gapped BLAST and PSI-BLAST: A new generation of protein database search programs. *Nucleic Acids Res., 25*, 3389-3402.

Brigle, K. E., Weiss, M. C., Newton, W. E., and Dean, D. R. (1987). Products of the iron-molybdenum cofactor-specific biosynthetic genes, *nifE* and *nifN*, are structurally homologous to the products of the nitrogenase molybdenum-iron protein genes, *nifD* and *nifK*. *J. Bacteriol., 169*, 1547-1553.

Burke, D. H., Hearst, J. E., and Sidow, A. (1993). Early evolution of photosynthesis - clues from nitrogenase and chlorophyll iron proteins. *Proc. Natl. Acad. Sci. USA, 90*, 7134-7138.

Chien, Y. T., Auerbuch, V., Brabban, A. D., and Zinder, S. H. (2000). Analysis of genes encoding an alternative nitrogenase in the archaeon *Methanosarcina barkeri* 227. *J. Bacteriol., 182*, 3247-3253.

Cole, J. R., Chai, B., Marsh, T. L., Farris, R. J., Wang, Q., Kulam, S. A., *et al.* (2003). The Ribosomal Database Project (RDP-II): Previewing a new autoaligner that allows regular updates and the new prokaryotic taxonomy. *Nucleic Acids Res., 31*, 442-443.

Daubin, V., Moran, N. A., and Ochman, H. (2003). Phylogenetics and the cohesion of bacterial genomes. *Science, 301*, 829-832.

Dean, D. R., and Brigle, K. E. (1985). *Azotobacter vinelandii nifD*-encoded and *nifE*-encoded polypeptides share structural homology. *Proc. Natl. Acad. Sci. USA, 82*, 5720-5723.

Dean, D. R., and Jacobson, M. R. (1992). Biochemical genetics of nitrogenase. In G. Stacey and R. H. Burris and H. J. Evans (Eds.), *Biological nitrogen fixation.* (pp. 763-834). New York, NY: Chapman and Hall.

Deppenmeier, U., Johann, A., Hartsch, T., Merkl, R., Schmitz, R. A., Martinez-Arias, R., *et al.* (2002). The genome of *Methanosarcina mazei*: Evidence for lateral gene transfer between bacteria and archaea. *J. Mol. Microbiol. Biotechnol., 4*, 453-461.

Doolittle, W. F., Boucher, Y., Nesbø, C. L., Douady, C. J., Andersson, J. O., and Roger, A. J. (2003). How big is the iceberg of which organellar genes in nuclear genomes are but the tip? *Phil. Trans. Royal Soc. Series B, 358*, 39-57.

Eisen, J. A., Nelson, K. E., Paulsen, I. T., Heidelberg, J. F., Wu, M., Dodson, R. J., *et al.* (2002). The complete genome sequence of *Chlorobium tepidum* TLS, a photosynthetic, anaerobic, green-sulfur bacterium. *Proc. Natl. Acad. Sci. USA, 99*, 9509-9514.

Fani, R., Gallo, R., and Lio, P. (2000). Molecular evolution of nitrogen fixation: The evolutionary history of the *nifD*, *nifK*, *nifE*, and *nifN* genes. *J. Mol. Evol., 51*, 1-11.

Galagan, J. E., Nusbaum, C., Roy, A., Endrizzi, M. G., Macdonald, P., FitzHugh, W., *et al.* (2002). The genome of *M. acetivorans* reveals extensive metabolic and physiological diversity. *Genome Res., 12*, 532-542.

Galibert, F., Finan, T. M., Long, S. R., Pühler, A., Abola, P., Ampe, F., *et al.* (2001). The composite genome of the legume symbiont *Sinorhizobium meliloti*. *Science, 293*, 668-672.

Golden, J. W., Robinson, S. J., and Haselkorn, R. (1985). Rearrangement of nitrogen fixation genes during heterocyst differentiation in the cyanobacterium *Anabaena*. *Nature, 314*, 419-423.

Haukka, K., Lindström, K., and Young, J. P. W. (1998). Three phylogenetic groups of *nodA* and *nifH* genes in *Sinorhizobium* and *Mesorhizobium* isolates from leguminous trees growing in Africa and Latin America. *Appl. Environ. Microbiol., 64*, 419-426.

Hennecke, H., Kaluza, K., Thöny, B., Fuhrmann, M., Ludwig, W., and Stackebrandt, E. (1985). Concurrent evolution of nitrogenase genes and 16S ribosomal RNA in *Rhizobium* species and other nitrogen-fixing bacteria. *Arch. Microbiol., 142*, 342-348.

Holland, D., Zilberstein, A., Zamir, A., and Sussman, J. L. (1987). A quantitative approach to sequence comparisons of nitrogenase MoFe protein alpha subunits and beta subunits including the newly sequenced *nifK* gene from *Klebsiella pneumoniae*. *Biochem. J., 247*, 277-285.

Hurek, T., Egener, T., and Reinhold-Hurek, B. (1997). Divergence in nitrogenases of *Azoarcus* spp., *Proteobacteria* of the beta subclass. *J. Bacteriol., 179*, 4172-4178.

Joerger, R. D., Loveless, T. M., Pau, R. N., Mitchenall, L. A., Simon, B. H., and Bishop, P. E. (1990). Nucleotide sequences and mutational analysis of the structural genes for nitrogenase 2 of *Azotobacter vinelandii*. *J. Bacteriol., 172*, 3400-3408.

Kaminski, P. A., Batut, J., and Boistard, P. (1998). A survey of symbiotic nitrogen fixation by rhizobia. In H. P. Spaink, A. Kondorosi and P. J. J. Hooykaas (Eds.), *The Rhizobiaceae* (pp. 431-460). Dordrecht, The Netherlands: Kluwer Academic Publishers.

Kaneko, T., Nakamura, Y., Sato, S., Asamizu, E., Kato, T., Sasamoto, S., *et al.* (2000). Complete genome structure of the nitrogen-fixing symbiotic bacterium *Mesorhizobium loti*. *DNA Res., 7*, 331-338.

Kaneko, T., Nakamura, Y., Sato, S., Minamisawa, K., Uchiumi, T., Sasamoto, S., *et al.* (2002). Complete genomic sequence of nitrogen-fixing symbiotic bacterium *Bradyrhizobium japonicum* USDA110. *DNA Res., 9*, 189-197.

Kaneko, T., Nakamura, Y., Wolk, C. P., Kuritz, T., Sasamoto, S., Watanabe, A., *et al.* (2001). Complete genomic sequence of the filamentous nitrogen-fixing cyanobacterium *Anabaena* sp. strain PCC 7120. *DNA Res., 8*, 205-213.

Kessler, P. S., and Leigh, J. A. (1999). Genetics of nitrogen regulation in *Methanococcus maripaludis*. *Genetics, 152*, 1343-1351.

Kessler, P. S., McLarnan, J., and Leigh, J. A. (1997). Nitrogenase phylogeny and the molybdenum dependence of nitrogen fixation in *Methanococcus maripaludis*. *J. Bacteriol., 179*, 541-543.

Kumar, S., Tamura, K., Jakobsen, I. B., and Nei, M. (2001). MEGA2: Molecular evolutionary genetics analysis software. *Bioinformatics, 17*, 1244-1245.

Loveless, T. M., and Bishop, P. E. (1999). Identification of genes unique to Mo-independent nitrogenase systems in diverse diazotrophs. *Can. J. Microbiol., 45*, 312-317.

Loveless, T. M., Saah, J. R., and Bishop, P. E. (1999). Isolation of nitrogen-fixing bacteria containing molybdenum-independent nitrogenases from natural environments. *Appl. Environ. Microbiol., 65*, 4223-4226.

Masepohl, B., and Klipp, W. (1996). Organization and regulation of genes encoding the molybdenum nitrogenase and the alternative nitrogenase in *Rhodobacter capsulatus*. *Arch. Microbiol., 165*, 80-90.

Moulin, L., Munive, A., Dreyfus, B., and Boivin-Masson, C. (2001). Nodulation of legumes by members of the beta-subclass of *Proteobacteria*. *Nature, 411*, 948-950.

Nölling, J., Breton, G., Omelchenko, M. V., Makarova, K. S., Zeng, Q. D., Gibson, R., *et al.* (2001). Genome sequence and comparative analysis of the solvent-producing bacterium *Clostridium acetobutylicum*. *J. Bacteriol., 183*, 4823-4838.

Normand, P., and Bousquet, J. (1989). Phylogeny of nitrogenase sequences in *Frankia* and other nitrogen-fixing microorganisms. *J. Mol. Evol., 29*, 436-447.

Postgate, J. R. (1974). Evolution within nitrogen-fixing systems. In M. J. Carlile and J. J. Skehel (Eds.), *SGM Symposium 24: Evolution in the microbial world* (pp. 263-292). Cambridge, UK: Cambridge University Press.

Postgate, J. R. (1982). *The fundamentals of nitrogen fixation*. Cambridge, UK: Cambridge University Press.

Postgate, J. R., and Eady, R. R. (1988). The evolution of biological nitrogen fixation. In H. Bothe, F. J. De Bruijn and W. E. Newton (Eds.), *Nitrogen fixation: Hundred years after* (pp. 31-40). Stuttgart, Germany: Gustav Fischer.

Raymond, J., Zhaxybayeva, O., Gogarten, J. P., Gerdes, S. Y., and Blankenship, R. E. (2002). Whole-genome analysis of photosynthetic prokaryotes. *Science, 298*, 1616-1620.

Ribbe, M., Gadkari, D., and Meyer, O. (1997). N_2 fixation by *Streptomyces thermoautotrophicus* involves a molybdenum dinitrogenase and a manganese superoxide oxidoreductase that couple N_2 reduction to the oxidation of superoxide produced from O_2 by a molybdenum-CO dehydrogenase. *J. Biol. Chem., 272*, 26627-26633.

Smith, D. R., Doucette-Stamm, L. A., Deloughery, C., Lee, H. M., Dubois, J., Aldredge, T., *et al.* (1997). Complete genome sequence of *Methanobacterium thermoautotrophicum* Delta H: Functional analysis and comparative genomics. *J. Bacteriol., 179*, 7135-7155.

Thiel, T. (1993). Characterization of genes for an alternative nitrogenase in the cyanobacterium *Anabaena variabilis*. *J. Bacteriol., 175*, 6276-6286.

Thiel, T. (1996). Isolation and characterization of the *vnfEN* genes of the cyanobacterium *Anabaena variabilis*. *J. Bacteriol., 178*, 4493-4499.

Thompson, J. D., Gibson, T. J., Plewniak, F., Jeanmougin, F., and Higgins, D. G. (1997). The CLUSTAL_X windows interface: Flexible strategies for multiple sequence alignment aided by quality analysis tools. *Nucleic Acids Res., 25*, 4876-4882.

Tiedje, J. M., and Davis, J. K. (2002). *Co-occurrence of dehalogenation and nitrogen fixation in reductively dechlorinating bacteria.* Paper presented at the US-Japan Joint Workshop on Systems Biology of Useful Microorganisms, Keio University, Japan.

Wolfinger, E. D., and Bishop, P. E. (1991). Nucleotide sequence and mutational analysis of the *vnfENX* Region of *Azotobacter vinelandii*. *J. Bacteriol., 173*, 7565-7572.

Young, J. P. W. (1990). The phylogeny of *nifH*: Gene duplication or lateral transfer? In P. M. Gresshoff, L. E. Roth, G. Stacey and W. E. Newton (Eds.), *Nitrogen fixation: Achievements and objectives* (pp. 840). New York, NY: Chapman and Hall.

Young, J. P. W. (1992). Phylogenetic classification of nitrogen-fixing organisms. In G. Stacey, R. H. Burris and H. J. Evans (Eds.), *Biological nitrogen fixation.* (pp. 43-86). New York, NY: Chapman and Hall.

Young, J. P. W. (2000). Molecular evolution in diazotrophs: Do the genes agree? In F. O. Pedrosa, M. Hungria, M. G. Yates and W. E. Newton (Eds.), *Nitrogen fixation: From molecules to crop productivity* (pp. 161-164). Dordrecht, The Netheralnds: Kluwer.

Zinoni, F., Robson, R. M., and Robson, R. L. (1993). Organization of potential alternative nitrogenase genes from *Clostridium pasteurianum*. *Biochim. Biophys. Acta, 1174*, 83-86.

SUBJECT INDEX

activating enzyme, see *draG/draT*
adenylylation/deadenylylation 24,
ADP-ribosylation, see covalent
    modification
Agrobacterium 86-87, 92, 102, 114,
    116-117, 213-,
alanine dehydrogenase 9,
Allorhizobium 83,
alternative nitrogenases, see also
    anf and *vnf*, 9, 61, 72-73, 78,
amino-acid transport 61,
ammonium assimilation 61-63,
ammonium transport, see *amt*
amt 11, 61-62,
Anabaena 2, 4, 39-40,
anfA 73-74, 76-78,
anfDK 8-9, 23, 78,
anfG 8-9, 23, 78, 233,
anf genes 8-9, 23, 73-78,
anfH 8-9, 23, 73-74, 78, 226-227,
anfI 8-9,
Anf nitrogenases (history) 236-237,
angiosperm associations 32,
Anthroceros 32,
Archaea 7-11,
A. vinelandii, see *Azotobacter
    vinelandii*
Azorhizobium 83, 92,
Azospirillum 75,
Azotobacter vinelandii 5, 23, 73, 78,
    223, 225,
bioinformatics 160-162,
Blasia 32,
Bradyrhizobium 2-3, 83, 85-86, 92, 99-
108, 116, 153-154, 186, 191-192, 223,
    fix genes 103-105,
    G+C 100,
    IS/MS 101,
    nif genes 223,
    nod genes 102-104,
    proteomics 153-154,
    respiratory chains 106-107,
    symbiotic island 102-103, 186,
    transporters 105-106,
bryophyte associations 32,
Burkholderia 83, 223,
cellulose synthase 41,
Chlorobium 2, 4, 223,
Chloroflexus 80,
circadian rhythm 41,
clostridia, see *Clostridium*
Clostridium 2, 4, 13-24, 223,
    genome 15-16,
covalent modification 22, 73-75,
cox/Cox 55, 57, 104, 106-107, 122,
cyanoglobin (*glbN*) 57-58,
cycad associations 32,
cytochrome oxidase, see *cox*/Cox
Desulfitobacteriuim 223,
dev/Dev 54-55,
dinitrogenase, see MoFe protein
dinitrogenase reductase, see Fe
    protein
DNA arrays 145-150, 170-178,
DNA supercoiling 74,
drat/DraT/*drag*/DraG 74-75,
dynamics 183-198,
E. coli, see *Escherichia coli*
electron transfer flavoprotein 121,
electrophoresis 151,
Escherichia coli 10-11, 20,
evolution 9, 201-214, 221-238,
FeMoco 9, 18,
Fe protein 8,
fdx/ferredoxin 18, 20-21, 59-60, 121,
fix genes 86-87, 103-105, 120-122,
formate dehydrogenase 122,
functional genomics 144-163, 169-179,
G+C content 23,
gene amplification 188-191,
gene expression 144-150,
genome architecture 183-187, 191-194,
genome amplification 183-184, 96,
genome co-integration 183-184, 193-
    198,
genome deletions 183-184,
genome dynamics 183-198,
genome instability 186-188,

genome islands 209-210,
genome rearrangements 183-184, 193-198,
Geosiphon 32,
glnB/K 10-11, 19, 22, 49, 73-75, 79,
glucose-6-phosphate dehydrogenase 55-56,
glutamate dehydrogenase 9, 23,
glutamate synthase 9, 23,
glutamine synthetase 9, 23, 55-56,
Gunnera 32,
glycolipid synthesis 54,
han/Han 48-50,
Haemophilus 1,
hep/Hep 54,
het/Het 36, 48-52, 54,
heterocyst differentiation 4, 29-30, 46-52,
heterocyst maturation 53-57,
hgl/Hgl 54,
histidine kinases 42, 74,
homocitrate 9,
horizontal gene transfer 205-209,
host specificity 84-86,
hox/Hox 103,
hup/Hup 55, 57, 60, 103, 106,
hvr/Hvr 73-74, 79-80,
hydrogenase, see *hup*/Hup
insertion sequences (IS/MS) 86-88, 101, 117-119, 183-186,
iron-molybdenum cofactor, see FeMoco
iron protein, see *nifH*/NifH
isocitrate dehydrogenase 55, 57,
invertase 56,
Klebsiella pneumoniae 2, 5, 73, 75, 146, 148,
K. pneumoniae, see *Klebsiella pneumoniae*
Lablab 84,
liverwort associations 32,
Lotus 149,
Macrozamia 32,
Magnetococcus 223,
mass spectrometry 151-152,
Medicago 84, 113, 147, 149, 157, 176, 178,
Melilotus 84, 113, 155,
Mesorhizobium 2-3, 83, 86, 92, 102, 116, 191-192, 209-210, 213, 223, 225,
Methanobacterium 7, 11,
Methanococcus 7-8, 10-11, 225,
Methanopyrus 7,
Methanosarcina 2, 4-5, 7-11, 223, 225,
Methanospirillum 7-11,
Methanothermobacter 2-5, 8-9, 223,
Methanothermococcus 2-5, 7,
Methylobacterium 83,
mod/Mod 8, 18, 20-21, 75-76,
MoFe protein 5, 8-9,
molybdenum transport 9, 18,
molybdenum-iron cofactor, see FeMoco
molybdenum-iron protein, see MoFe protein
mop/Mop 73-74,
mycorrhizal associations 32,
nap/Nap 62, 104,
nar/Nar 62,
nifA/NifA 22, 72-77, 79,
nifB 9, 58-60, 76, 79,
nifDK 14, 17-21, 24, 58-60, 74-76,
NifDK, see MoFe protein
nifEN /NifEN 8, 17-21, 58-60, 79,
nif gene cluster/organization 8-9, 17-21, 75-80, 86, 99, 120-121, 222-225,
nifH/NifH 8-9, 14, 17-21, 24, 58-60, 73-76, 225-227,
nifI/NifI 8, 11, 17-22, 24,
nifL/NifL 22, 75,
nifN-B 17-21,
nifS 9,58-59, 79,
nifU 9, 58-59, 74, 79,
nifV/NifV 59-60, 79,
nifVα/*nifVω* 17-21,
nifW 79,
nifX/NifX 8, 58-60, 79,
nir/Nir 18, 20, 62, 105,
nitrate assimilation/transport 61-63,
nitrate reductase 62, 104, 122,
nitric oxide reductase 105, 122,

nitrite reductase 62, 105, 122,
nitrogen (fixed) regulation, see
 regulation by fixed nitrogen
nitrous oxide reductase 122,
Nod (*nod*) factors 84-87, 99, 102-103, 119-120, 147,
nodulation 83-85,
Nostoc 4-5, 27-64, 223,
 chemotaxis 33,
 fixed nitrogen assimilation 61-63,
 free-living state 29-31, 36,
 genome 37-46,
 heterocysts 29-30, 36-37, 40, 46-57,
 hormogonia 32-34, 40,
 infectivity 31-34,
 nif genes 57-61, 223,
 photosynthesis 29, 34-35,
 rhizobia comparison 32, 37,
 symbiotic state 31-32, 34-37,
nrt/Nrt 62,
ntc/Ntc 33, 36, 48-50, 52,
ntr/Ntr 22, 24, 49, 73-74, 77-78,
oxygen (O_2) control 22, 74, 121,
oxygen (O_2) protection/tolerance 4, 14, 32,
P_{II} proteins 8, 10-11, 19, 22, 49, 74-75, 79,
Paenibacillus 13,
Parasponia 84,
pat/Pat 48-52,
6-phosphogluconate dehydrogenase 55-56,
photosynthesis 4, 162-163,
phylogeny (organism) 202-205,
phylogeny (nitrogenases) 221-238,
 organism-phylogeny comparison 231-238,
 nifDKEN 227-232,
 nifH 225-227,
phytochrome 41,
Prochlorococcus 40,
proteomics 150-160,
 bacterial 152-156,
 in symbiosis 155-157,
 peribacteroid membrane 156-157,
 protein-protein interactions 160,
Ralstonia 83,
rearrangements (genomic) 183-198,
reg/Reg 73-74, 79,
regulation by ammonia, see
 regulation by fixed nitrogen
regulation by fixed nitrogen 21-22, 73-74,
reiterated sequences 183-186,
response regulator proteins 43,
Rhizobium 2-3, 5, 83-93, 99, 146-147,
 NGR234 83-93, 116, 147, 173, 185, 191-194,
  chromosome 87-88,
  genome structure 85, 191-192,
  IS/MS 86-88, 185,
  megaplasmid 87-88,
  replicon plasticity 88-90,
  symbiotic plasmid 85-87,
  transcriptomics 91-92, 173,
 R. etli 133-140, 153-154, 185-198,
  evolution 138-140,
  genome plasticity 136-138, 185-198,
  genome structure 134-136,
  IS/MS 185-186,
  proteomics 153-154, 158-160,
  taxonomy 138-140,
 R. leguminosarum 152-156,
  proteomics 152-156,
Rhodobacter 5, 71-80,
Rhodopseudomonas 71-80, 107-108, 148, 223,
Rhodospirillum 5, 75, 223, 225,
rnf genes 79,
serine/threonine protein kinases 45,
Sesbania 84,
sigma^{54} (σ^{54})/*rpoN*/RpoN 49, 56, 73-74, 77-78, 86, 104-105,
Sinorhizobium 2-3, 83-87, 92, 102, 113-128, 146-148, 169-179, 186, 191-192, 213, 223,
 pSymA 114-128,
  carbon metabolism 121-123,
  chemotaxis 123,
  genome structure 114-115,

IS/MS 117-119,
nif/Nif 120-121, 223,
nitrogen metabolism 121-123,
nod/Nod 119-120,
plasmid biology 117,
proteomics 153-155,
regulation 124-125,
stress responses 125-126,
sulfur metabo,ism 126,
transporters 123-124,
pSymB 114-115,
transcriptomics 169-179,
symbiotic island 3, 85, 102-103, 209-210,
symbiotic plasmid 3, 85-90, 114-116,
symbiosome 156-157,
Synechococcus 41,
Synechocystis 4, 37-39,
taxonomy 201-214,
transcriptomics 144-150, 157-160, 169-179,
bacterial 144-150,
nodulation 148-150,
transport of fixed nitrogen, see *amt*/ammonium transport
Trichodesmium 62, 223,
Trigonella 84, 113,
urease 61,
vnfDK 8-9,
vnfEN 8-9, 227-232,
vnfG 8-9, 233,
vnf genes 8-9, 23,
vnfH 8-9, 226-227,
vnfI 8-9,
Vnf nitrogenases (history) 235-236,
xis 59-60,